Dynamic Optimization

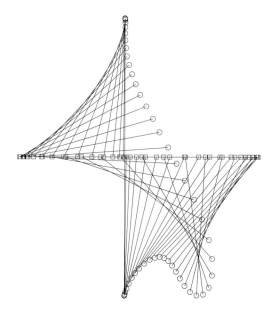

Arthur E. Bryson, Jr.

Pigott Professor of Engineering Emeritus
Stanford University

 **ADDISON-WESLEY**

An imprint of Addison Wesley Longman, Inc.

Menlo Park, California • Reading, Massachusetts • Harlow, England
Berkeley, California • Don Mills, Ontario • Sydney • Bonn • Amsterdam • Tokyo • Mexico City

Acquisitions Editor: Paul Becker
Assistant Editor: Anna Friedlander
Editorial Assistant: Royden Tonomura
Production Manager: Pattie Myers
Senior Production Editor: Teri Hyde
Art and Design Supervisor: Kevin Berry
Composition Art: Interactive Composition Corporation
Cover Design: Karl Miyajima
Cover Image: Arthur E. Bryson, Jr.
Text Design: Eigentype Compositors
Text Printer and Binder: Maple Vail
Cover Printer: Phoenix Color

Library of Congress Cataloging-in-Publication Data

    Bryson, Arthur E. (Arthur Earl)
    Dynamic optimization / Arthur E. Bryson, Jr.
       p.   cm.
    Includes bibliographical refernces and index.
    ISBN 0-201-59790-X
    1. Mathematical optimization.   2. Control theory.   3. Calculus of variations. I. Title.
  QA402.5.B784   1999
  519.3–dc21

Instructional Material Disclaimer: The programs presented in this book have been included for their instructional value. They have been tested with care but are not guaranteed for any particular purpose. Neither the publisher or the authors offer any warranties or representations, nor do they accept any liabilities with respect to the programs.

The full complement of supplement teaching materials is available to qualified instructors.

ISBN 0-201-59790-X (Book)
ISBN 0-201-36187-6 (Package)
1 2 3 4 5 6 7 8 9 10—MV—02 01 00 99 98

Addison Wesley Longman, Inc.
2725 Sand Hill Road
Menlo Park, California 94025

Contents

The DYNOPT Toolbox
(Mᴀᴛʟᴀʙ Codes)

Acronyms and Abbreviations

- A/C—Aircraft
- AMP—Accessory Minimum Problem
- CMG—Control Moment Gyro
- CHTC—Continuous with Hard Terminal Constraints
- CSTC—Continuous with Soft Terminal Constraints
- CV—Calculus of Variations
- DHTC—Discrete with Hard Terminal Constraints
- DLQR—Discrete Linear-Quadratic Regulator
- DO—Dynamic Optimization
- DP—Dynamic Programming
- DSTC—Discrete with Soft Terminal Constraints
- DTDP—Discrete Thrust Direction Programming
- DVDP—Discrete Velocity Direction Programming
- EL—Euler-Lagrange
- EOM—Equations of Motion
- GEP—Generalized Eigenvalue Problem
- HJB—Hamilton-Jacobi-Bellman
- HTC—Hard Terminal Constraints
- ILOS—Initial Line-of-Sight
- LHP—Left Half-Plane
- LOS—Line-of-Sight
- LP—Linear Programming
- LQ—Linear-Quadratic
- LQR—Linear-Quadratic Regulator
- Max—Maximum
- MIMO—Multiple Input/Multiple Output

- MISC—Minimum Integral-Square Control
- Min—Minimum
- MISO—Multiple Input/Single Output
- MOI—Moment of Inertia
- NFP—Normal Force Programming
- NLP—Nonlinear Programming
- NMP—Nonminimum Phase
- NOC—Neighboring Optimum Control
- NR—Newton-Raphson
- PI—Performance Index
- POIC—Parameter Optimization problems with Inequality Constraints
- QPI—Quadratic Performance Index
- RHP—Right Half Plane
- RK—Runge-Kutta
- RRCE—Reciprocal Root Characteristic Equation
- RRL—Reciprocal Root Locus
- S/C—Spacecraft
- SFB—State Feedback
- SS—Steady-State
- SIMO—Single Input/Multiple Output
- SISO—Single Input/Single Output
- SQP—Sequential Quadratic Programming
- SRCE—Symmetric Root Characteristic Equation
- SRL—Symmetric Root Locus
- STC—Soft Terminal Constraints
- SVD—Singular Value Decomposition
- TDP—Thrust Direction Programming
- TI—Time Invariant
- TZ—Transmission Zero
- VDP—Velocity Direction Programming
- ZOH—Zero-Order-Hold
- 2D—Two-Dimensional
- 3D—Three-Dimensional

Preface

Dynamic optimization is the process of determining control and state histories for a dynamic system over a finite time period to minimize a performance index. There may be constraints on the final states of the system and on the 'in-flight' states and controls. An important special case is when the dynamic system is linear and the performance index is a quadratic functional of the states and controls. The main tools of dynamic optimization are the *calculus of variations* and *dynamic programming*. Dynamic optimization is used to determine efficient maneuvers of aircraft, spacecraft, and robots, and in the design of structures where the independent variable is distance along the structure instead of time.

Computational algorithms are developed for solving practical problems and many examples and problems are presented. The algorithms are coded in MATLAB which is a popular software package for engineers interested in dynamics and control. Solutions for the examples and the many problems are provided on a disc that goes with the book. The problems are difficult but doing them is the only way to really learn the subject.

The book starts with a review of parameter optimization (ordinary calculus) and then treats dynamic optimization (the calculus of variations), first with fixed final time and no constraints, then with terminal constraints, and then with terminal constraints and open final time. This is followed by chapters on linear-quadratic problems which are of practical interest in themselves but also develop the theory needed to consider the second variation and neighboring-optimal feedback control. Next is a chapter on dynamic programming, an interesting but not very practical method of nonlinear feedback control, and then a chapter on neighboring-optimal feedback control which is practical. The next to last chapter deals with inequality constraints, first for static systems (nonlinear programming) and then for dynamic systems using inverse dynamic optimization. The last chapter covers singular problems, i.e., problems where the second variation is identically zero for a finite time; these problems occur when the system equations are linear in one or more of the controls. There is an appendix giving a short history of dynamic optimization.

Chapter 5 describes dynamic optimization for linear systems with time-varying feedback gains. With fast computers having large memory storage, this is now an attractive alternative to constant-gain feedback control, since it cuts the time to reach a desired state almost in half. In Chapters 9 and 10 we describe an inverse dynamic optimization method due to Seywald (1994) that uses nonlinear programming software to solve dynamic optimization problems with inequality constraints or singular arcs.

Before the advent of the digital computer (about 1950) only rather simple dynamic optimization problems could be solved (in terms of tabulated functions). Now, with powerful digital computers, numerical solutions can be found for realistic problems. Some current aircraft contain flight management computers that find optimal flight paths in real time and send them to the autopilot for implementation. Digital control is now commonplace, where a digital computer is the logic element in a feedback control system involving sensors and actuators. Spaceflight would not have been possible without digital control. Microprocessors have made it possible to use digital control in cars, home appliances, robots, and even toys.

This book updates and extends the first half of *Applied Optimal Control*.[1] An update and extension of the second half is under preparation with the tentative title 'Optimal Control with Uncertainty'; it will deal with optimal linear feedback control in the presence of uncertain inputs and an uncertain dynamic model. In the intervening 29 years the development and spread of personal computers has made it possible to do more interesting problems while learning the subject. Hence this book contains more examples and problems than its predecessor.

The codes presented here were prepared for use on personal computers. Several aerospace companies have developed codes for very large problems which require supercomputers. For example Boeing (Ref. HP) has developed OTIS (Optimal Trajectories by Implicit Simulation) and Lockheed-Martin (Ref. BCS) has developed POST (Program to Optimize Simulated Trajectories) which use collocation techniques with NLP software. Collocation techniques are very effective but are not discussed in this text in an effort to limit the size of the book.

The discrete algorithms presented in Chapters 2 to 4 are largely there for pedagogical reasons since they are simpler and lead into the continuous algorithms. However they also lead into the discrete algorithms for the linear-quadratic problems of Chapters 5 and 6 which are used more than the continuous algorithms in current practice.

I should like to thank Carolyn Edwards for her patient work in putting the text on the computer. Using LATEX I made changes and additions on my personal computer. Special thanks go to Sun Hur Diaz and Paul M. Montgomery for showing me how to use MATLAB to integrate differential equations with ease, and to the Mathworks for creating MATLAB which has made easy work out of things that were very tedious only a few years ago.

I am very appreciative to the following individuals for their careful reviews of the manuscript and for their constructive suggestions: Bruce Conway, University of Illinois; Mark L. Psiaki, Cornell University; and Yiyuan Zhao, University of Minnesota.

Arthur E. Bryson
Stanford, California

---

[1]Written with my esteemed colleague Yu-Chi Ho, published in 1969 by the Blaisdell Publishing Co., and reprinted by the Hemisphere Publishing Co. in 1975. Hemisphere later became a member of the Taylor and Francis Group of Bristol, Pa.

Static Optimization

1.1 Problems without Constraints

A simple class of static optimization problems involves finding the values of p parameters $y_1, \ldots, y_p$ that minimize a performance index that is a function of these parameters,

$$L(y_1, \ldots, y_p) \ .$$

For convenience, we shall use a more compact nomenclature; let

$$y = \begin{bmatrix} y_1 \\ \vdots \\ y_p \end{bmatrix} = \text{parameter vector} \ , \tag{1.1}$$

and write the performance index as

$$L(y) \ . \tag{1.2}$$

If there are no constraints on possible values of y and if the function $L(y)$ has first and second partial derivatives everywhere, then the function can be approximated in the neighborhood of y^o by the first three terms of the Taylor series

$$L(y) \approx L(y^o) + L_y(y - y^o) + \frac{1}{2}(y - y^o)^T L_{yy}(y - y^o) \ , \tag{1.3}$$

where the derivatives L_y and L_{yy} are evaluated at y^o and

$$\frac{\partial L}{\partial y} \equiv L_y = [L_{y_1} \ L_{y_2} \cdots L_{y_p}] \ , \tag{1.4}$$

$$\frac{\partial^2 L}{\partial y^2} \equiv L_{yy} = \begin{bmatrix} L_{y_1 y_1} & \cdots & L_{y_1 y_p} \\ \vdots & \vdots & \vdots \\ L_{y_p y_1} & \cdots & L_{y_p y_p} \end{bmatrix}. \tag{1.5}$$

L_y is a row matrix of first derivatives that is called *the gradient of L* and L_{yy} is a square symmetric matrix of second derivatives that is called *the Hessian of L*.

From the approximation (1.3), it is clear that *necessary conditions for a minimum* are

$$L_y = 0 , \tag{1.6}$$

and

$$L_{yy} \geq 0 , \tag{1.7}$$

by which we mean that the Hessian must be positive semidefinite (all eigenvalues either positive or zero). Points that satisfy the first order necessary condition (1.6) are called *stationary* points.

Sufficient conditions for a local minimum are (1.6) and the strong form of (1.7)

$$L_{yy} > 0 , \tag{1.8}$$

i.e., the Hessian must be positive definite (all eigenvalues positive). If the point is stationary but the Hessian is positive semidefinite, additional information is needed to establish whether or not the point is a minimum. Such a point is called a *singular point*. If L is a linear function of y, then all of the eigenvalues of the Hessian matrix are zero, and no minimum exists.

Example 1.1.1—Problems in Two Dimensions (2D)

(a) *Minimum*: Both eigenvalues of the Hessian are positive at $y_1 = y_2 = 0$ (see Figure 1.1):

$$L = \begin{bmatrix} y_1 & y_2 \end{bmatrix} \begin{bmatrix} 1 & -1 \\ -1 & 4 \end{bmatrix} \begin{bmatrix} y_1 \\ y_2 \end{bmatrix}.$$

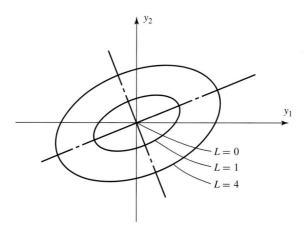

FIGURE 1.1   A Minimum Point

(b) *Saddle Point*: One positive and one negative eigenvalue of the Hessian (see Figure 1.2):

$$L = [\,y_1 \quad y_2\,] \begin{bmatrix} -1 & 1 \\ 1 & 3 \end{bmatrix} \begin{bmatrix} y_1 \\ y_2 \end{bmatrix}.$$

(c) *Singular Point*: One positive and one zero eigenvalue of the Hessian (see Figure 1.3):

$$L = (y_1 - y_2^2)(y_1 - 3y_2^2) \, .$$

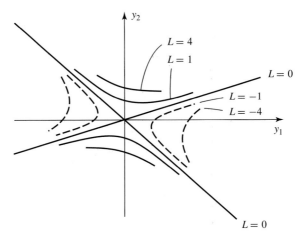

FIGURE 1.2   A Saddle Point

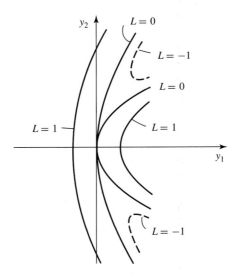

FIGURE 1.3   A Singular Saddle Point

1.2 Problems with Equality Constraints

A more general class of static optimization problems involves finding the p-vector y to minimize a function $L(y)$ with a set of n *constraint relations*

$$f^{(i)}(y) = 0 , \quad i = 1, \cdots, n ,$$

where $n < p$.

For convenience, we shall again use a more compact nomenclature. Let

$$f = \begin{bmatrix} f^{(1)} \\ \vdots \\ f^{(n)} \end{bmatrix} = \text{constraint vector} . \tag{1.9}$$

In this nomenclature, the problem may be stated as follows: Find the parameter vector y that minimizes

$$L(y) \tag{1.10}$$

with the n constraint relations

$$f(y) = 0 , \tag{1.11}$$

where $n < p$.

In this chapter we discuss only cases where the systems (1.10) and (1.11) are *nonlinear* so that a minimum can exist without adding inequality constraints (inequality constraints are discussed in Chapter 9). Of course, nonlinearity does not, in itself, ensure the existence of a minimum.

A Geometric Approach to the First-Order Necessary Conditions

Suppose that we have a feasible y, i.e., $f(y) = 0$, and we wish to see if it also provides a minimum of L. We look for an infinitesimal change in y that will decrease L while keeping $f = 0$ to first order in dy

$$dL = L_y dy < 0 , \tag{1.12}$$

$$df = f_y dy = 0 . \tag{1.13}$$

Now L_y is the *gradient* of L with respect to y, and the rows of f_y are the gradients of $f^{(i)}$ with respect to y, $i = 1, \cdots, n$. These latter gradients are all perpendicular to the hypercurve $f = 0$ since they are perpendicular to the hypersurfaces $f^{(i)} = 0$ whose intersection forms the hypercurve.

At a minimum, L_y *must have no component parallel to the hypercurve* $f = 0$; otherwise a dy could be found to make $dL < 0$ while keeping $df = 0$ to first order in dy. Thus, L_y

must also be perpendicular to $f = 0$ at a minimum; hence the rows of f_y must constitute *a basis* for L_y, i.e., L_y must be a linear combination of the n constraint gradients f_y

$$L_y = -\sum_{i=1}^{n} \lambda_i f_y^{(i)} \; ,$$

where λ_i is a set of n constants. In our more compact notation, this can be written as

$$L_y = -\lambda^T f_y \; , \tag{1.14}$$

where

$$\lambda^T = [\lambda_1, \cdots, \lambda_n] \; . \tag{1.15}$$

Thus, first-order necessary conditions for a stationary point (a minimum, maximum, or saddle point) are

$$f(y) = 0 \; , \tag{1.16}$$

$$L_y + \lambda^T f_y = 0 \; , \tag{1.17}$$

which may be regarded as $n + p$ equations for the $n + p$ unknowns λ and y. The λ_i are called *Lagrange multipliers*. Equation (1.17) is often written as

$$H_y = 0 \; , \tag{1.18}$$

where

$$H \stackrel{\Delta}{=} L + \lambda^T f \; . \tag{1.19}$$

Interpretation of the Lagrange Multipliers

Suppose we had the constraints $f = c$ instead of the constraints $f = 0$, where c is a vector of constants. Then we would define the H function of (1.19) to be

$$H = L + \lambda^T (f - c) \; . \tag{1.20}$$

It follows that

$$dH = (L_y + \lambda^T f_y)dy - \lambda^T dc \; . \tag{1.21}$$

At a stationary point, the coefficient of dy vanishes, leaving

$$dH = -\lambda^T dc \; . \tag{1.22}$$

From this, it is clear that

$$\lambda^T = -L_c^{min} \equiv \frac{\partial L^{min}}{\partial c} \tag{1.23}$$

where $L = L(y, c)$, i.e., $-\lambda$ is a sensitivity vector of the min value of L to the level of the constraints, c.

Adjoining Approach to First-Order Necessary Conditions

Another (equivalent) way of arriving at (1.17) is to "adjoin" the constraints (1.11) to the performance index (1.10) with a set of n "undetermined multipliers" $\lambda_1, \ldots, \lambda_n$:

$$H(y, \lambda) \triangleq L(y) + \sum_{i=1}^{n} \lambda_i f^{(i)}(y) \equiv L(y) + \lambda^T f(y) . \tag{1.24}$$

Suppose we have guessed some values for the parameters y; differential changes in H due to differential changes in y are given by

$$dH = H_y dy , \tag{1.25}$$

where

$$H_y = L_y + \lambda^T f_y . \tag{1.26}$$

Now $H \equiv L$ if the constraints $f = 0$ are met, and a necessary condition for a minimum of H is that $H_y = 0$, which leads directly to (1.16) and (1.17).

Note that (1.26) has the interpretation

$$H_y \equiv (L_y)_f , \tag{1.27}$$

i.e., H_{y_i} is the derivative of L with respect to y_i, letting y_j change, $j \neq i$, so that $f = $ constant, whereas L_{y_i} is the derivative of L with respect to y_i, holding $y_j = $ constant, $j \neq i$. Still another way to say this is that H_y is the component of L_y parallel to $f = 0$.

Decomposition into a Decision Vector and a Dependent Vector

It is sometimes convenient to decompose y into two vectors, x of dimension n and u of dimension $m = p - n$, in such a way that u determines x from the n constraint equations (1.11)

$$f(x, u) = 0 . \tag{1.28}$$

In this case (1.17) becomes two vector equations:

$$L_x + \lambda^T f_x = 0 , \tag{1.29}$$

$$L_u + \lambda^T f_u = 0 , \tag{1.30}$$

Since u determines x it follows that f_x must be nonsingular; hence, from (1.29):

$$\lambda^T = -L_x f_x^{-1} . \tag{1.31}$$

Substituting this into (1.30) allows λ to be eliminated:

$$L_u - L_x f_x^{-1} f_u = 0 . \tag{1.32}$$

Thus necessary conditions for a stationary point are simply (1.28) and (1.32), which constitute $n + m = p$ equations for the $n + m = p$ unknowns x and u. Note that (1.32) is simply $H_u \equiv (L_u)_f = 0$.

Stationary Points in 2D and in Three Dimensions (3D)

Suppose $p = 2$ and $n = 1$, i.e., we have two parameters and one constraint. Then $y = [\, y_1 \ y_2\,]^T$, and λ is a scalar; for this case, (1.17) becomes

$$L_y = -\lambda f_y \,, \tag{1.33}$$

which says that the gradient of L is in the same or the opposite direction of the gradient of f at a stationary point (see Figure 1.4), or L_y has no component along $f = 0$ at a stationary point.

Suppose $p = 3$ and $n = 2$, i.e., there are three parameters and two constraints. Then $y = [\, y_1 \ y_2 \ y_3\,]^T$ and $\lambda = [\, \lambda_1 \ \lambda_2\,]^T$; for this case (1.17) becomes

$$L_y = -\lambda_1 f_y^{(1)} - \lambda_2 f_y^{(2)} \,, \tag{1.34}$$

which says that L_y must be coplanar with $f_y^{(1)}$ and $f_y^{(2)}$ at a stationary point. Now $f^{(1)} = f^{(2)} = 0$ is a space curve (the intersection of two surfaces) in the three-dimensional y-space, so this also means that L_y must have no component along this space curve at a stationary point (see Figure 1.5), or the surface $L = $ constant must be tangent to the space curve at a stationary point.

Example 1.2.1—A Linear-Quadratic (LQ) Problem with Two Parameters and One Constraint

Find the two-parameter vector y that yields a stationary value of the quadratic performance index

$$L = \frac{1}{2} \left(\frac{y_1^2}{a^2} + \frac{y_2^2}{b^2} \right) \,,$$

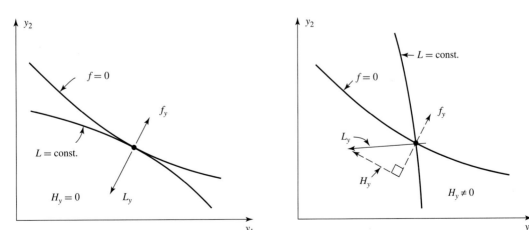

FIGURE 1.4   Stationary and Non-Stationary Points in Two Dimensions

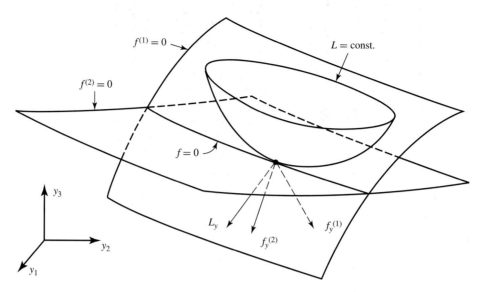

FIGURE 1.5   Stationary Point in Three Dimensions; $L = $ Constant is Tangent to $f = 0 \iff L_y$ is Coplanar with $f_y^{(1)}$ and $f_y^{(2)}$

with the linear constraint

$$f(y) = y_1 + my_2 - c = 0 \, ,$$

where $a, b, m,$ and c are scalar constants.

The curves of constant L are ellipses, with L increasing with the size of the ellipse, whereas $y_1 + my_2 - c = 0$ is a fixed straight line (see Figure 1.6).

Clearly, the min value of L satisfying the constraint is obtained when the ellipse is just tangent to the straight line. The H function is

$$H = \frac{1}{2} \left(\frac{y_1^2}{a^2} + \frac{y_2^2}{b^2} \right) + \lambda(y_1 + my_2 - c) \, ,$$

and necessary conditions for a stationary point are

$$f = y_1 + my_2 - c = 0 \, , \quad H_{y_1} = \frac{y_1}{a^2} + \lambda = 0 \, , \quad H_{y_2} = \frac{y_2}{b^2} + \lambda m = 0 \, .$$

These three equations for the three unknowns $y_1, y_2,$ and λ have a simple solution

$$\lambda = \frac{-c}{a^2 + m^2b^2} \, , \quad y_1 = -\lambda a^2 \, , \quad y_2 = -\lambda mb^2 \, ,$$

and the min value of L is

$$L_{\min} = \frac{c^2}{2(a^2 + m^2b^2)} \, .$$

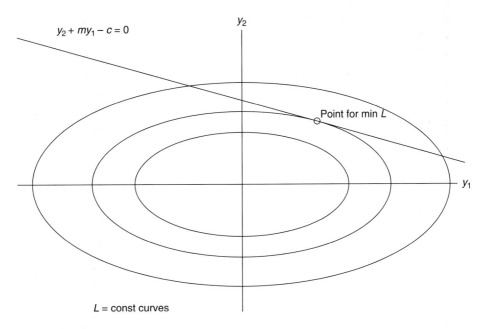

$y_2 + my_1 - c = 0$

y_2

Point for min L

y_1

L = const curves

FIGURE 1.6   Example of Minimization with Respect to Two Parameters and One Constraint

Note that

$$\lambda = - \left(L_c^{min} \right)_f , \quad y_1 + my_2 - c = 0 .$$

Example 1.2.2—A Problem with Three Parameters and Two Constraints; Max Climb Rate of an Aircraft (A/C)

The net force on an aircraft maintaining a steady rate of climb must be zero. If we choose force components parallel and perpendicular to the flight path (see Figure 1.7), this yields the two constraint equations

$$f^{(1)}(V, \gamma, \alpha) = T(V) \cos(\alpha + \epsilon) - D(V, \alpha) - mg \sin \gamma = 0 ,$$
$$f^{(2)}(V, \gamma, \alpha) = T(V) \sin(\alpha + \epsilon) + L(V, \alpha) - mg \cos \gamma = 0 ,$$

where V = velocity, γ = flight path angle, α = angle-of-attack, m = mass of aircraft, g = gravitational force per unit mass, ϵ = angle between thrust axis and zero-lift axis (a given constant), L = aerodynamic lift, D = aerodynamic drag, and T = engine thrust.

The performance index is rate of climb

$$J = V \sin \gamma ,$$

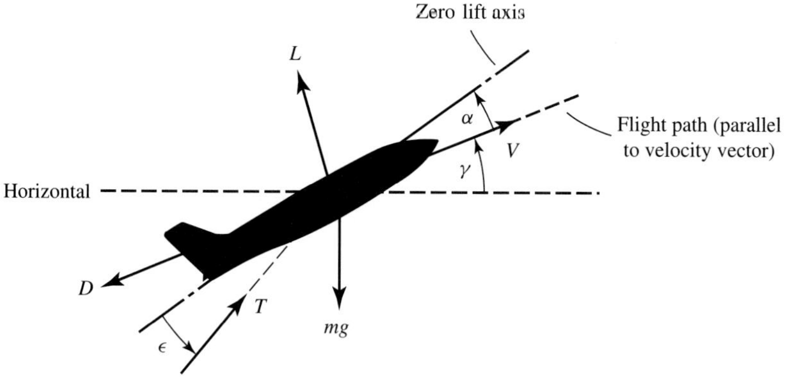

FIGURE 1.7   Force Equilibrium for an Aircraft in Steady Climb

The three parameters are $y = [\alpha, V, \gamma]^T$; the H function is

$$H = V \sin\gamma + \lambda_1[T\cos(\alpha + \epsilon) - D - mg\sin\gamma] + \lambda_2[T\sin(\alpha + \epsilon) + L - mg\cos\gamma] ,$$

and necessary conditions for a stationary value of climb rate are $f^{(1)} = f^{(2)} = 0$ and

$$H_V = \sin\gamma + \lambda_1[T_V\cos(\alpha + \epsilon) - D_V] + \lambda_2[T_V\sin(\alpha + \epsilon) + L_V] = 0 ,$$

$$H_\gamma = V\cos\gamma - \lambda_1 mg\cos\gamma + \lambda_2 mg\sin\gamma = 0 ,$$

$$H_\alpha = \lambda_1[-T\sin(\alpha + \epsilon) - D_\alpha] + \lambda_2[T\cos(\alpha + \epsilon) + L_\alpha] = 0 .$$

These five equations determine the five unknowns V, γ, α, λ_1, λ_2. For realistic lift, drag, and thrust functions these equations must be solved numerically (Sections 1.3 and 1.5 present numerical algorithms.) Data for a specific aircraft are given in Problem 1.3.21.

Problems

1.2.1 Rectangle of Max Perimeter Inscribed in an Ellipse:
(a) Find x, y to max $P = 4(x + y)$ subject to the constraint

$$\frac{x^2}{a^2} + \frac{y^2}{b^2} = 1 .$$

(b) Find x, y for the case $a = 1, b = 2$.

1.2.2 Rectangular Parallelepiped of Max Volume Contained in an Ellipsoid:
(a) Find x, y, z to max $V = 8xyz$, subject to the constraint

$$\frac{x^2}{a^2} + \frac{y^2}{b^2} + \frac{z^2}{c^2} = 1 .$$

(b) Find x, y, z for the case $a = 1, b = 2, c = 3$.

1.2.3 Pseudo-Inverse Solution to an Indeterminate Set of Linear Equations: Consider an indeterminate set of linear equations $Ax = b$ where A is a rectangular matrix with fewer rows than columns (i.e., there are fewer equations than elements of x).

(a) Show that the min length vector x that satisfies $Ax = b$ is

$$x = A^T (AA^T)^{-1} b \ .$$

The matrix $A^T (AA^T)^{-1}$ is called the *pseudo-inverse* of A.

(b) Find x using (a) for the case

$$A = \begin{bmatrix} 1 & 2 & 3 \\ 1 & -1 & 2 \end{bmatrix}, \quad b = \begin{bmatrix} 10 \\ 10 \end{bmatrix}.$$

1.2.4 Min Distance between Two Lines in 3D: A straight line in a three-dimensional space is the intersection of two planes, which can be represented in vector matrix notation as $Ax = b$ where x is a 3×1 vector of cartesian coordinates, A is a 2×3 matrix, and b is a 2×1 vector. Consider *two lines* $A_1 x^{(1)} = b_1$, $A_2 x^{(2)} = b_2$.

(a) Using parameter optimization with Lagrange multipliers, show that the points on the two lines that are closest to each other are

$$x^{(1)} = D_1^{-1}(d_1 + d_2 - C_1 d_2) , \quad x^{(2)} = D_2^{-1}(d_1 + d_2 - C_2 d_1) ,$$

where

$$C_1 = A_1^T (A_1 A_1^T)^{-1} A_1 , \quad C_2 = A_2^T (A_2 A_2^T)^{-1} A_2 , \quad d_1 = A_1^T (A_1 A_1^T)^{-1} b_1 ,$$

$$d_2 = A_2^T (A_2 A_2^T)^{-1} b_2 , \quad D_1 = C_1 + C_2 - C_1 C_2 , \quad D_2 = C_1 + C_2 - C_2 C_1 .$$

(b) Give a geometrical interpretation of the Lagrange multipliers.

(c) Find $x^{(1)}$, $x^{(2)}$ for the case

$$A_1 = \begin{bmatrix} 1 & 2 & 3 \\ 1 & -1 & 2 \end{bmatrix} , \quad b_1 = \begin{bmatrix} 10 \\ 10 \end{bmatrix} ,$$

$$A_2 = \begin{bmatrix} 3 & 2 & 1 \\ 2 & -1 & 1 \end{bmatrix} , \quad b_2 = \begin{bmatrix} 3 \\ 4 \end{bmatrix} .$$

1.2.5 Min Time Path Through a Region with Two Layers of Constant Velocity Magnitude: A lifeguard can run at velocity v_1 and can swim at velocity v_2. Let $y > y_1$ be the water and $y < y_1$ be the land; he is located at $x = 0$, $y = 0$. We wish to find the min-time path from his location to a point (x_2, y_2) where $y_2 > y_1$. We assume that the path consists of two straight lines, changing direction at the water's edge (see Figure 1.8), so the problem may be stated as

$$\min_{\theta_1, \theta_2} L = \frac{y_1 \sec \theta_1}{v_1} + \frac{(y_2 - y_1) \sec \theta_2}{v_2}$$

subject to

$$f = x_2 - y_1 \tan \theta_1 - (y_2 - y_1) \tan \theta_2 = 0 \ .$$

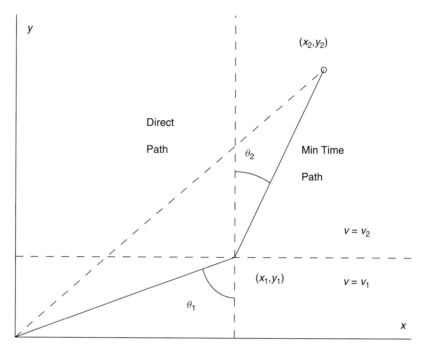

FIGURE 1.8   Min Time Path Through a Region with Two Layers of Constant Velocity Magnitude

(a) Use a Lagrange multiplier to show that

$$\frac{\sin \theta_1}{v_1} = \frac{\sin \theta_2}{v_2},$$

which is Snell's law.

(b) Snell's law and the constraint $f = 0$ are two nonlinear equations in the two unknowns θ_1, θ_2. For $x_2 = 300$, $y_2 = 300$, $y_1 = 100$, $v_1 = 25$, $v_2 = 6$ (units ft, sec) verify the min-time path shown in Figure 1.8 and the min time of 45.1 sec. Verify that the time for the direct (straight-line) path is 52.8 sec.

1.2.6 General Quadratic Performance Index with Linear Equality Constraints: Consider the problem of finding the parameter vectors x and u to minimize

$$L = \frac{1}{2}(x^T Q x + u^T R u)$$

subject to the constraints

$$x + Gu - c = 0 \; .$$

(a) Adjoin the constraints with the Lagrange multiplier vector λ^T and show that

$$u = Kc \; , \quad L^{\min} = \frac{1}{2} c^T Sc \; , \quad -\lambda = Sc \equiv \left[\frac{\partial L^{\min}}{\partial c} \right]^T \; ,$$

where

$$K = (R + G^T QG)^{-1} G^T Q \equiv R^{-1} G^T S \text{ if } R^{-1} \text{ exists} \; ,$$

$$S = Q - QG(R + G^T QG)^{-1} G^T Q \equiv (Q^{-1} + GR^{-1}G^T)^{-1} \text{ if } Q^{-1}, \ R^{-1} \text{ exist} \; .$$

The equivalence of the two expressions for K and S are a special case of the *matrix inversion lemma*. We shall use them frequently in the remainder of the book.

(b) Use (a) to find the solution for the case $Q = \mathrm{diag}[1 \ 2 \ 3]$, $R = \mathrm{diag}[4 \ 5]$ and

$$G = \begin{bmatrix} 3 & 2 \\ 1 & -1 \\ -4 & -3 \end{bmatrix}, \quad c = \begin{bmatrix} -2 \\ 3 \\ -1 \end{bmatrix}.$$

1.2.7 Optimal Performance of a Sailboat: An approximate model of a sailboat moving at constant velocity V relative to the water is shown in Figure 1.9.

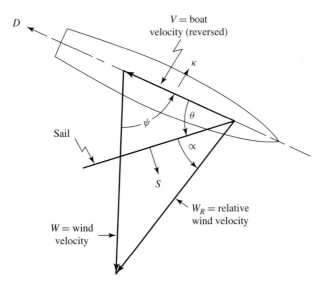

FIGURE 1.9   Force Equilibrium for a Sailboat Moving at Constant Velocity

The boat is moving at an angle ψ relative to the wind, which has velocity W relative to the water. The sail is set at an angle θ to the centerline of the boat, and the aerodynamic force on the sail is approximated as acting normal to the sail with magnitude

$$S = c_1 W_r^2 \sin \alpha ,$$

where c_1 is a constant, W_r = the magnitude of the relative wind, and α = the angle-of-attack of the sail (see Figure 1.9).

The hydrodynamic force on the keel is resolved into components parallel and perpendicular to the centerline, namely K and D, where D is given approximately by

$$D = c_2 V^2 ,$$

where c_2 is a constant.

In equilibrium (at constant V) the force parallel to the boat centerline must vanish, which implies that

$$D = S \sin \theta .$$

(a) Using the law of sines and the law of cosines in the wind triangle, show that

$$W_R \sin(\alpha + \theta) = W \sin \psi , \quad W_R^2 = V^2 + W^2 - 2VW \cos \psi .$$

These two relations and the equilibrium of forces parallel to the centerline in the form

$$V^2 = \mu^2 W_R^2 \sin \alpha \sin \theta ,$$

where $\mu^2 \overset{\Delta}{=} c_1/c_2$, determine (V, W_R, α) given (θ, ψ).

(b) Show that max velocity for given ψ is obtained when $\alpha = \theta$ and $\xi = \cos \theta$ is a solution of the quartic equation

$$4\xi^4 - (2\mu \sin 2\psi)\xi^3 - (4 - \mu^2 \sin^2 \psi)\xi^2 + (2\mu \sin 2\psi)\xi + (1 - \mu^2) \sin^2 \psi = 0 .$$

(c) Show that the max velocity over all possible ψ is obtained when

$$\alpha = \theta , \quad \tan \psi = -\frac{\cos \theta}{2\mu \sin^4 \theta} ,$$

which implies the following cubic equation for $\sin \theta$:

$$2\mu \sin^3 \theta + \mu \sin \theta - 1 = 0 .$$

(d) Show that the maximum upwind velocity, $-V \cos \psi$, is obtained when

$$\alpha = \theta , \quad \tan \psi = -\frac{\cos \theta}{(\cos 2\theta + \mu^2 \sin^4 \theta)^{1/2}} ,$$

which implies the following quartic equation for $\sin \theta$:

$$2\mu^2 \sin^4 \theta + (\mu^2 - 6) \sin^2 \theta - 2\mu \sin \theta + 1 = 0.$$

(e) Verify the polar plot of V/W vs. ψ shown in Figure 1.10 for the case $\mu = 1$. Max $V/W = .593$ and occurs at $\psi = 73.3$ deg, with $W_R/W = 1.006$, $\alpha = \theta = 36.1$ deg, while max $-V \cos \psi = .2610$ and occurs at $\psi = 133.7$ deg with $W_R/W = 1.291$, $\alpha = \theta = 17.0$ deg.

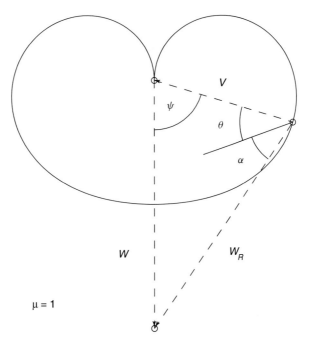

FIGURE 1.10   Max Velocity of a Sailboat
vs. Heading—Note $\theta = \alpha$

1.2.8 Min Glide Angle and Min Descent Rate: Consider a glider in a steady descent; the equations of equilibrium are those in Example 1.2.2 with thrust $T = 0$ and γ replaced by $-\gamma$. Approximate expressions for lift and drag are

$$L = C_{L\alpha}\alpha q S , \quad D = (C_{D0} + \eta C_{L\alpha}\alpha^2)q S ,$$

where α = angle-of-attack, $C_{L\alpha} = dC_L/d\alpha$ = lift coefficient curve slope, C_{D0} = zero-lift drag coefficient, $q = \rho V^2/2$ = dynamic pressure, ρ = air density, S = reference area (usually wing planform area), and η = efficiency factor ($0 < \eta < 1$).

(a) Show that the equations of equilibrium above may be written as

$$\eta V^2(\alpha^2 + \alpha_m^2) = \sin \gamma , \quad V^2\alpha = \cos \gamma ,$$

where V^2 is in units of $g\ell$, $\ell \triangleq 2m/\rho SC_{L\alpha}$ = characteristic length, and α_m is defined as $\sqrt{C_{D0}/\eta C_{L\alpha}}$.

(b) Using the results of (a), show that

$$\gamma_{\min} = \tan^{-1}(2\eta\alpha_m) \, ,$$

which is obtained when $\alpha = \alpha_m$.

(c) Show that the min value of $\dot{z} = V \sin \gamma$ is obtained when

$$\alpha = \alpha_m \sqrt{\zeta^2 - 1 - \zeta\sqrt{\zeta^2 - 8}} \, ,$$

where $\zeta \overset{\Delta}{=} 1/(2\eta\alpha_m)$.

(d) For $\alpha_m = 1/12$, $\eta = 1/2$, find the min glide angle and corresponding V, α, and descent rate.

Answer: $\gamma = .0831$, $V = 3.458$, $\alpha = .0833$, $V \sin \gamma = .2872$.

(e) For $\alpha_m = 1/12$, $\eta = 1/2$, find the min descent rate and corresponding V, α, γ.

Answer: $\gamma = .0964$, $V = 2.614$, $\alpha = .1457$, $V \sin \gamma = .2515$. Note this condition is 'on the back side' of the lift curve ($\alpha > \alpha_m$) and occurs at a much slower velocity than the condition for min γ.

1.2.9 Min Gliding Turn Radius and Max Gliding Turn Rate: In a steady gliding turn, the equations of equilibrium are (see Figure 1.11)

$$D = mg \sin \gamma \, , \quad L \cos \sigma = mg \cos \gamma \, , \quad L \sin \sigma = \frac{m(V \cos \gamma)^2}{r} \, ,$$

where γ = flight-path-angle below the horizontal $\equiv$ helix angle of flight path, σ = bank angle of glider, r = turn radius $\equiv$ radius of the helix, and the other quantities were defined in Problem 1.2.8.

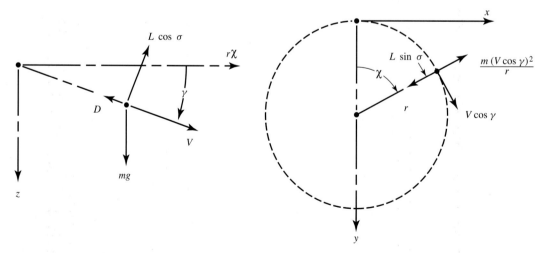

FIGURE 1.11   Aircraft in a Steady Gliding Turn; On Left—Helix Unwrapped; On Right—Top View of Helix

(a) Using the normalizations of Problem 1.2.8, show that the three relations may be written as

$$\eta V^2(\alpha^2 + \alpha_m^2) = \sin\gamma \ , \quad V^2\alpha\cos\sigma = \cos\gamma \ , \quad r = \frac{\cos^2\gamma}{\alpha\sin\sigma} \ .$$

where r is in units of ℓ.

(b) Using the results of (a), show that min r, given γ, is

$$r_{\min} = \frac{\cos^2\gamma}{\alpha_m\sqrt{\mu^2 - 1}} \ .$$

which is obtained when

$$\sec 2\sigma = 2\mu^2 - 1 \ , \quad \alpha = \alpha_m\sqrt{2\mu^2 - 1} \ ,$$

where

$$\mu \overset{\triangle}{=} \frac{\tan\gamma}{2\eta\alpha_m} \equiv \frac{\tan\gamma}{\tan\gamma_{\min}} \ .$$

Note that μ must be ≥ 1 which implies $\alpha \geq \alpha_m$, $\tan\gamma \geq 2\eta\alpha_m$. Note also that $\sigma \leq 45$ deg.

(c) Turn rate is given by $\dot\chi = V\cos\gamma/r \equiv \tan\sigma/V$, where time is in units of $\sqrt{\ell/g}$. Show that the maximum turn rate, given γ, is

$$\dot\chi_{\max} = \sqrt{\frac{2\alpha_m(\mu - 1)}{\cos\gamma}} \ .$$

which is obtained when $\sec\sigma = \sqrt{2\mu - 1}$, $\alpha = \alpha_m\sqrt{2\mu - 1}$, where μ was defined in (b). Again μ must be ≥ 1, which implies $\alpha \geq \alpha_m$, $\tan\gamma \geq 2\eta\alpha_m$, but in this case, $\sigma \leq 90$ deg.

(d) For $\alpha_m = 1/12$, $\eta = 1/2$, plot $1/r_{\min}$, V, α, σ vs. γ.

(e) For $\alpha_m = 1/12$, $\eta = 1/2$, plot $\dot\chi_{\max}$, V, α, σ vs. γ.

1.2.10 Reachable Ground Area for a Glider: A glider moving along the line $y = 0$ is at altitude z and at $x = 0$. We wish to find the largest area on the ground ($z = 0$) that the glider can reach from this position using constant α and σ. This area can be found by considering min radius helical paths for several glide angles, which were found in Problem 1.2.9. From Figure 1.11 the final values of (x, y) are

$$x = r\sin\chi \ , \quad y = r(1 - \cos\chi) \ ,$$

where the initial altitude above the ground is $z = r\chi\tan\gamma$. Using this knowledge, verify the max reachable ground area (using constant α and σ) for a space shuttle orbiter with $\alpha_m = 1/3$, $\eta = 1/2$, from initial altitudes $z/\ell = 2, 4, 6, 8, 10, 12$ shown in Figure 1.12. Typical values of ℓ for this type of vehicle are from 500 to 1000 ft.

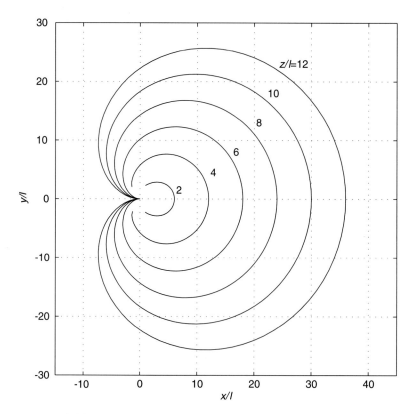

FIGURE 1.12   Reachable Landing Area from Various Altitudes for a Space Shuttle Orbiter with $\alpha_m = 1/3$, $\eta = 1/2$

1.2.11 Max Climb Angle and Max Climb Rate:

(a) The equations of equilibrium are those in Example 1.2.2. Using (L, D) from Problem 1.2.8, show that these relations may be written as

$$T\cos(\alpha + \epsilon) - \eta V^2(\alpha^2 + \alpha_m^2) = \sin\gamma \ , \quad T\sin(\alpha + \epsilon) + V^2\alpha = \cos\gamma \ ,$$

where T is in units of mg, V is in units of $\sqrt{g\ell}$, $\ell \stackrel{\Delta}{=} 2m/C_{L\alpha}\rho S$, and $\alpha_m \stackrel{\Delta}{=} \sqrt{C_{D0}/\eta C_{L\alpha}}$ = angle of attack for maximum L/D.

(b) For T and ϵ constant, show that max γ occurs when

$$\eta V^2(\alpha^2 - \alpha_m^2) + \begin{vmatrix} T\sin(\alpha + \epsilon) & \eta(\alpha^2 + \alpha_m^2) \\ T\cos(\alpha + \epsilon) & \alpha \end{vmatrix} = 0 \ .$$

(c) Climb rate is given by

$$\dot{h} = V\sin\gamma \ .$$

For T and ϵ constant, show that max $\dot{h}$ occurs when

$$\begin{vmatrix} V\cos\gamma & 0 & \sin\gamma \\ \cos\gamma & T\sin(\alpha + \epsilon) + 2\eta V^2\alpha & 2V\eta(\alpha^2 + \alpha_m^2) \\ \sin\gamma & T\cos(\alpha + \epsilon) + V^2 & 2V\alpha \end{vmatrix} = 0 \ .$$

(d) For $\alpha_m = 1/12$, $\eta = .5$, $\epsilon = 2\,\text{deg}$, verify the solid curves of $\gamma_{\max}$ and the corresponding values of α and V vs. T shown in Figure 1.13. (*Hint*: Use FSOLVE in MATLAB.)

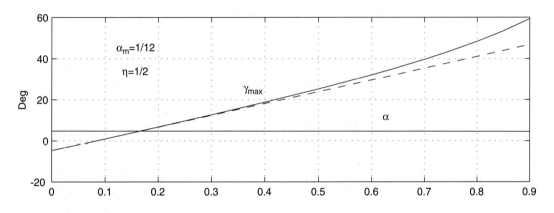

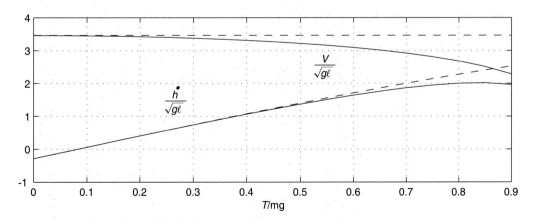

FIGURE 1.13 Max Climb Angle vs. T/mg for $\alpha_m = 1/12$, $\eta = 1/2$; Dashed Curves Calculated Assuming $T/mg \ll 1$

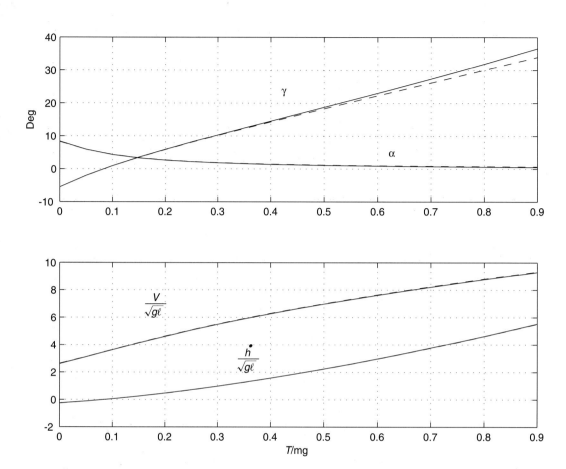

FIGURE 1.14　Max Climb Rate vs. T/mg for $\alpha_m = 1/12$, $\eta = 1/2$; Dashed Curves Calculated Assuming $T/mg \ll 1$

(e) For $\alpha_m = 1/12$, $\eta = .5$, $\epsilon = 2$ deg, verify the solid curves of maximum climb rate and the corresponding values of α and V vs. T shown in Figure 1.14. (*Hint*: Use FSOLVE in MATLAB.)

1.2.12 Max Climb Angle and Max Climb Rate for $T/mg \ll 1$:　In Problem 1.2.11, if $|T|$, $|\alpha|$, $|\epsilon|$, $|\gamma|$ are all small compared to one, the term $T\sin(\alpha + \epsilon)$ is negligible compared to $V^2\alpha$ and $T\cos(\alpha + \epsilon) \approx T$, $\sin\gamma \approx \gamma$, $\cos\gamma \approx 1$. The normalized equations of equilibrium then become

$$T - \eta V^2(\alpha^2 + \alpha_m^2) \approx \gamma \ , \quad V^2\alpha \approx 1; \ ,$$

where T is in units of mg, V is in units of $\sqrt{g\ell}$, $\ell \triangleq 2m/C_{L\alpha}\rho S$, and $\alpha_m \triangleq \sqrt{C_{D0}/\eta C_{L\alpha}} =$ angle of attack for maximum L/D.

(a) For this approximation, show that the maximum climb angle is

$$\gamma_{\max} \approx T - 2\eta\alpha_m \ ,$$

and the corresponding values of α and V are $\alpha = \alpha_m$, $V = 1/\sqrt{\alpha_m}$.

(b) For this approximation, show that the maximum climb rate ($\dot{h} \approx V \cdot \gamma$) is obtained when

$$\alpha \approx \sqrt{3\alpha_m^2 + \frac{T^2}{(2\eta)^2}} - \frac{T}{2\eta} \ .$$

(c) For $\alpha_m = 1/12$, $\eta = 1/2$, verify the dashed curves of $\gamma_{\max}$ and the corresponding values of α_m, V, and $\dot{h}$ vs. T shown in Figure 1.13.

(d) For $\alpha_m = 1/12$, $\eta = 1/2$, verify the dashed curves of $\dot{h}_{\max}$ and the corresponding values of α_m, γ, and V vs. T shown in Figure 1.14.

1.2.13 Min Climbing Turn Radius and Max Climbing Turn Rate: An A/C in a steady climbing turn (see Figure 1.15) is in equilibrium when

$$T \cos(\alpha + \epsilon) - D = mg \sin\gamma \ , \quad [T \sin(\alpha + \epsilon) + L]\cos\sigma = mg \cos\gamma \ ,$$

$$[T \sin(\alpha + \epsilon) + L]\sin\sigma = \frac{m(V \cos\gamma)^2}{r} \ , \quad r\dot{\chi} = V \cos\gamma \ ,$$

where γ = flight path climb angle = helix angle of flight path, $\dot{\chi}$ = turn rate of the velocity vector, r = turn radius, α = angle-of-attack, σ = bank angle, T = thrust, ϵ = thrust alignment angle, V = velocity, m = aircraft mass, and g = gravitational force per unit mass.

(a) Using (L, D) from Problem 1.2.8, show that these relations may be written as

$$T \cos(\alpha + \epsilon) - \eta V^2(\alpha^2 + \alpha_m^2) = \sin\gamma \ , \quad \left[T \sin(\alpha + \epsilon) + V^2\alpha\right]\cos\sigma = \cos\gamma \ ,$$

$$\frac{1}{r} = \frac{\tan\sigma}{V^2 \cos\gamma} \ , \quad \dot{\chi} = \frac{\tan\sigma}{V} \ ,$$

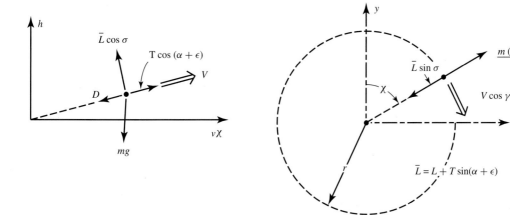

FIGURE 1.15 Aircraft in a Steady Climbing Turn; On Left—Helix Unwrapped; On Right—Top View of the Helix

where $\alpha_m \triangleq \sqrt{C_{D0}/\eta C_{L\alpha}}$, T is in units of mg, r in $\ell \triangleq 2m/C_{L\alpha}\rho S$, V in $\sqrt{g\ell}$, and time in $\sqrt{\ell/g}$. Note that the first two equations determine α and V given σ, T, γ, which then determine r and $\dot\chi$.

(b) For specified T, γ show that a necessary condition for min climbing turn radius r is

$$\cos^2\sigma = \eta\left(\alpha + \frac{\alpha_m^2}{\alpha}\right)\frac{T\cos(\alpha+\epsilon)+V^2}{T\sin(\alpha+\epsilon)+2\eta V^2\alpha}\,.$$

(c) For specified T, γ show that a necessary condition for max climbing turn rate $\dot\chi$ is

$$1 + \cos^2\sigma = 2\eta\left(\alpha + \frac{\alpha_m^2}{\alpha}\right)\frac{T\cos(\alpha+\epsilon)+V^2}{T\sin(\alpha+\epsilon)+2\eta V^2\alpha}\,.$$

(d) For $T = .2$, $\alpha_m = 1/12$, $\eta = .5$ verify the solid curves in the plots of $1/r_{min}$, α, σ, V vs. γ shown in Figure 1.16.

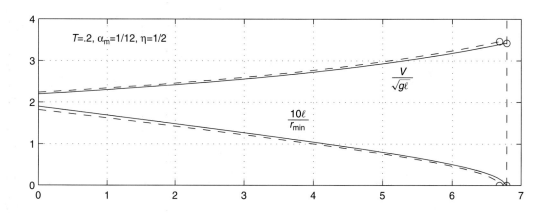

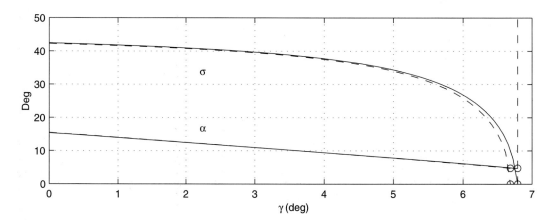

FIGURE 1.16   Min Climbing Turn Radius vs. γ for $T/mg = .2$, $\alpha_m = 1/12$, $\eta = 1/2$; Dashed Curves Calculated Assuming $T/mg \ll 1$

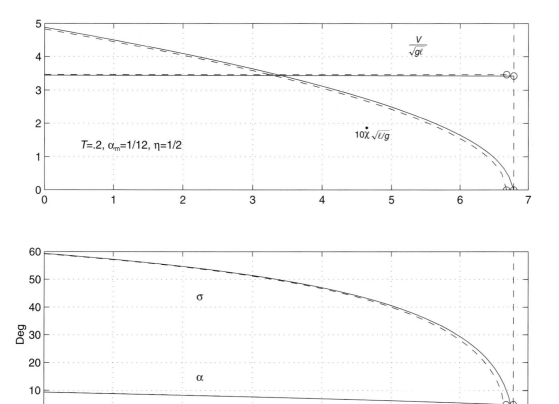

FIGURE 1.17 Max Climbing Turn Rate vs. γ for $T/mg = .2$, $\alpha_m = 1/12$, $\eta = 1/2$; Dashed Curves Calculated Assuming $T/mg \ll 1$

(e) For $T = .2$, $\alpha_m = 1/12$, $\eta = .5$ verify the solid curves in the plots of $\dot{\chi}_{max}$, α, σ, V vs. γ shown in Figure 1.17.

1.2.14 Min Climbing Turn Radius and Max Climbing Turn Rate for $T/mg \ll 1$: If α, ϵ, γ, and T are all $\ll 1$ then $T\sin(\alpha + \epsilon$ is negligible compared to $V^2\alpha$ and the normalized equations may be written as

$$T \approx \gamma + \frac{\eta(\alpha + \alpha_m^2/\alpha)}{\cos\sigma} \quad , \quad \frac{1}{r} \approx \alpha\sin\sigma, \quad \dot{\chi} = \sin\sigma\sqrt{\alpha/\cos\sigma} \ ,$$

where $V^2 \approx 1/(\alpha\cos\sigma)$.

(a) For T and γ specified, show that

$$\frac{1}{r_{min}} = \alpha_m\sqrt{2\mu^2 - 1} \ ,$$

which is obtained when $\sec 2\sigma = 2\mu^2 - 1$, $\alpha = \alpha_m\sqrt{2\mu^2 - 1}$, where $\mu \overset{\Delta}{=}$ $T - \gamma/(2\eta\alpha_m)$.

(b) For T and γ specified, show that

$$\dot{\chi}_{max} = \sqrt{2\alpha_m(\mu - 1)},$$

which is obtained when $\sec\sigma = \sqrt{2\mu - 1}$, $\alpha = \alpha_m\sqrt{2\mu - 1}$, and μ was defined in (a).

(c) For $T = .2$, $\alpha_m = 1/12$, $\eta = .5$ verify the dashed curves in the plots of $1/r_{min}$, α, σ, V vs. γ shown in Figure 1.16.

(d) For $T = .2$, $\alpha_m = 1/12$, $\eta = .5$ verify the dashed curves in the plots of $\dot{\chi}_{max}$, α, σ, V vs. γ shown in Figure 1.17.

1.2.15 Min Weight Cantilever Truss: A cantilever truss supports a load W (see Figure 1.18). If both members are made of the same material, the weight of the truss is proportional to the volume of material

$$J = L_1 A_1 + L_2 A_2,$$

where $(L_i, A_i) = $ (length, cross-sectional area) of the ith member.

From elementary mechanics, the loads in the members are $P_1 = W/\tan\theta$ (compression) and $P_2 = W/\sin\theta$ (tension). For min weight, the tension member will obviously be at full allowable stress σ, so that $A_2 = P_2/\sigma$. The stress in the compression member must be less than σ to avoid buckling; it depends on a slenderness parameter z proportional to the square of the member length divided by some characteristic cross-section dimension. If the compression member has a square cross-section of side b, we shall take $z = (6/\pi^2)(\sigma/E)(L/b)^2$ where $E = $ Young's modulus of the material and $L = $ length of the member. For slender members ($z > 1$) the allowable stress is the Euler buckling stress

$$\frac{P_1}{\sigma b^2} = \frac{1}{2z},$$

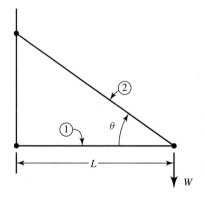

FIGURE 1.18   Cantilever Truss
Supporting a Weight W

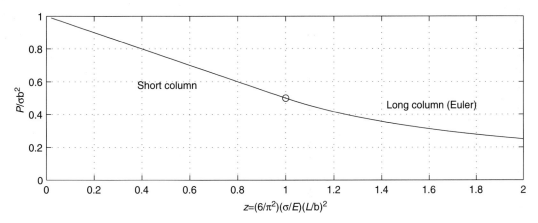

FIGURE 1.19   Allowable Axial Stress vs. Slenderness Parameter $z \triangleq (6/\pi^2)(\sigma/E)(L/b)^2$ for a Column of Length L and Square Cross-Section of Side b

For short members ($z < 1$) we shall take the allowable stress to be the empirical "short column" stress

$$\frac{P_1}{\sigma b^2} = 1 - \frac{z}{2} \ .$$

Figure 1.19 is a plot of this allowable stress vs. z.

It follows that the min weight truss will be attained by choosing θ and b to minimize

$$J = Lb^2 + \frac{LW}{\sigma \sin \theta \cos \theta} \ ,$$

subject to the column stress constraint of Figure 1.19.

If W is measured in units of $(6/\pi^2)L^2\sigma^2/E$ and J in units of $(6/\pi^2)L^3\sigma/E$, the problem is to find θ and z to minimize

$$J = \frac{1}{z} + \frac{W}{\sin \theta \cos \theta}$$

subject to

$$zW - h(z) \tan \theta = 0 \ ,$$

where

$$h(z) = 1 - z/2 \ , \quad z < 1 \quad \text{and} \quad h(z) = 1/(2z) \ , \quad z > 1 \ .$$

(a) For $z > 1$ show that the optimality condition may be written as

$$\tan \theta = \sqrt{1 + z} \ .$$

(b) For $z < 1$ show that

$$\tan \theta = \sqrt{2} = \text{constant} \ ,$$

and verify Figure 1.20.

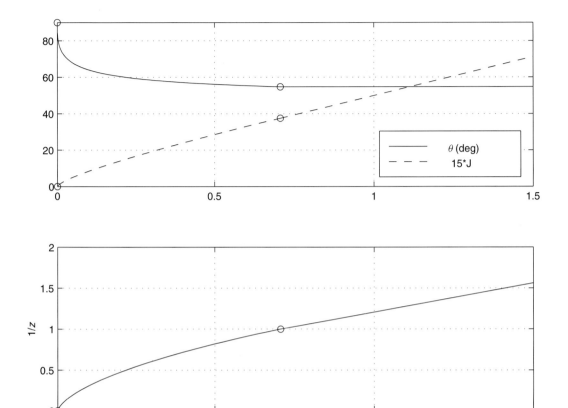

FIGURE 1.20   Min Normalized Volume J and Optimal (θ, z) vs. Normalized Weight W

1.2.16 Min Weight Inverted Cantilever Truss: Do Problem 1.2.15 with the truss inverted (flat side up) so that the tension and compression members are switched.

1.2.17 Min Weight Simple Truss: The truss shown in Figure 1.21 supports a load W at its center. If all members are made of the same material, the weight of the truss is proportional to the volume of material

$$J = \sum_i L_i A_i \ ,$$

where $(L_i, A_i) = $ (length, cross-sectional area) of the ith member.

Let $P_i = $ the load in member i. From elementary mechanics $P_1 = P_2 = W/(2\sin\theta)$ (compression), $P_3 = P_4 = W\cos\theta/(2\sin\theta)$ (tension), and $P_5 = W$ (tension).

For min weight, the tension members will obviously be at full allowable stress σ, so that $A_i = P_i/\sigma$ for $i = 3, 4, 5$. The stress in the compression members must be less than σ to avoid buckling. If the compression members have square cross-sections of side b, the axial stress is given by the relations in Problem 1.2.15 (replacing L by $L\sec\theta$).

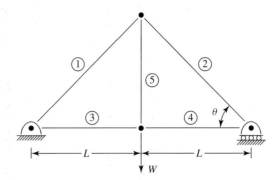

FIGURE 1.21 Simple Truss Supporting a Weight W

(a) Using the same definition of z and the same normalization of variables as in Problem 1.2.15, show that the problem may be stated as finding θ and z to minimize

$$J = W \left(\frac{1}{\tan \theta} + \tan \theta \right) + \frac{2}{z \cos \theta} \, ,$$

subject to the constraint

$$W = \frac{\sin \theta \cos^2 \theta}{z^2} \, , \quad z > \cos^2 \theta$$

$$= \sin \theta \left(\frac{2}{z} - \frac{1}{\cos^2 \theta} \right) \, , \quad z \le \cos^2 \theta \, .$$

(b) Find min J as a function W and make plots like those of Figure 1.20 of min J and the corresponding values of (θ, z) versus W.

(c) For a wooden truss ($E = 1{,}760{,}000$ lb/in², $\sigma = 1200$ lb/in²) with $L = 20$ ft, determine θ for the min weight truss to support a 1000-lb load, and find b for the compression and tension members.

1.2.18 Min Weight Inverted Simple Truss: Do Problem 1.2.17 with the truss inverted (flat side up) and the load in the center on the flat side so that the tension and compression members are switched.

1.2.19 Gradient Projection: Given a p-row vector a and the n by p-matrix B, $n < p$, we wish to find the n-column vector λ to minimize $J = (a - \lambda^T B)(a^T - B^T \lambda)$. If $J_{min} = 0$, the n rows of B form a *basis* for a. If $J_{min} \ne 0$, then $a - \lambda^T B$ is orthogonal to each row of B.

(a) Show that the solution is

$$\lambda^T = aB^* \Rightarrow a - \lambda^T B = a(I - B^* B) \, ,$$

where

$$B^* \overset{\Delta}{=} B^T(BB^T)^{-1} \overset{\Delta}{=} \text{generalized inverse of } B \; .$$

(b) Apply (a) to the case where $a = L_y$ and $B = -f_y$ to show that

$$\lambda^T = -L_y f_y^* \Rightarrow H_y \overset{\Delta}{=} L_y + \lambda^T f_y = L_y(I - f_y^* f_y) \; ,$$

so that $H_y \equiv (L_y)_f = $ the component of L_y that is tangential to $f = 0$ (see Figs. 1.4 and 1.5).

1.2.20 Max and Min Singular Values of a Matrix: Given a matrix A with fewer rows than columns, show that the maximum and min values of the magnitude of the vector $y = Ax$ with unit magnitude of the vector x are the square roots of the largest and smallest eigenvalues of $A^T A$. *Hint*: Maximize and minimize $y^T y$ with respect to x subject to the constraint $x^T x = 1$ using a Lagrange multiplier.

The square roots of the eigenvalues of $A^T A$ are called the *singular values* of the matrix A. MATLAB has a command SVD (for Singular Value Decomposition) that finds these singular values using a numerical method that does not require forming the matrix AA^T.

1.3 Numerical Solution with Gradient Methods

Unless the relations for $L(y)$ and $f(y)$ in Section 1.2 are quite simple, numerical methods must be used to determine the parameter vector y that minimizes $L(y)$ with $f(y) = 0$.

POP—A Gradient Algorithm for Parameter OPtimization with Equality Constraints

A straightforward numerical method is to guess an initial y and then take small steps Δy in the direction of $-H_y^T = $ the direction of *steepest descent*. This will lead to a local minimum if one exists. If there are several local minima, this method may not find the global minimum. Choosing the step size requires some experience. If the steps are too large, the minimum may be overshot; if the steps are too small, it takes many steps to reach the minimum.

We first state the algorithm and then give a brief derivation.

Enter data
- Guess y.
- Choose $k > 0$ for a minimum, $k < 0$ for a maximum.
- Choose η so that $|\Delta y|$ is not too large, $0 < \eta \le 1$.
- Choose stopping tolerance *tol*.

Find L, f, L_y, f_y, f_y^*, and H_y

$(*)L = L(y),\ f = f(y),\ L_y = L_y(y),\ f_y = f_y(y).$

- $f_y^* = f_y^T (f_y f_y^T)^{-1}.$
- $\lambda^T = -L_y f_y^*.$
- $H_y = L_y + \lambda^T f_y.$

Find improved value of y

- $dy = -\eta f_y^* f - k H_y^T.$
- If $\max(|f|, |dy|/\sqrt{p}) < tol$, then end.
- Replace y by $y + dy$ and go to $(*)$.

This algorithm may be interpreted as finding a small change Δy, to minimize a linear approximation to ΔL subject to a quadratic penalty on Δy and a linear approximation to coming closer to satisfying the constraints $f(y) = 0$:

$$\min_{\Delta y} \Delta J = L_y \Delta y + \frac{\Delta y^T \Delta y}{2k} \ , \tag{1.35}$$

subject to

$$f_y \Delta y = -\eta f \ , \tag{1.36}$$

where $L_y,\ f,\ f_y$ are evaluated at the current estimate of y. y *should be normalized so that a change of one unit of each element of y is approximately of equal significance.* $\eta = 1$ asks for the constraints to be satisfied in one step, so $\eta < 1$ asks only to move part way toward satisfying the constraints. k is a scalar step-size parameter; we shall say more about choosing k below.

To solve this linear-quadratic problem, we adjoin the constraint (1.36) to the performance index with a Lagrange multiplier vector $\bar\lambda$:

$$\Delta J \triangleq L_y \Delta y + \frac{\Delta y^T \Delta y}{2k} + \bar\lambda^T (\eta f + f_y \Delta y) \ , \tag{1.37}$$

Assuming that $L_y,\ f,\ f_y$ are fixed matrices, take a differential of ΔJ. For $d(\Delta J) = 0$ with arbitrary $d(\Delta y)$, it is clearly necessary that

$$L_y + \bar\lambda^T f_y + \frac{\Delta y^T}{k} = 0 \ , \tag{1.38}$$

from which

$$\Delta y^T = -k(L_y + \bar\lambda^T f_y) \ . \tag{1.39}$$

Substituting (1.39) into (1.36), we may solve for $\bar\lambda^T$:

$$\bar\lambda^T = \lambda^T + \frac{\eta}{k} f^T Q^{-1} \ , \tag{1.40}$$

where

$$Q = f_y f_y^T , \quad \lambda^T = -L_y f_y^* , \quad f_y^* \overset{\Delta}{=} f_y^T Q^{-1} . \tag{1.41}$$

f_y^* is the weighted *generalized inverse* of f_y.

Finally, substituting (1.40) and (1.41) into (1.39) gives

$$\Delta y = -\eta f_y^* f - k H_y^T , \tag{1.42}$$

where

$$H_y \overset{\Delta}{=} L_y + \lambda^T f_y . \tag{1.43}$$

When starting with a possibly poor guess, one can put $k = 0$ (i.e., ask for no improvement in L) and pick $\eta < 1$ to limit the size of $|dy|$ so that the linear approximation is reasonable. Over a few steps, gradually increase η to 1, and a feasible solution ($f \approx 0$) should be obtained, after which $\eta = 1$ can be used with $k \neq 0$ to minimize L.

k must be chosen by interpolation so that L decreases at a reasonable rate with each iteration. If $|k|$ is too large, each step will "overshoot" the minimum and L will actually increase; if $|k|$ is too small, many iterations will be required to reach the vicinity of the minimum.

Other gradient algorithms are available that do not require the user to interpolate k, and that converge more rapidly near the minimum (e.g., Refs. Gr, GMSW, and Ln). They use the successive values of H_y to approximate the Hessian H_{yy}; they are called *quasi-Newton* algorithms since they tend toward Newton-Raphson (NR) algorithms as the number of iterations increases. NR algorithms are second-order gradient algorithms (covered in the next section). These algorithms are necessarily more complicated than POP; obviously there is a trade-off between algorithm complexity and speed of convergence.

Table 1.1 lists a MATLAB function file that implements the POP algorithm.

TABLE 1.1   A MATLAB Code for POP

```
function [L,y,f]=pop(name,y,k,tol,eta,mxit)
% Parameter OPtimization using a generalized gradient algorithm; outputs L
% and y (p by 1) are optimum performance index & parameter vector; con-
% straints are f=0 where f is (n by 1) with n < p; user must supply a sub-
% routine 'name' that computes L,f,Ly,fy; input y is a guess; y should be
% normalized so that a change of one unit in each element of y has roughly
% the same significance; k is a scalar step size parameter; stopping cri-
% terion is max(fn,dyn)<tol, where fn=norm(f); dyn=norm(dy)/sqrt(p); eta=
% fraction of constraint violation to be removed; mxit=max no. iterations.
%
it=0; dyn=1; fn=1; p=length(y);
disp('    it        L       fn          dyn')
while max(fn,dyn) > tol
   [L,f,Ly,fy]=feval(name,y); fn=norm(f); fyi=fy'/(fy*fy'); lat=-Ly*fyi;
   Hy=Ly+lat*fy; dy=-eta*fyi*f-k*Hy'; dyn=norm(dy)/sqrt(p);
   disp([it L fn dyn]); y=y+dy; if it> mxit, break, end; it=it+1;
end
```

Example 1.3.1—Max Velocity of a Sailboat

This is a numerical solution of Problem 1.2.7. The subroutine containing L, f, L_y, f_y) is listed below.

```
function [L,f,Ly,fy]=slbt(y)
% Sailboat max velocity; L = V; y = [V,Wr,al,th,ps]',
%
V=y(1); Wr=y(2); al=y(3); th=y(4); ps=y(5); sa=sin(al); ca=cos(al);
st=sin(th); ct=cos(th); sp=sin(ps); cp=cos(ps); sat=sin(al+th);
cat=cos(al+th); L=V; Ly=[1 0 0 0 0];
f=[V^2-Wr^2*sa*st; Wr^2-V^2-1+2*V*cp; Wr*sat-sp];
fy=[2*V -2*Wr*sa*st -Wr^2*ca*st -Wr^2*sa*ct 0; -2*(V-cp) 2*Wr 0 0
    -2*V*sp; ... 0 sat Wr*cat Wr*cat -cp];
```

An edited diary is listed below that solves the problem using POP. It took 28 iterations to bring the norms of f and dy below .00005.

```
% Script e01_3_1.m for gradient solution of Ex. 1.3.1;
%
y=[.5 1 .5 .5 1.5]'; k=-1.2; tol=1e-5; eta=1; mxit=50;
[L,y,f]=pop('slbt',y,k,tol,eta,mxit);
   it      L        fn       dyn
    0     0.5000   0.2385   0.0733
    1     0.5793   0.0149   0.0263

    -       -        -        -
   27     0.5931   0.0000   0.0000
   28     0.5931   0.0000   0.0000
y'=[.5931 1.0057 .6307 .6307 1.2798]; f=1e-9*[.1183 .0627
      .0065]'
```

The MATLAB Code CONSTR

CONSTR is a quasi-Newton code in the MATLAB Optimization Toolbox (Ref. Gr) that minimizes a scalar performance index $f(y)$, where y is a parameter vector, with a vector of equality and/or inequality constraints $g(y) \leq 0$. It is easier to use than POP since the user does not have to guess the scalar step-size parameter k. Also it handles inequality constraints that POP does not (see Chapter 9). CONSTR uses a Sequential Quadratic Programming algorithm where the second-derivative matrices (the Hessians) are estimated after each iteration using the differences between the current gradients and the previous gradients (see POPN in Section 1.5, where the Hessians are entered explicitly).

The user must generate a MATLAB function file 'not name' that calculates $f(y)$ and $g(y)$ and must supply an initial guess for y. Generating a function file to calculate the gradients f_y and g_y is optional; if it is not provided, CONSTR calculates the gradients numerically by finite differencing; convergence is more rapid if gradients are supplied. The user must also enter the number of equality constraints, that must form the first part of the constraint vector g. The user should, in most cases, also supply lower and upper bounds on the values of the parameters, although this is not required. Other options include specifying the accuracy of (f, g, y), and specifying the number of function evaluations before stopping.

The syntax of the CONSTR command is y=constr('name', y, optn, vlb, vub, 'nameg', p), where output y is the optimal value of y, 'name' is the .m file that calculates f and g, input y is the initial guess, 'optn' is the vector of options (see example below), 'vlb' is a vector of lower bounds for y, 'vub' is a vector of upper bounds for y, 'nameg' is another .m file containing the analytical gradients of f and g (this is optional; if not used, replace 'nameg' by [ ]), and p is a vector of fixed parameters needed by the codes 'name' and 'nameg'.

Example 1.3.2

This is the same example solved above with the POP code (max velocity of a sailboat with $\mu = 1$). The MATLAB code for computing the performance index f and the constraints g is listed below, followed by a diary that performs the optimization using CONSTR.

```
function [f,g]=slbt_fg(y,mu)
% Sailboat max velocity; y = [V Wr al th ps]'.
%
V=y(1); Wr=y(2); al=y(3); th=y(4); ps=y(5); sa=sin(al); st=sin(th);
cp=cos(ps); sp=sin(ps); sat=sin(al+th);
%
f=-V; g(1)=V^2-mu*Wr^2*sa*st; g(2)=Wr^2-V^2-1+2*V*cp;
   g(3)=Wr*sat-sp;
```

```
% Diary 'slbt1.dia'
%
y=[.5 1 .5 .5 1.5]; optn(1)=1; optn(13)=3; optn(14)=200; mu=1;
y=constr('slbt_fg',y,optn,vlb,vub,[],mu)
```

f-COUNT	FUNCTION	MAX{g}	STEP	Procedures
6	-0.5	0.179263	1	
12	-0.579115	0.0110093	1	mod Hess
-	-	-	-	
60	-0.593087	1.9529e-008	1	mod Hess
61	-0.593087	1.00598e-008	1	mod Hess

```
Optimization Converged Successfully; Active Constraints:
     1     2      3
y = [0.5931    1.0057  0.6307  0.6307     1.2797]
```

Problems

1.3.1 to 1.3.18:   Solve Problems 1.2.1 to 1.2.18 numerically using either the POP or the CONSTR code. For analytical problems, assume some reasonable numerical data and compare the analytical answer with the numerical answer for those data.

1.3.19 A Quadratic-Quadratic Problem: Find $(x, \ y, \ z)$ to maximize

$$L = xy + 2xz + 3yz$$

subject to

$$f = x^2 + 2y^2 + 3z^2 - 1 = 0 \ .$$

Answer: $x = \pm.54098, \ y = \pm.39376, \ z = \pm.36389; \ L_{max} = 1.03659.$

1.3.20 Min Distance between Two Ellipses: Find the min distance between ellipse 1 and ellipse 2. Ellipse 1 is

$$\frac{\bar{x}^2}{2^2} + \frac{\bar{y}^2}{1} = 1 \ ,$$

where

$$\begin{bmatrix} \bar{x} \\ \bar{y} \end{bmatrix} = \begin{bmatrix} \cos\theta & \sin\theta \\ -\sin\theta & \cos\theta \end{bmatrix} \begin{bmatrix} x \\ y \end{bmatrix} \ , \ \theta = 30 \ \text{deg} \ .$$

Ellipse 2 is

$$\frac{x^2}{4^2} + \frac{y^2}{2^2} = 1 \ .$$

Answer: Min distance $= .592$ between $(1.223, \ 1.302)$ on ellipse 1 and $(1.336, \ 1.885)$ on ellipse 2.

1.3.21 Max Climb Angle and Max Rate of Climb for the 727 A/C: For a 727 A/C at take-off, the thrust, drag, and lift (T, D, L), in units of $W = $ A/C weight (180,000 lb) are given approximately by

$$T = A_0 + A_1 V + A_2 V^2 \ ,$$
$$D = C_D V^2, \ \ C_D = B_0 + B_1\alpha + B_2\alpha^2 \ ,$$
$$L = C_L V^2, \ \ C_L = C_0 + C_1\alpha + C_2(\alpha - \alpha_1)^2 \ ,$$

where $V = $ velocity in units of $\sqrt{g\ell}$, $g = $ gravitational force per unit mass, $\ell = 2W/(\rho g S)$ $= $ a characteristic length, $\rho = $ mass density of the air, $S = $ wing area (here 1560 ft^2). From curve fits to real data (Ref. MWM), $A_0 = .2476$, $A_1 = -.04312$, $A_2 = .008392$, $B_0 = .07351$, $B_1 = -.08617$, $B_2 = 1.996$, $C_0 = .1667$, $C_1 = 6.231$, $C_2 = 0$ if $\alpha < \alpha_1$, $C_2 = -21.65$ if $\alpha > \alpha_1$, $\alpha_1 = 12\pi/180$, $\epsilon = 2\pi/180$. Here we took $\rho = .002203$ slugs/ft^3 corresponding to sea level.

(a) Show that max L/D is 10.52 and occurs at $\alpha = 9.14$ deg.

(b) With reference to Example 1.2.2, show that the max climb angle, γ, is 6.99 deg and occurs at $V = 276$ ft/sec, $\alpha = 10.43$ deg $\Rightarrow \dot{h} = 33.6$ ft/sec. Note that this is on the "back side" of the L/D curve.

(c) Show that the max rate of climb, $\dot{h} = 37.6$ ft/sec and occurs at $V = 342$ ft/sec, $\alpha = 6.39$ deg, $\gamma = 6.31$ deg. Note that this is on the "front side" of the L/D curve.

1.3.22 Min Turn Radius and Max Turn Rate; 727 A/C Climb-out: The equations of equilibrium for a steady climbing turn were given in Problem 1.2.13.

(a) Using the data of Problem 1.3.21, show that the min turning radius for the 727 A/C at sea level, using full thrust T, with $\gamma = 0$, is 2059 ft and occurs with $\sigma = 55.2$ deg, $\alpha = 16.9$ deg, $V = 309$ ft/sec.

(b) Find the min turning radius at sea level for the 727 A/C, using full thrust T, with $\gamma = 2, 4, 6$ deg (note that $r \to \infty$ as $\gamma \to 6.99$ deg = the max climb angle with $\sigma = 0$).

(c) Show that the max turn rate at sea level for the 727 A/C, using full thrust T, with $\gamma = 0$, is 9.22 deg/sec and occurs with $\sigma = 60.7$ deg, $\alpha = 13.7$ deg, V = 356 ft/sec.

(d) Find the max turn rate at sea level for the 727 A/C, using full thrust T, with $\gamma = 2, 4, 6$ deg (note that $\dot{\chi} \to 0$ as $\gamma \to 6.99$ deg = the max climb angle with $\sigma = 0$).

1.3.23 F4 A/C Cruise Condition for Min Fuel Consumption:

(a) For the F4 A/C described in Example 4.4.3, use CONSTR to find the steady level-flight ($\gamma = 0$) condition for M fuel consumption per unit distance. Use $T = \eta T_{max}(V, h)$, where $0 \le \eta \le 1$. Assume constant specific fuel consumption, $\sigma = .29 \times 10^{-3}$ lb per sec of fuel per lb of thrust, so that fuel consumption per unit distance is given by

$$J = \frac{\sigma \eta T}{V} ,$$

where $T_{max}(V, h)$ is given Example 4.4.3. The constraint equations are

$$L - mg + T \sin(\alpha + \epsilon) = 0 , \quad D - T \cos(\alpha + \epsilon) = 0 ,$$

where $L = L(V, h, \alpha)$, $D = D(V, h, \alpha)$ are given in Example 4.4.3.
  Answer for $W = 38{,}000$ lb: $J_{min} = 6.62$ lb/mile at $V = 822$ ft/sec, $h = 32{,}250$ ft, $\alpha = .0744$ rad, $\eta = .268$.

(b) Find the steady level-flight ($\gamma = 0$) condition for min fuel consumption per unit time, where

$$J = \sigma \eta T ,$$

  Answer for $W = 38{,}000$ lb: $J_{min} = .997$ lb/sec at $V = 757$ ft/sec, $h = 32{,}250$ ft, $\alpha = .0881$ rad, $\eta = .272$.

1.4 Sufficient Conditions for a Minimum

A stationary point may not be a minimum, as demonstrated in the examples without constraints, Figs. 1.2 and 1.3.

To calculate the Hessian for a problem with equality constraints, we approximate the H function quadratically in the neighborhood of a test point (see Ref. Ln):

$$\Delta H \approx H_y \Delta y + H_\lambda \Delta \lambda + \frac{1}{2} \left[(\Delta y)^T \quad (\Delta \lambda)^T \right] \begin{bmatrix} H_{yy} & H_{y\lambda} \\ H_{\lambda y} & H_{\lambda\lambda} \end{bmatrix} \begin{bmatrix} \Delta y \\ \Delta \lambda \end{bmatrix} .$$

Now $H = L + \lambda^T f$, so $H_\lambda = f^T$, $H_{\lambda\lambda} = 0$, and $H_{\lambda y} = f_y$. Thus ΔH can be written as

$$\Delta H \approx H_y \Delta y + f^T \Delta\lambda + \frac{1}{2} \left[(\Delta y)^T \quad (\Delta\lambda)^T \right] \begin{bmatrix} H_{yy} & f_y^T \\ f_y & 0 \end{bmatrix} \begin{bmatrix} \Delta y \\ \Delta\lambda \end{bmatrix}. \tag{1.44}$$

Examining (1.44), we see that a sufficient condition for a local minimum at a stationary point $(H_y = 0, \ f = 0)$ is that

$$(\Delta y)^T H_{yy} \Delta y > 0 \ , \tag{1.45}$$

for all values of Δy such that

$$\Delta f \approx f_y \Delta y = 0 \ . \tag{1.46}$$

Note that investigating H to second order with $f = 0$ to first order is equivalent to investigating L to second order with $f = 0$ to second order (Ref. Ln).

From (1.45) $H_{yy} > 0$ is sufficient but not necessary for a local minimum. To obtain a less restrictive condition we investigate how ΔH varies on a hypersphere of radius $\Delta\sigma$ with the stationary point at its center, i.e., on

$$(\Delta y)^T \Delta y = (\Delta\sigma)^2 \ . \tag{1.47}$$

We adjoin the constraint (1.47) to (1.44) with a scalar Lagrange multiplier $\mu/2$

$$\Delta H \approx \frac{1}{2} \Delta y^T H_{yy} \Delta y + \Delta\lambda^T f_y \Delta y + \frac{\mu}{2} (\Delta\sigma^2 - \Delta y^T \Delta y) \ . \tag{1.48}$$

The differential change of ΔH caused by differential changes in Δy and $\Delta\lambda$ is then

$$d(\Delta H) \approx \left[\Delta y^T \quad \Delta\lambda^T \right] \left\{ \begin{bmatrix} H_{yy} & f_y^T \\ f_y & 0 \end{bmatrix} - \begin{bmatrix} \mu I & 0 \\ 0 & 0 \end{bmatrix} \right\} \begin{bmatrix} d(\Delta y) \\ d(\Delta\lambda) \end{bmatrix} . \tag{1.49}$$

If the min value of ΔH on the hypersphere is positive then the stationary point is a local minimum. The min points on the hypersphere are stationary points, and a necessary condition that $d(\Delta H) = 0$ for arbitrary $d(\Delta y)$ and $d(\Delta\lambda)$ is that

$$\begin{bmatrix} H_{yy} - \mu I & f_y^T \\ f_y & 0 \end{bmatrix} \begin{bmatrix} \Delta y \\ \Delta\lambda \end{bmatrix} = 0 \ . \tag{1.50}$$

This is a *generalized eigenvalue problem* for determining the eigenvectors Δy, $\Delta\lambda$ and the eigenvalues μ. For nontrivial solutions, the determinant of the matrix must vanish, which yields a polynomial equation in μ of order $m = p - n$. Now, provided Δy and $\Delta\lambda$ are eigenvectors of (1.50), it follows from (1.44) and (1.48), that

$$\Delta H_{\min} = \frac{1}{2} \Delta y^T (\mu \Delta y - f_y^T \Delta\lambda) \equiv \frac{\mu}{2} \Delta y^T \Delta y \equiv \frac{\mu}{2} (\Delta\sigma)^2 \ . \tag{1.51}$$

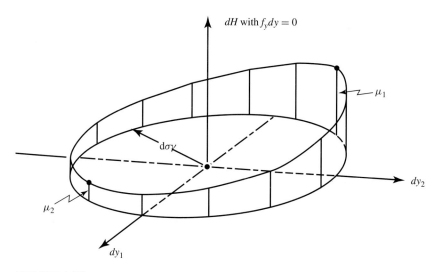

FIGURE 1.22   Sufficient Conditions for a Constrained Local Min in 2D—Ensuring that ΔH is Positive in an Infinitesimal Region Around a Stationary Point

Thus if all m of the eigenvalues μ_i of (1.50) are positive, ΔH must increase to second order in Δy in all directions away from the stationary point, so it is a local minimum of H holding $f = 0$. Figure 1.22 shows the case where y has only two components.

MATLAB has a command for finding the generalized eigenvalues and eigenvectors of $A - \mu B$ where B is singular, namely, 'eig(A,B)'. The A, B matrices corresponding to (1.50) are

$$A \triangleq \begin{bmatrix} H_{yy} & f_y^T \\ f_y & 0 \end{bmatrix}, \quad B \triangleq \begin{bmatrix} I & 0 \\ 0 & 0 \end{bmatrix}. \tag{1.52}$$

The output lists $n + p$ "eigenvalues," but there are only $m = p - n$ true eigenvalues; the others are listed as "NaN", i.e., 'not a number' (see the MATLAB file POPN in the next section that truncates out the NaN eigenvalues and corresponding eigenvectors).

Sufficient Conditions in Terms of a Decision Vector *u* and a Dependent Vector *x*

Sufficient conditions can also be expressed, in a less symmetric way, by decomposing y into a decision vector u and a dependent vector x. Equation (1.46) becomes

$$f_x \Delta x + f_u \Delta u = 0 , \tag{1.53}$$

and, since f_x is nonsingular, we may solve (1.53) for Δx

$$\Delta x = -(f_x)^{-1} f_u \Delta u . \tag{1.54}$$

Equation (1.45) may then be written as

$$\begin{bmatrix} (\Delta x)^T & (\Delta u)^T \end{bmatrix} \begin{bmatrix} H_{xx} & H_{xu} \\ H_{ux} & H_{uu} \end{bmatrix} \begin{bmatrix} \Delta x \\ \Delta u \end{bmatrix} > 0 . \tag{1.55}$$

Substituting (1.54) into (1.55), the sufficient condition becomes

$$(L_{uu})_f > 0 , \tag{1.56}$$

where

$$(L_{uu})_f = H_{uu} + H_{xu}^T (x_u)_f + (x_u)_f^T H_{xu} + (x_u)_f^T H_{xx} (x_u)_f , \tag{1.57}$$

and

$$(x_u)_f = -(f_x)^{-1} f_u . \tag{1.58}$$

Example 1.4.1—A Quadratic-Quadratic Problem

Find the parameters y_1 and y_2 that minimize

$$L = \frac{1}{2} \left(\frac{y_1^2}{a^2} + \frac{y_2^2}{b^2} \right)$$

subject to

$$f = c - y_1 y_2 = 0 ,$$

where (a, b, c) are positive constants.

The curves of constant L are ellipses with L increasing with the size of the ellipse, whereas $f = 0$ is a hyperbola with two branches (see Figure 1.23). Min L occurs when the ellipse is just tangent to the hyperbola.

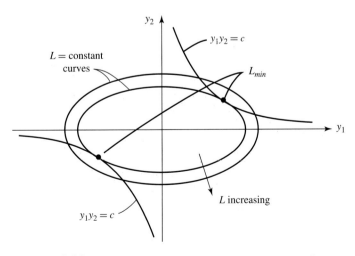

FIGURE 1.23   Example of a Minimization Subject to a Nonlinear Constraint

The H function is

$$H = \frac{1}{2}\left(\frac{y_1^2}{a^2} + \frac{y_2^2}{b^2}\right) + \lambda(c - y_1 y_2) .$$

Hence necessary conditions for a stationary point are $f = 0$ and

$$H_{y_1} = \frac{y_1}{a^2} - \lambda y_2 , \quad H_{y_2} = \frac{y_2}{b^2} - \lambda y_1 .$$

These equations have the simple solution

$$y_1 = \pm\sqrt{ac/b} , \quad y_2 = \pm\sqrt{bc/a}, \quad \lambda = 1/ab, \quad L_{min} = c/ab .$$

Using the *generalized eigenvalue approach*, we determine that

$$H_{yy} = \begin{bmatrix} 1/a^2 & -\lambda \\ -\lambda & 1/b^2 \end{bmatrix} , \quad f_y = \begin{bmatrix} -y_2 & -y_1 \end{bmatrix} .$$

Solving the determinant (1.50) gives the eigenvalue

$$\mu = \frac{y_1^2/a^2 + y_2^2/b^2 + 2c/ab}{y_1^2 + y_2^2} ,$$

which is positive, so both stationary points are minima.

Of course, it would have been simpler in this case to eliminate y_2 using $f = 0$, so that

$$L = \frac{1}{2}\left(\frac{y_1^2}{a^2} + \frac{c^2}{b^2 y_1^2}\right) ,$$

since this is a minimization problem without constraints. However, this type of simplification is not possible in more complicated problems.

Problems

1.4.1 Min and Max Distance from a Circle to an Ellipse: Given that $a > b > 1$, use a Lagrange multiplier and the second-derivative test to show that the min distance between the circle $x^2 + y^2 = 1$ and the ellipse $x^2/a^2 + y^2/b^2 = 1$ occurs with $(x, y) = (\pm 1, 0)$ on the circle and $(x, y) = (\pm a, 0)$ on the ellipse, while the max distance occurs with $(x, y) = (\mp 1, 0)$ on the circle and $(x, y) = (\pm a, 0)$ on the ellipse.

1.5 Numerical Solution with Newton-Raphson Methods

First-order gradient algorithms like those in Section 1.3, approximate the L and f functions locally with osculating hyperplanes. Newton-Raphson (NR) algorithms approximate the L and f functions locally with osculating *quadric surfaces,* i.e., they use the Hessian as well

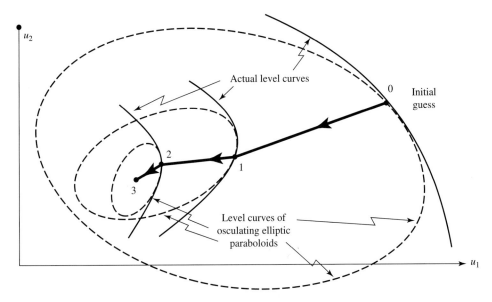

FIGURE 1.24   Typical Convergence of an NR Algorithm with Two Decision Parameters

as the gradient at the test point. Figure 1.24 is a sketch of the typical convergence of an NR algorithm with two decision parameters u_1 and u_2. Starting from an initial guess, each step fits an elliptic paraboloid to the local data and goes to the minimum of that surface (in this case, the center of the elliptical cross-section) for the next step.

Simple NR algorithms work only if the Hessian is positive definite (so that the quadratic hypersurface is a hyperelliptic paraboloid). This is a serious drawback, since, in many problems, it is difficult to find an initial guess where the Hessian is positive definite.

Another drawback is the extra time spent in determining and entering the second-derivative data into the algorithm, and the extra time necessary to compute the Hessian at each step.

However, the big advantage of NR algorithms over gradient algorithms is their rapid convergence to a precise solution *if the initial guess is such that the Hessian is positive-definite.*

Many algorithms are now available that are more efficient and more user friendly than the simple versions presented above and below, that combine the advantages of gradient and NR algorithms, e.g., NPSOL (Ref. GMSW) and the MATLAB Optimization Toolbox (Ref. Gr).

POPN—Parameter Optimization with Equality Constraints; a Newton-Raphson Algorithm

The NR derived here finds the parameter vector y to minimize or maximize $L(y)$ subject to $f(y) = 0$, where the dimension of f is less than the dimension of y.

Starting with an initial guess of y, a differential of (1.44) gives

$$d(\Delta H) \approx \left\{ \begin{bmatrix} H_y & f^T \end{bmatrix} + \begin{bmatrix} \Delta y^T & \Delta \lambda^T \end{bmatrix} \begin{bmatrix} H_{yy} & f_y^T \\ f_y & 0 \end{bmatrix} \right\} \begin{bmatrix} d(\Delta y) \\ d(\Delta \lambda) \end{bmatrix}. \tag{1.59}$$

At a minimum, $d(\Delta H) = 0$ for arbitrary $d(\Delta y)$ and $d(\Delta \lambda)$, which requires that

$$\begin{bmatrix} H_{yy} & f_y^T \\ f_y & 0 \end{bmatrix} \begin{bmatrix} \Delta y \\ \Delta \lambda \end{bmatrix} = - \begin{bmatrix} H_y^T \\ f \end{bmatrix}. \tag{1.60}$$

Equation (1.60) is easily solved for Δy and $\Delta \lambda$.

The NR algorithm is summarized below.

- Guess y and set $it = 0$.
  (∗) $f = f(y)$, $L = L(y)$.
- $f_y = f_y(y)$, $L_y = L_y(y)$.
- If $it = 0$ then

$$\lambda^T = -L_y \cdot f_y^* ,$$

where

$$f_y^* \triangleq f_y^T (f_y f_y^T)^{-1} ,$$

which is the weighted *generalized inverse* of f_y.
- $H_y = L_y + \lambda^T f_y$.
- If norm(g) < *tol* then end, where

$$g \triangleq \begin{bmatrix} H_y^T \\ f \end{bmatrix}.$$

- $H_{yy} = H_{yy}(y) \triangleq L_{yy} + \sum_{i=1}^n \lambda_i \cdot f_{yy}^i$.
-

$$\begin{bmatrix} \Delta y \\ \Delta \lambda \end{bmatrix} = - \begin{bmatrix} H_{yy} & f_y^T \\ f_y & 0 \end{bmatrix}^{-1} \begin{bmatrix} H_y^T \\ f \end{bmatrix}.$$

- Replace y by $y + \Delta y$ and λ by $\lambda + \Delta \lambda$. Set $it = it + 1$.
- Go to (∗).

In the POPN algorithm above, if we replace H_{yy} by I/k, where k is a positive scalar, we obtain the POP algorithm of Section 1.3.

A MATLAB file is listed in Table 1.2 that implements this algorithm.

Example 1.5.1—Max Velocity of a Sailboat Using POPN

This is a numerical solution of Example 1.3.1 using POPN. The subroutine containing L, f and their derivatives with respect to y is listed below. The first part of this subroutine is identical to the subroutine used in the example for POP, so it could also be used with POP to get a convex approximation for use with POPN.

TABLE 1.2   POPN—A MATLAB Newton-Raphson Code

```
function [L,y,ev,evec]=popn(name,y,tol,mxit)
% Parameter OPtimization with equality constraints using a
% Newton-raphson algorithm. L is optimum performance index, output y is
% optimum parameter vector; fn = norm(f); Hyn=norm(Hy); (ev,evec)=
% (eigvals, eigvecs) of Hessian; mxit=max no. iterations input y is
% guess of parameter vector; stopping criterion is max(fn,Hyn) < tol;
% subroutine 'name' contains L,f,Ly,fy,Lyy,fyy.
%
format compact; it=0; fn=1; Hyn=1; p=length(y);
disp('       it       L       fn       Hyn')
while max([fn Hyn]) > tol
 [L,f,Ly,fy,Lyy,fyy]=feval(name,y); fn=norm(f); n=length(f);
 if it==0, lat=-Ly*fy'/(fy*fy'); end; Hy=Ly+lat*fy; Hyn=norm(Hy);
 Hyy=Lyy; for i=1:n, Hyy=Hyy+lat(i)*fyy([1+p*(i-1):i*p],:); end
 A=[Hyy fy'; fy zeros(n)]; dv=-A\[Hy';f]; y=y+dv([1:p]);
 lat=lat+dv([p+1:n+p])'; disp([it L fn Hyn]);
 if it>mxit, break, end; it=it+1;
end
%
% Finds generalized eigenvalues and eigenvectors of Hessian:
B=diag([ones(1,p) zeros(1,n)]); [X,D]=eig(A,B); ev=diag(D)';
ev=ev([1+2*n:n+p]); evec=real(X([1:p],[1+2*n:n+p]));
```

```
function [L,f,Ly,fy,Lyy,fyy]=slbt2(y)
% Subroutine for Example 1.5.1; sailboat max velocity using POPN,
%   an NR
% code; L=V; y=[V Wr al th ps]'.
%
V=y(1); Wr=y(2); al=y(3); th=y(4); ps=y(5); sa=sin(al);
  ca=cos(al);
st=sin(th); ct=cos(th); p=sin(ps); cp=cos(ps); sat=sin(al+th);
cat=cos(al+th);
L=V; Ly=[1 0 0 0 0];
f=[V^2-Wr^2*sa*st; Wr^2-V^2-1+2*V*cp; Wr*sat-sp];
fy=[2*V -2*Wr*sa*st -Wr^2*ca*st -Wr^2*sa*ct 0; -2*(V-cp) 2*Wr 0
    0 ... -2*V*sp; 0 sat Wr*cat Wr*cat -cp];
Lyy=zeros(5);
f1=[2 0 0 0 0; 0 -2*st*sa -2*Wr*ca*st -2*Wr*sa*ct 0; 0 0 Wr^2*
    sa*st ... -Wr^2*ca*ct 0; 0 0 0 Wr^2*sa*st 0; 0 0 0 0 0];
fyy=f1+f1'-diag(diag(f1));
f2=[-2 0 0 0 -2*sp; 0 2 0 0 0; zeros(2,5); 0 0 0 0 -2*V*cp];
fyy([6:10],:)=f2+f2'-diag(diag(f2));
```

```
f3=[0 0 0 0 0; 0 0 cat cat 0; 0 0 -Wr*sat -Wr*sat 0; 0 0 0 -Wr*sat
   0; ... 0 0 0 0 sp];
fyy([11:15],:)=f3+f3'-diag(diag(f3));
```

An edited MATLAB diary solving the problem using POPN is shown below. It took only four iterations to bring the norms of f and of H_y below 10^{-8}. Note the two generalized eigenvalues are negative (indicating a maximum). The difficult part of using this algorithm is in selecting an initial guess for which the eigenvalues are negative. In an unfamiliar problem, it may be necessary to first use POP to get close to the maximum.

```
y=[.6 1 .6 .6 1.3]'; tol=1e-8; mxit=10;
[L,y,ev]=popn('slbt2',y,tol,mxit);
      it        L         fn        Hyn
       0     0.6000    0.0375     0.0050
  1.0000     0.5915    0.0007     0.0039
  2.0000     0.5931    0.0000     0.0005
  3.0000     0.5931    0.0000     0.0000
  4.0000     0.5931    0.0000     0.0000
y'=[.5931 1.0057 .6308 .6308 1.2797]; ev=-.2081 -.7075
```

Using FSOLVE with the Stationarity Conditions

Another approach, that avoids calculating the second derivatives, is to use the MATLAB Optimization Toolbox code FSOLVE with the first-order necessary conditions for a stationary solution:

$$f = 0 , \quad H_y \overset{\Delta}{=} L_y + \lambda^T f_y = 0 . \tag{1.61}$$

These represent $n + p$ equations for the $n + p$ unknowns λ, y. However, this apparently straightforward method may not converge unless very good initial guesses of λ and y are supplied; such guesses can be obtained using codes like POP or CONSTR.

This is an NR algorithm since FSOLVE numerically finds the gradient of the first-order necessary conditions, thus implicitly finding the second derivatives of both L and f.

Example 1.5.2—Max Velocity of a Sailboat Using FSOLVE

Below is an edited diary of the numerical solution of Example 1.3.1 using FSOLVE including the required subroutine SLBT-F. In this case a good initial guess of λ was not required.

```
% Script e01_5_2.m for FSOLVE solution of the sailboat of
   Example 1.3.1;
%
y1=[.5 1 .5 .5 1.5]; la=[0 0 0]; y=[y1 la]; optn(1)=1; optn(2)=1e-8;
y=fsolve('slbt_f',y,optn)
```

f-COUNT	RESID	STEP-SIZE	GRAD/SD	LINE-SEARCH
9	1.05688	1	-0.535	
-	-	-	-	
20	0.524986	1	0.0285	int_step
86	3.82673e-022	1	-2.36e-020	incstep

Calculation Converged Successfully
y=[.5931 1.0057 .6308 .6308 1.2797 -.6996 .2780 -1.1008]; f=slbt_f(y)
f=1e-10*[.0589 .0597 -.0946 .0609 -.1326 -.0206 -.0206 .0118]

```
function f=slbt_f(y)
% Subroutine for Example 1.5.2; FSOLVE solution of sailboat
  Example 1.3.1;
%
y1=y([1:5]); la=y([6:8]); [L,f1,Ly,fy]=feval('slbt',y1); f=[f1'
  Ly+la*fy];
```

Problems

1.5.1 to 1.5.22   Solve Problems 1.3.1 to 1.3.22 numerically using POPN or FSOLVE.

1.6   Chapter Summary

Parameter Optimization with Equality Constraints

Choose y to minimize $L(y)$ with $f(y) = 0$, where y is $(p \times 1)$, f is $(n \times 1)$, and $n < p$.

Necessary Conditions for a Stationary Point

y and λ must satisfy

$$f = 0 \quad (n \text{ equations}), \quad H_y \overset{\Delta}{=} L_y + \lambda^T f_y = 0 \quad (p \text{ equations}),$$

which simply says that $L_y dy = 0$ must be true for all dy such that $f_y dy = 0$.

A Gradient Algorithm

1. Guess y where y has been normalized so that one unit of each element is approximately of the same significance.
2. Evaluate f, f_y, L, L_y.
3. $f_y^* = f_y^T (f_y f_y^T)^{-1}$.
4. $\lambda^T = -L_y f_y^*$.
5. $H_y = L_y + \lambda^T f_y = $ component of L_y parallel to $f = 0$.
6. $dy = -\eta f_y^* f - k H_y^T$, $0 < \eta \leq 1$, $k > 0$.
7. If $|dy| < \epsilon$ then stop.
8. $y = y + dy$.
9. Go to (2).

An Explicit NR Algorithm

Omit (4) in the gradient algorithm after the first iteration and replace (6) through (8) by the following:

6. $H_{yy} = H_{yy}(y) \triangleq L_{yy} + \sum_{i=1}^{n} \lambda_i f_{yy}^i.$

7. (a) $\begin{bmatrix} dy \\ d\lambda \end{bmatrix} = -\begin{bmatrix} H_{yy} & f_y^T \\ f_y & 0 \end{bmatrix}^{-1} \begin{bmatrix} H_y^T \\ f \end{bmatrix}.$

   (b) If $\max(|dy|, |d\lambda|) < \epsilon$ then stop.

8. $y = y + dy$ and $\lambda = \lambda + d\lambda.$

An Implicit NR Algorithm

Use a nonlinear equation solver like MATLAB's FSOLVE to find y, λ that satisfy the first-order necessary conditions.

Dynamic Optimization

2.1 Discrete Dynamic Systems

A discrete dynamic system is described by an n-dimensional *state vector* $x(i)$ at step i. Choice of an m-dimensional *control vector* $u(i)$ determines a transition of the system to state $x(i + 1)$ through the relations

$$x(i + 1) = f[x(i), u(i), i] \,, \tag{2.1}$$

where

$$x(0) = x_0 \,. \tag{2.2}$$

A fairly general optimization problem for such systems is to find the sequence of control vectors $u(i)$ for $i = 0, \ldots, N - 1$ to minimize a performance index of the form

$$J = \phi[x(N)] + \sum_{i=0}^{N-1} L[x(i), u(i), i] \,, \tag{2.3}$$

subject to (2.1) and (2.2) with N, x_0, and the functions f specified. This is a parameter optimization problem with equality constraints, so it can be solved using the methods of Chapter 1, by treating the control vector histories $u(i)$ as the unknown parameters. Given an initial guess of $u(i)$, the state histories $x(i)$ can be calculated to determine J. $J_{u(i)}$, the gradient of J with respect to $u(i)$, may be calculated numerically by calculating the state histories $N \times m$ times changing only one element of $u(i)$ each time; the MATLAB code FMINU will do this, and does it quite rapidly for small problems on current personal computers. The code FMINU then iterates to find the optimal $u(i)$ (see Example 2.2.2).

However, all the components of $J_{u(i)}$ can be calculated by *one backward sequencing of a set of equations* that are adjoint to the state equations (2.1). This saves considerable time in solving large problems. This section develops these adjoint equations that take advantage of the sequential form of (2.1).

Necessary Conditions for a Stationary Solution

Adjoin the constraints (2.1) and (2.2) to the performance index (2.3) with a sequence of Lagrange multiplier vectors $\lambda(i)$ as follows:

$$\bar{J} = \phi[x(N)] + \sum_{i=0}^{N-1} \left\{ L(i) + \lambda^T(i+1)[f(i) - x(i+1)] \right\} + \lambda^T(0)[x_0 - x(0)] . \quad (2.4)$$

Define the *discrete Hamiltonian* as $H(i) \overset{\Delta}{=} H[x(i), u(i), \lambda(i+1), i]$ as follows:

$$H(i) = L[x(i), u(i), i] + \lambda^T(i+1)f[x(i), u(i), i] . \quad (2.5)$$

Using (2.5) in (2.4) and changing indices of summation on the last term in (2.4) yields

$$\bar{J} = \phi[x(N)] - \lambda^T(N)x(N) + \lambda^T(0)x_0 + \sum_{i=0}^{N-1} [H(i) - \lambda^T(i)x(i)] . \quad (2.6)$$

Now consider differential changes in $\bar{J}$ due to differential changes in $u(i)$ and x_0:

$$d\bar{J} = \left[\phi_x - \lambda^T(N) \right] dx(N) + \lambda^T(0)dx_0 + \sum_{i=0}^{N-1} \left\{ \left[H_x(i) - \lambda^T(i) \right] dx(i) + H_u(i)du(i) \right\} . \quad (2.7)$$

To avoid having to determine the differential changes $dx(i)$ produced by a given $du(i)$ sequence, we choose the multiplier sequence $\lambda(i)$ so that the coefficients of $dx(i)$ and $dx(N)$ in (2.7) vanish, i.e., we choose

$$\lambda^T(i) = H_x(i) \equiv L_x(i) + \lambda^T(i+1)f_x(i) , \quad (2.8)$$

for $i = 0, \ldots, N-1$, with boundary conditions

$$\lambda^T(N) = \phi_x . \quad (2.9)$$

Equation (2.7) then becomes

$$d\bar{J} = \lambda^T(0)dx_0 + \sum_{i=0}^{N-1} H_u(i)du(i) . \quad (2.10)$$

Thus $H_u(i)$ is the gradient of J with respect to $u(i)$ while holding x_0 constant and satisfying

(2.1), and $\lambda^T(0)$ is the gradient of J with respect to x_0 while holding $u(i)$ constant and satisfying (2.1). If x_0 is specified, then $dx_0 = 0$.

For a stationary solution $d\bar{J} = 0$ for arbitrary $du(i)$; this can happen only if

$$H_u(i) = 0 \ , \ i = 0, \ldots, N - 1 \ . \tag{2.11}$$

$H_u(i)$ is called the *pulse response sequence*.

Hence, to find a control vector sequence $u(i)$ that produces a stationary value of J, we must solve the following difference equations, *the discrete Euler-Lagrange EL equations*:

$$x(i + 1) = f[x(i), u(i), i], \tag{2.12}$$

$$\lambda(i) = H_x^T(i) \equiv L_x^T(i) + f_x^T(i)\lambda(i + 1) \ , \tag{2.13}$$

where $u(i)$ is determined from (2.11), which may be written as

$$H_u(i) \equiv L_u(i) + \lambda^T(i + 1)f_u(i) = 0 \ . \tag{2.14}$$

The boundary conditions for these difference equations are split, n at $i = 0$ and n at $i = N$:

$$x(0) = x_0, \tag{2.15}$$

$$\lambda(N) = \phi_x^T \ . \tag{2.16}$$

This is a *two-point boundary value problem (TPBVP)*. Note that (2.12) and (2.13) are *coupled* since $u(i)$ depends on $\lambda(i + 1)$ through (2.14) and the coefficients in (2.13) depend on $x(i)$ and $u(i)$. The Appendix at the end of the chapter gives an alternative derivation using the concept of consistent linear equations.

Since both $x(i)$ and $\lambda(i)$ have dimension n, $u(i)$ has dimension m, and $i = 0, \cdots, N-1$, the two-point boundary value problem has $N(2n + m)$ unknowns and an equal number of equations. It can be solved as a large set of simultaneous nonlinear algebraic equations (for example, using the command FSOLVE in MATLAB). However, a more efficient method of solution will be described later, exploiting the sequential nature of the problem.

Example 2.1.1—Discrete Velocity Direction Programming (DVDP) with Gravity; Max Range in a Given Time

A bead slides on a wire without friction in a gravitational field. We wish to find the shape of the wire to maximize the horizontal range in a given time t_f. We shall do this by programming the direction of the velocity, i.e., the angle of the wire below the horizontal $\gamma(t)$ as a function of time t (see Figure 2.1).

This is the dual problem to the famous *brachistochrone problem* of finding the shape of a wire to minimize the time t_f to cover a given horizontal distance ("brachistochrone" means "shortest time" in Greek). It was posed and solved by Jakob Bernoulli in the seventeenth century (Ref. Go).

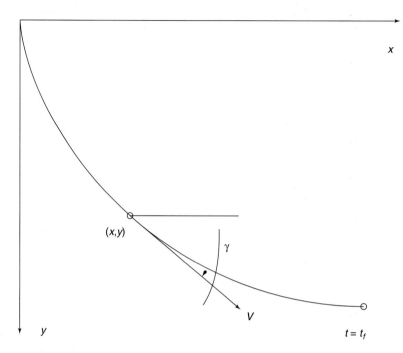

FIGURE 2.1   Nomenclature for Velocity Direction Programming Problem (Brachistochrone Problem)

To discretize the problem we shall change γ only at time intervals of length ΔT. Thus we wish to find a control sequence $\gamma(i)$ for $i = 0$ to $N - 1$ to maximize x at time t_f where $\Delta T = t_f/N$, $t = i\Delta T$. Figure 2.2 shows the case where $N = 5$ and the limiting case $N \to \infty$.

From elementary mechanics, the velocity of the bead at time $(i + 1)\Delta T$ is

$$v(i + 1) = v(i) + g\Delta T \cdot \sin \gamma(i) , \quad v(0) = 0 , \tag{2.17}$$

From simple kinematics, the x-coordinate of the bead at time $(i + 1)\Delta T$ is

$$x(i + 1) = x(i) + \Delta \ell(i) \cos \gamma(i) , \quad x(0) = 0 , \tag{2.18}$$

where

$$\Delta \ell(i) = \Delta T v(i) + \tfrac{1}{2} g(\Delta T)^2 \sin \gamma(i) , \tag{2.19}$$

is the length of the wire between the ith corner and the $(i + 1)^{st}$ corner.

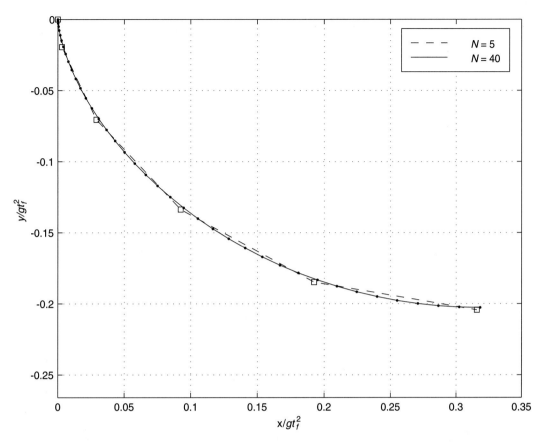

FIGURE 2.2   DVDP for Max Range with Gravity for $N = 5$ and $N = 40$; x vs. y

Let $s = [v \quad x]^T$ = state vector. Then $L = 0$, $\phi[s(N)] = x(N)$, and the Hamiltonian is

$$H(i) = \lambda_v(i + 1)[v(i) + g\Delta T \sin \gamma(i)] + \lambda_x(i + 1)[x(i) + \Delta \ell(i) \cos \gamma(i)] . \quad (2.20)$$

The adjoint equations (2.13) become

$$\lambda_v(i) = H_v(i) = \lambda_v(i + 1) + \Delta T \cos \gamma(i)\lambda_x(i + 1) , \quad \lambda_v(N) = 0 , \quad (2.21)$$

$$\lambda_x(i) = H_x(i) = \lambda_x(i + 1) , \quad \lambda_x(N) = 1 . \quad (2.22)$$

Clearly

$$\lambda_x(i) = 1 \text{ for all } i . \quad (2.23)$$

Using (2.23) and $\Delta = t_f/N$), the optimality condition (2.14) can be written as

$$0 = \cos \gamma(i)\frac{\lambda_v(i+1)}{t_f} - \frac{v(i)}{gt_f}\sin \gamma(i) + \frac{1}{2N}\cos[2\gamma(i)] , \qquad (2.24)$$

which determines $\gamma(i)$ given $v(i)$ and $\lambda_v(i+1)$.

Defining $\alpha \triangleq \pi/2N$, the solution to this TPBVP (see Problem 2.1.1) is

$$\gamma(i) = \frac{\pi}{2} - \alpha(i + \tfrac{1}{2}) , \quad i = 0, \ldots, N-1 \qquad (2.25)$$

$$\frac{v(i)}{gt_f} = \frac{\sin(\alpha i)}{2N \sin(\alpha/2)} , \quad i = 0, \ldots, N \qquad (2.26)$$

$$\frac{x(i)}{gt_f^2} = \frac{\cos(\alpha/2)}{4N^2 \sin(\alpha/2)}\left[i - \frac{\sin(2\alpha i)}{2 \sin \alpha}\right] , \quad i = 0, \ldots, N \qquad (2.27)$$

$$\frac{\lambda_v(i)}{t_f} = \frac{\cos(\alpha i)}{2N \sin(\alpha/2)} , \quad i = 1, \ldots, N \qquad (2.28)$$

Figure 2.3 shows γ vs. t/t_f for $N = 5$ and $N = 40$. As $N \to \infty$, $\gamma \to (\pi/2)(1-t/t_f)$, which is the continuous solution.

Note that the y-coordinate of the bead does not enter the problem. However, from simple kinematics

$$y(i+1) = y(i) + \Delta\ell(i)\sin \gamma(i) , \quad y(0) = 0 . \qquad (2.29)$$

From this relation and the expression for $\gamma(i)$ above, it follows that

$$\frac{y(i)}{gt_f^2} = \frac{\cos(\alpha/2)}{8N^2 \sin(\alpha/2) \sin \alpha}[1 - \cos(2\alpha i)] . \qquad (2.30)$$

Note that

$$\frac{x(N)}{gt_f^2} = \frac{1}{\pi}\frac{\alpha/2}{\tan(\alpha/2)} \to \frac{1}{\pi} \text{ as } N \to \infty , \qquad (2.31)$$

$$\frac{y(N)}{gt_f^2} = \frac{2}{\pi^2}\frac{\alpha/2}{\tan(\alpha/2)}\frac{\alpha}{\sin \alpha} \to \frac{2}{\pi^2} \text{ as } N \to \infty . \qquad (2.32)$$

Example 2.1.2—Discrete LQ Problems

Here we wish to find the control vector sequence $u(i)$, $i = 0, \ldots, N-1$ that minimizes the quadratic performance index

$$J = \frac{1}{2}x^T(N)Q_f x(N) + \frac{1}{2}\sum_{i=0}^{N-1}[x^T(i)Q(i)x(i) + u^T(i)R(i)u(i)] , \qquad (2.33)$$

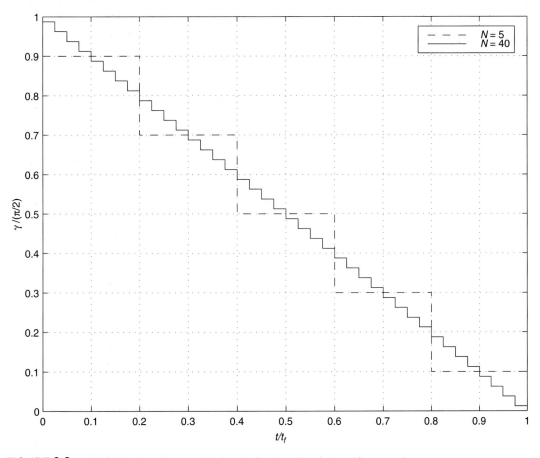

FIGURE 2.3   DVDP for Max Range with Gravity for $N = 5$ and $N = 40$ γ vs. t/t_f

where $Q(i)$, $R(i)$, and Q_f are positive definite matrices, with the linear difference equation constraints

$$x(i + 1) = \Phi(i)x(i) + \gamma(i)u(i) \tag{2.34}$$

for $i = 0, \ldots, N - 1$ and $x(0)$ specified.

The $H(i)$ sequence for this problem is

$$\tfrac{1}{2}[x^T(i)Q(i)x(i) + u^T(i)R(i)u(i)] + \lambda^T(i + 1)[\Phi(i)x(i) + \gamma(i)u(i)] , \tag{2.35}$$

where

$$\lambda^T(i) = \lambda^T(i + 1)\Phi(i) + x^T(i)Q(i) , \tag{2.36}$$

$$\lambda^T(N) = x^T(N)Q_f . \tag{2.37}$$

The stationarity condition (2.14) is

$$H_u(i) \equiv u^T(i)R(i) + \lambda^T(i+1)\gamma(i) = 0 , \qquad (2.38)$$

or

$$u(i) = -[R(i)]^{-1}\gamma^T(i)\lambda(i+1) . \qquad (2.39)$$

This yields the following *linear* TPBVP:

$$x(i+1) = \Phi(i)x(i) - \gamma(i)[R(i)]^{-1}\gamma^T(i)\lambda(i+1) , \qquad (2.40)$$

$$\lambda(i) = Q(i)x(i) + \Phi^T(i)\lambda(i+1) , \qquad (2.41)$$

with boundary conditions

$$x(0) \quad \text{specified} , \qquad (2.42)$$

$$\lambda(N) = Q_f x(N) . \qquad (2.43)$$

See Chapter 5 for solutions using transition matrices or discrete Riccati equations.

Problems

2.1.1 DVDP for Max Range with Gravity; a Brachistochrone Problem:

(a) Verify the solution given in Example 2.1.1, i.e., show that the stated solution satisfies the stationarity conditions.
(b) Verify Figs. 2.2 and 2.3.

2.1.2 DVDP for Max Range with $u_c = Vy/h$; a Zermelo Problem: A ship travels with constant velocity V with respect to the water through a region where the velocity of the current is parallel to the x-axis but varies with y, so that

$$\dot{x} = V\cos\theta + u_c(y) ,$$
$$\dot{y} = V\sin\theta ,$$

where θ is the ship's heading relative to the x-axis. E. Zermelo (Ref. Ze) was the first to treat such problems.

(a) For $u_c = Vy/h$, i.e., linear variation of the current with y, and θ held constant for time intervals of length ΔT, show that the x and y coordinates at the ends of these intervals are given by the following difference equations:

$$y(i+1) = y(i) + \Delta T \sin\theta(i) ,$$
$$x(i+1) = x(i) + \Delta T [\cos\theta(i) + y(i)] + \tfrac{1}{2}(\Delta T)^2 \sin\theta(i) ,$$

where ΔT is in units of h/V and (x, y) are in units of h.

(b) For $x(0) = y(0) = 0$, and given t_f, show that the $\theta(i)$ sequence that maximizes $x(t_f)$ in N steps is such that:

$$\tan \theta(i) = \left(N - i - \tfrac{1}{2}\right) \Delta T \ ,$$

where

$$\Delta T = \frac{t_f}{N} \ , \quad \text{and} \ \ i = 0 , \cdots, \ N - 1 \ .$$

(c) For $Vt_f/h = 2$, calculate and plot optimal $x(i)$ vs. $y(i)$ for $N = 5$ and $N = 40$ using the optimality condition of (b) with the discretized equations of motion of (a). Compare your solution with Figure 2.4.

2.1.3 DVDP for Min Distance to a Meridian on a Sphere: We wish to find the heading angle sequence $\beta(i)$ (β is positive counterclockwise from east) as a function of longitude ϕ to minimize the distance traveled in going along the surface of a unit sphere from a point

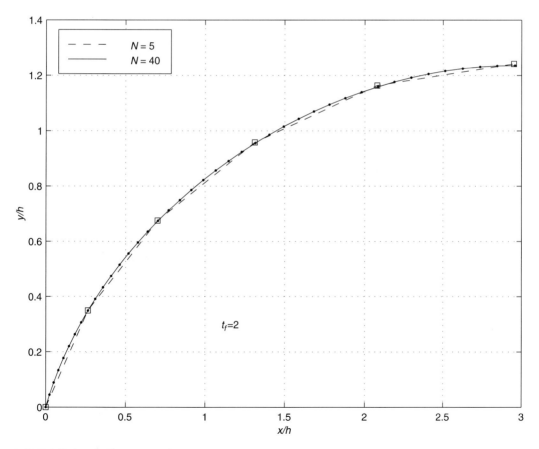

FIGURE 2.4 DVDP for Max Range with $u_c = Vy/h$; a Zermelo Problem

with (latitude, longitude) $= (\theta_0, 0)$ to the meridian where $\phi = \phi_f$. The continuous system equations are

$$\frac{ds}{d\phi} = \sec \beta \cos \theta, \quad \frac{d\theta}{d\phi} = \tan \beta \cos \theta.$$

where $s =$ distance along the path.

(a) For β held constant for longitude intervals of $\Delta\phi$ show that the discrete EOM are

$$\theta(i+1) = g(z) \triangleq 2 \tan^{-1} z - \pi/2 , \quad s(i+1) = \frac{\theta(i+1) - \theta(i)}{\sin \beta(i)} ,$$

where $z \triangleq b \tan[\theta(i)/2 + \pi/4]$, $b \triangleq \exp[\tan \beta(i)\Delta\phi]$.

(b) Show that the discrete EL equations are the two equations in (a) with $\theta(0) = \theta_0$, and

$$\lambda(i) = -\frac{1}{s} + \left[\lambda(i+1) + \frac{1}{s}\right] g_\theta , \quad \lambda(N) = 0 ,$$

$$0 = H_{\beta(i)} = -\frac{c}{s^2}[g(z) - \theta(i)] + \left[\lambda(i+1) + \frac{1}{s}\right] g_\beta ,$$

where $s = \sin \beta(i)$, $c = \cos \beta(i)$, and

$$g_\theta = \frac{b^2 + z^2}{b(1 + z^2)} , \quad g_\beta = \frac{2z\Delta\phi}{c^2(1 + z^2)} .$$

and $i = N - 1, \ldots, 0$.

(c) Using the FSOLVE command in the MATLAB Optimization Toolbox, find the solution to this TPBVP with $\theta_0 = 40$ deg, $\phi_f = 50$ deg for $N = 5$ and $N = 40$, and plot θ vs. ϕ. Compare your solutions with Figure 2.5. *Hint:* Sequence the system equations forward using a guess of $\beta(i)$ and saving $\theta(i)$; then sequence the $\lambda(i)$ equation backward, calculating $f(i) \triangleq H_{\beta(i)}$ as you go; use FSOLVE to make $f(i) = 0$.

2.1.4 DVDP for Max Range with $V = V_0(1+y/h)$; a Fermat Problem: A particle moves through a region in which its instantaneous velocity magnitude is a function of position, i.e., $V = V(x, y)$, where (x, y) are rectangular coordinates. The velocity direction makes an angle θ with the x-axis. The equations of motion are

$$\dot{x} = V(x, y)\cos \theta , \quad \dot{y} = V(x, y)\sin \theta .$$

(a) For θ held constant for time intervals of length ΔT, and $V = V_0(1 + y/h)$, show that

$$y(i+1) = [1 + y(i)]a(i) - 1 ,$$
$$x(i+1) = x(i) + [1 + y(i)]\cot \theta(i)[a(i) - 1] ,$$

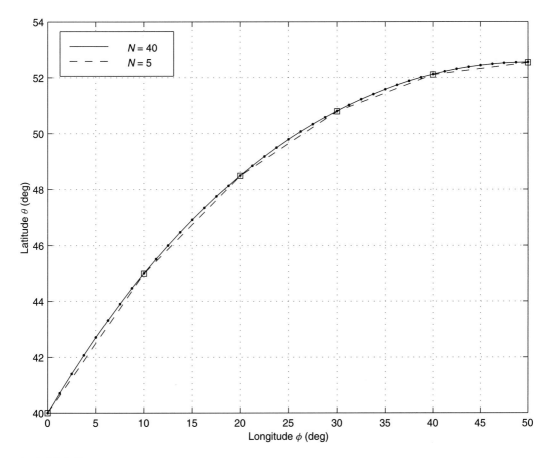

FIGURE 2.5 DVDP for Min Distance to a Meridian on a Sphere

where $a(i) \triangleq \exp[\Delta T \sin \theta(i)]$ and (x, y) are in units of h, $\Delta T = t_f/N$, t_f is in units of h/V_0, and $i = 0, \ldots, N - 1$.

(b) For $x(0) = y(0) = 0$, show that the TPBVP for determining the path that maximizes $x(N)$ requires solving the two equations in (a) and

$$\lambda_y(i) = \frac{a(i) - 1}{\tan \theta(i)} + a(i)\lambda_y(i + 1) , \quad \lambda_y(N) = 0 ,$$

$$0 = \frac{a(i)\Delta T \cos^2 \theta(i)}{\sin \theta(i)} - \frac{[a(i) - 1]}{\sin^2 \theta(i)} + \lambda_y(i + 1)a(i)\Delta T \cos \theta(i) ,$$

where $i = N - 1, \ldots, 0$.

(c) Using the FSOLVE command in the MATLAB Optimization Toolbox, find the solution to this TPBVP with $x(0) = y(0) = 0$ for normalized $t_f = 2$, with $N = 5$ and $N = 40$ and plot x vs. y. Compare your solution with Figure 2.6. See the hint in Problem 2.1.3(c).

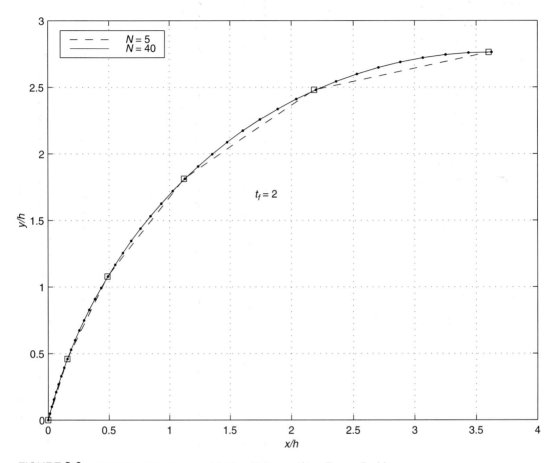

FIGURE 2.6   DVDP for Max Range with $V = V_0(1 + y/h)$; a Fermat Problem

2.2 Numerical Solution with Gradient Methods

Only very simple problems can be solved in terms of tabulated functions. Hence, in this section, we consider algorithms for numerical solution. We shall do this using the *Mayer formulation* of the problem instead of the *Bolza formulation* of (2.1) to (2.3). These two formulations are exactly equivalent, but the Mayer form yields simpler expressions, that makes it easier to code for numerical solutions.

In the Mayer formulation the state vector of (2.1) and (2.2) is augmented by one state $x_{n+1}(i)$ that is the cumulative sum of L to step i, i.e.,

$$x_{n+1}(i + 1) = x_{n+1}(i) + L[x(i), u(i), i] , \quad x_{n+1}(0) = 0 .$$

Thus the performance index (2.3) becomes

$$J = \phi[x(N)] + x_{n+1}(N) \stackrel{\Delta}{=} \bar{\phi}[\bar{x}(N)] ,$$

where

$$\bar{x} \triangleq \begin{bmatrix} x \\ x_{n+1} \end{bmatrix} .$$

Dropping the bar on x and ϕ, the problem may be stated as finding a vector sequence $u(i), i = 0, \cdots, N-1$, to minimize (or maximize)

$$\phi[x(N)] , \qquad (2.44)$$

subject to

$$x(i+1) = f[x(i), u(i)] , \qquad (2.45)$$

where u is $(n_c \times N)$, both f and x are $(n \times 1)$, and $x(0)$, N are specified.

We first describe an iterative gradient algorithm DOP0 (for Discrete OPtimization with 0 terminal constraints). It requires a subroutine that calculates f, ϕ, ϕ_x, f_x, and f_u and selection of a step-size parameter k. A MATLAB implementation of DOP0 is listed in Table 2.1.

Then we describe the use of the MATLAB command FMINU (for Function MINimization Unconstrained) to solve this problem. This code does *not* require the selection of a step-size parameter. Furthermore it does not require the analytical gradients ϕ_x, f_x, f_u since it will calculate them numerically. However the code will run faster for big problems if these gradients are provided.

Finally, in Section 2.5, we describe a *shooting algorithm,* which is a multiple interpolation algorithm for sequencing the EL equations backward to match the specified initial conditions. This algorithm yields a precise solution, but it converges only if the initial guess of the final state is very good; such a guess can be provided by DOP0 or FMINU.

DOP0—Discrete OPtimization with 0 Terminal Constraints

This is a simplified version of the POP algorithm of Section 1.3 since there are "no constraints" in the sense of that algorithm; simply replace L by ϕ and y by $u \triangleq [u(0) \cdots u(N-1)]^T$. This is a first-order gradient algorithm, which starts with an initial guess of $u(i)$, finds J and the gradient of J with respect to $u(i)$, then changes $u(i)$ by a small amount $du(i)$ in the direction of the negative gradient, and iterates until the magnitude of $u(i)$ becomes small. A more detailed description is given below.

- Enter $x(0)$, t_f, k, tol, and a guess of $u(i)$, $i = 0, \cdots, N-1$; choose $k > 0$ if minimizing, $k < 0$ if maximizing.
- (∗) Sequence forward. Compute and store $x(i)$.
- Evaluate $\phi[x(N)]$ and set $\lambda^T(N) = \phi_x$.
- Sequence backward. Compute and store the pulse response sequence $H_u(i)$.
  For $i = N-1, \cdots, 0$:

$$H_u(i) = \lambda^T(i+1)f_u(i) , \qquad (2.46)$$

$$\lambda^T(i) = \lambda^T(i+1)f_x(i) . \qquad (2.47)$$

TABLE 2.1   A MATLAB Code for DOP0

```
function [u,s,la0]=dop0(name,u,s0,tf,k,tol,mxit)
% Discrete OPtim. with 0 term. Constraints, tf specified, nc controls.
% Inputs: name must be in single quotes; function file 'name' computes
% s(i+1)= f(s(i),u(i))for flg=1, (phi,phis) for flg=2, and (fs,fu)
% for flg=3; u(nc,N) = estimate of optimal u; s0(ns,1) = initial state;
% tf = final time; k = step size parameter; u should be normalized so
% that the elements of du are roughly the same size; stopping criterion
% is max(dua) < tol; mxit = max no iterations; outputs (u,s) = improved
% (u,s) histories; la0 = optimal lambda(0) for possible use with a
% shooting algorithm.
% BASIC version 1984; MATLAB version 1994; rev. 1/8/98
%
if nargin<7, mxit=10; end;
ns=length(s0); [nc,N]=size(u); s=zeros(ns,N+1); dum=zeros(nc,1);
la=zeros(ns,1); Hu=zeros(N,nc); dua=1; it=0; dt=tf/N; s(:,1)=s0;
disp('    Iter.      phi       dua');
while norm(dua)>tol,
 % Forward sequencing and store state histories x(:,i):
 for i=1:N, s(:,i+1)=feval(name,u(:,i),s(:,i),dt,(i-1)*dt,1); end
 % Performance index phi and b.c. for backward sequence phis:
 [phi,phis]=feval(name,dum,s(:,N+1),dt,N*dt,2); la=phis';
 % Backward sequencing and store Hu(i);
 for i=N:-1:1
    [fs,fu]=feval(name,u(:,i),s(:,i),dt,(i-1)*dt,3);
    Hu(i,:)=la'*fu; la=fs'*la;
 end; la0=la;
 % New u(i):
 for j=1:nc,
    du(j,:)=-k*Hu(:,j)'; dua(j)=norm(du(j,:))/sqrt(N);
 end; u=u+du;
 disp([it phi dua]);
 if it>mxit, break, end; it=it+1;
end
```

- Compute $\Delta u(i)$ and Δu_{avg}. For $i = 0, \cdots, N-1$

$$\Delta u(i) = -k H_u^T(i) , \tag{2.48}$$

$$\Delta u_{avg} = \sqrt{\frac{1}{N} \sum_{i=0}^{N-1} \Delta u^T(i) \Delta u(i)} . \tag{2.49}$$

- If $\Delta u_{avg} < tol$ then stop.

- Compute new $u(i)$:

$$u(i) \leftarrow u(i) + \Delta u(i) , \qquad (2.50)$$

- Go to ($*$).

A code for implementing the DOP0 algorithm in MATLAB is listed in Table 2.1. The data for the problem being solved are put into a subroutine "name" that gives the f functions, the performance index ϕ and its gradient ϕ_s, and the derivatives f_s, f_u. Note that MATLAB starts sequences with $i = 1$ where we have used $i = 0$ above.

Example 2.2.1—DVDP for Max Range with Gravity

This is the same example solved analytically in Example 2.1.1. The subroutine used is listed below.

```
function [f1,f2]=dvdp0(ga,s,dt,t,flg)
% s=[v x]'; ga=gamma; t in units of tf, v in g(tf), x in g(tf)^2.
%
v=s(1); x=s(2);
if flg==1,
  f1=s+dt*[sin(ga); v*cos(u)+dt*sin(2*ga)/4];          % f1 = f
elseif flg==2,
  f1=x; f2=[0 1];                        % f1 = phi, f2 = phis
elseif flg==3,
  f1=[1 0; dt*cos(ga) 1];                       % f1 = fs
  f2=dt*[cos(ga); -v*sin(ga)+dt*cos(2*ga)/2];          % f2 = fu
end
```

An edited MATLAB diary using DOP0 is shown below. It took only five iterations to bring du_a below .0001. It takes a little practice to learn how to select the step-size parameter k. Start with small values and put mxit=3 so that the code runs only three steps and stops; gradually increase k until dua and ϕ increase instead of decreasing (for a minimization problem); k is then too large and you will overshoot the minimum; decrease k by a small amount so that dua and ϕ decrease and dua is a reasonable size (i.e., one could expect linear prediction from one step to the next to be fairly accurate).

```
% Script e02_2_1.m; DVDP for max xf with gravity;
%
ga=[1:-.25:0]; s0=[0 0]'; tf=1; k=-7; tol=5e-5;
[ga,s]=dop0('dvdp0',ga,s0,tf,k,tol);
    Iter.      phi        dua
        0     0.2780     0.3553

        -        -          -

    6.0000     0.3157     0.0000
ga=[1.4137 1.0996 .7854 .4712 .1571]
s=[0 .3566 .9836 1.8709 2.9486; 0 .4341 .8246 1.1245 1.2457]
```

Use of the MATLAB Code FMINU

The FMINU command in the MATLAB Optimization Toolbox finds the minimum of a function 'f' with respect to a parameter vector 'u' (or parameter matrix 'u'), with or without the analytical gradient f_u. All that is required of the user is a MATLAB .m file that computes $f(u)$ given u, which, in this case, means sequencing (2.45) forward to calculate ϕ, which must be called 'f' when using FMINU (a source of possible confusion). For faster solution, the user can supply gradient information to FMINU; for small problems, this does not save much time when one takes into account the time necessary to generate the gradient subroutine. The MATLAB files below solve the same VDP example treated above. The first file DVDP0 is an operating code which generates an initial guess of $u(i)$ and supplies the options required by FMINU; the second file DVDP F calculates the performance index, and the third (optional) file DVDP GR supplies the analytical gradients.

Example 2.2.2—DVDP for Max Range with Gravity Using FMINU

```
% Script e02_2_2.m; DVDP for max range with gravity using MATLAB code FMINU
% with or without analytical gradient; s=[v x]';
%
% Optimal control history ga(i) using FMINU:
%
N=10; optn(1)=1; s0=[0 0]'; tf=1; ga=[1:-1/(N-1):0];    % Initial guess
% Without analytical gradient:
% ga=fminu('dvdp_f',ga,optn,[],s0,tf,N);
% With analytical gradient:
ga=fminu('dvdp_f',ga,optn,'dvdp_gr',s0,tf,N);
[f,v,x]=dvdp_f(ga,s0,tf,N);

function [f,v,x]=dvdp_f(u,s0,tf,N)
% Subroutine for e02_2_2.m; DVDP with gravity using FMINU; u = estimate
% of optimal u; s0 = initial state; tf = final time; N = number of steps.
%
dt=tf/N;     v(1)=s0(1); x(1)=s0(2);
% Forward sequencing:
for i=1:N
  x=x+dt*v*cos(u(i))+dt^2*sin(2*u(i))/4; v=v+dt*sin(u(i));
end
% Performance index f:
f=-x;

function df=dvdp_gr(u,s0,tf,N)
% u=estimate of optimal u; s0=initial state; tf=final time; N=no. steps.
%
Hu=zeros(1,N); dt=tf/N; v(1)=s0(1); x(1)=s0(2);
% Forward sequencing & store v(i) & x(i):
```

```
for i=1:N
  v(i+1)=v(i)+dt*sin(u(i));
  x(i+1)=x(i)+dt*v(i)*cos(u(i))+dt^2*sin(2*u(i))/4;
end;
% Gradient phix:
phis=[0 1]; la=phis';
% Backward sequencing and store Hu(i);
for i=N:-1:1
  fs=[1 0; dt*cos(u(i)) 1];
  fu=dt*[cos(u(i)); -v(i)*sin(u(i))+dt*cos(2*u(i))/2];
  Hu(i)=la'*fu; la=fs'*la;
end;
%
df=-Hu;
```

An edited diary of the solution with and without analytical gradients is given below. Note there are fewer function evaluations when the analytical gradients are supplied.

```
% Diary e02_2_2.dia; N=5;
%
% WITHOUT analytical gradients:
e02_2_2
f-COUNT    FUNCTION    STEP-SIZE      GRAD/SD  LINE-SEARCH
    7      -0.27798          1      -0.0129
   16      -0.315546   5.96982       0.000192  int_st
   25      -0.315612   2.56598       3.91e-008 int_st
   34      -0.315688   3.10319      -8.99e-009 incstep
Optimization Terminated Successfully;
ga=[1.4136    1.0993    0.7852    0.4714    0.1571]
[v;x]=0    0.1975    0.3757    0.5171    0.6079    0.6392
       0    0.0031    0.0291    0.0923    0.1925    0.3157
%
% WITH analytical gradients:
e02_2_2
f-COUNT    FUNCTION    STEP-SIZE      GRAD/SD  LINE-SEARCH
    2      -0.27798          1      -0.0129
    6      -0.315546   5.96982       0.000192  int_st
   10      -0.315612   2.56592       1.31e-009 int_st
   14      -0.315688   3.10666       2.09e-009 int_st
Optimization Terminated Successfully;
ga=[1.4138    1.0994    0.7853    0.4716    0.1569]
[v;x]=0    0.1975    0.3757    0.5171    0.6080    0.6392
       0    0.0031    0.0291    0.0923    0.1925    0.3157
```

Problems

2.2.1 to 2.2.4:   Do Problems 2.1.1 to 2.1.4 numerically using DOP0 or FMINU.

2.2.5 DVDP for Max Range Using Gravity and Thrust:

(a) Show that a discrete-step version of Problem 2.3.5 is to find the sequence $\gamma(i)$, $i = 0, \cdots, N - 1$, to maximize $x(N)$ subject to

$$V(i + 1) = V(i) + \Delta T[a + \sin \gamma(i)] \ ,$$

$$x(i + 1) = x(i) + \Delta \ell(i) \cos \gamma(i) \ ,$$

$$y(i + 1) = y(i) + \Delta \ell(i) \sin \gamma(i) \ ,$$

where $\Delta T =$ the normalized time step and $\Delta \ell(i) \triangleq V(i)\Delta T + [a + \sin \gamma(i)](\Delta T)^2/2$.

(b) Using a code like DOP0 or FMINU, solve the problem posed in (a) for $a = 1.5$ with zero initial state and $N = 10$. Compare your results with those for the continuous version of the problem shown in Figure 2.10 where, effectively, $N \to \infty$. Note that $y(i)$ is an ignorable state but is needed to plot the optimal path.

2.2.6 DVDP for Max Range with Gravity, Thrust, and Drag:

(a) Show that a discrete-step version of Problem 2.3.6 is to find the sequence $\gamma(i)$, $i = 0, \cdots, N - 1$, to maximize $x(N)$ subject to

$$V(i + 1) = b(i) \tanh g_1(i) \ ,$$

$$x(i + 1) = x(i) + \cos \gamma(i) \log \frac{\cosh g_1(i)}{\cosh g(i)} \ ,$$

$$y(i + 1) = y(i) + \sin \gamma(i) \log \frac{\cosh g_1(i)}{\cosh g(i)} \ ,$$

where

$$b(i) \triangleq [a - \sin \gamma(i)]^{\frac{1}{2}} \ , \quad g(i) \triangleq \tanh^{-1} \frac{V(i)}{b(i)} \ , \quad g_1(i) \triangleq g(i) + b(i)\Delta T \ ,$$

and $\Delta T =$ the normalized time step.

(b) Using a code like DOP0 or FMINU, solve the problem posed in (a) for $a = .05$, $t_f = 5$ with zero initial state and $N = 20$. Compare your results with those for the continuous version of the problem shown in Figure 2.11 where, effectively, $N \to \infty$. Note that $y(i)$ is an ignorable state but is needed to plot the optimal path.

2.2.7 Discrete-Step Calculation of Min Drag Nose Shape:

(a) Show that a discrete-step version of Example 2.3.3, approximating the nose shape as a sequence of conical segments, is to find the sequence $\theta(i)$, $i = 0, \cdots, N - 1$, to minimize

$$\frac{D}{\pi q} = C_p \left(\frac{\pi}{2}\right) [r(N)]^2 + \sum_{i=0}^{N-1} C_p[\theta(i)]\{[r(i)]^2 - [r(i + 1)]^2\} \ ,$$

where $C_p(\theta) = 2\sin^2\theta$ and

$$r(i+1) = r(i) - \frac{\ell}{N}\tan[\theta(i)] , \quad r(0) = a .$$

(b) Using a code like DOP0 or FMINU, solve the problem posed in (a) for $a/\ell = 1/4$ with $N = 10$. Compare the value of C_D that you obtain with the value given in Example 2.3.3 where, effectively, $N \to \infty$.

2.3 Continuous Dynamic Systems

Optimal programming problems for continuous systems are problems in the *calculus of variations*. They may be considered as limiting cases of optimal programming problems for discrete systems in which the time increment between steps becomes small compared to characteristic times of the continuous system. The reverse procedure is more common today; continuous systems are approximated by discrete systems for simulation on digital computers.

A continuous-step dynamic system is described by an n-dimensional *state vector* $x(t)$ at time t. Choice of an m-dimensional *control vector* $u(t)$ determines the time rate of change of the state vector through the relations

$$\dot{x} = f(x, u, t) , \tag{2.51}$$

A fairly general optimization problem for such a system is to find the time history of the control vector $u(t)$ for $t_0 \le t \le t_f$ to minimize a performance index of the form

$$J = \phi[x(t_f)] + \int_{t_0}^{t_f} L(x, u, t)dt , \tag{2.52}$$

subject to (2.51) with

$$t_0 , \quad t_f , \quad \text{and } x(t_0) \text{ specified} . \tag{2.53}$$

Necessary Conditions for a Stationary Solution

Adjoin the constraints (2.51) to the performance index (2.52) with a time-varying Lagrange multiplier vector $\lambda(t)$ as follows:

$$\bar{J} = \phi[x(t_f)] + \int_{t_0}^{t_f} \left\{ L[x(t), u(t), t] + \lambda^T(t)[f[x(t), u(t), t] - \dot{x}] \right\} dt . \tag{2.54}$$

Define the scalar *Hamiltonian* function $H[x(t), u(t), \lambda(t), t]$ that we shall call $H(t)$ for a shorter notation:

$$H(t) \stackrel{\Delta}{=} L[x(t), u(t), t] + \lambda^T(t)f[x(t), u(t), t] . \tag{2.55}$$

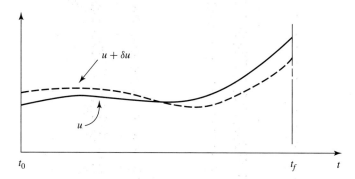

FIGURE 2.7   An Infinitesimal Variation in the Control History $\delta u(t)$

Also, let us integrate the $\lambda^T \dot{x}$ term in (2.54) by parts, yielding

$$\bar{J} = \phi[x(t_f)] - \lambda^T(t_f)x(t_f) + \lambda^T(t_0)x(t_0) + \int_{t_0}^{t_f} \left\{ H[x(t), u(t), \lambda(t), t] + \dot{\lambda}^T x(t) \right\} dt \;.$$
(2.56)

Now consider an infinitesimal variation in $u(t)$ that we shall call $\delta u(t)$ such as the one shown in Figure 2.7. Such a variation will produce variations in the state histories $\delta x(t)$ and a variation in the performance index $\delta \bar{J}$ that could be calculated from

$$\delta \bar{J} = \left[(\phi_x - \lambda^T) \delta x \right]_{t=t_f} + \left[\lambda^T \delta x \right]_{t=t_0} + \int_{t_0}^{t_f} \left[(H_x + \dot{\lambda}^T) \delta x + H_u \delta u \right] dt \;.$$
(2.57)

To avoid having to determine the functions $\delta x(t)$ produced by $\delta u(t)$, we choose the multiplier functions $\lambda(t)$ so that the coefficients of $\delta x(t)$ and $\delta x(t_f)$ in (2.57) vanish, i.e., we choose

$$\dot{\lambda}^T = -H_x \equiv -L_x - \lambda^T f_x \;,$$
(2.58)

with boundary conditions

$$\lambda^T(t_f) = \phi_x(t_f) \;.$$
(2.59)

Equation (2.57) then becomes

$$\delta \bar{J} = \lambda^T(t_0)\delta x(t_0) + \int_{t_0}^{t_f} H_u \delta u \, dt \;.$$
(2.60)

Thus $H_u(t)$ *is the impulse response function for* J, while holding $x(t_0)$ constant and satisfying (2.51), i.e., a unit impulse in δu at time t_1 will produce $\delta J = H_u(t_1)$. Also $\lambda^T(t_0) \equiv J_x(t_0)$, i.e., $\lambda^T(t_0)$ is the gradient of J with respect to $x(t_0)$, while holding $u(t)$ constant and satisfying (2.51). If $x(t_0)$ is specified, then $\delta x(t_0) = 0$.

For a stationary solution, $\delta \bar{J} = 0$ for arbitrary $\delta u(t)$; this can happen only if

$$H_u = 0 , \quad t_0 \leq t \leq t_f . \tag{2.61}$$

Equations (2.58), (2.59), and (2.61) are the EL equations in the calculus of variations.

Hence, to find a control vector function $u(t)$ that produces a stationary value of the performance index J, we must solve the following *differential equations*:

$$\dot{x} = f(x, u, t) , \tag{2.62}$$

$$\dot{\lambda} = -H_x^T \equiv -L_x^T - f_x^T \lambda , \tag{2.63}$$

where $u(t)$ is determined from (2.61), which may be written as

$$H_u \equiv L_u + \lambda^T f_u = 0 . \tag{2.64}$$

The boundary conditions for these differential equations are split, some at $t = t_0$ and some at $t = t_f$:

$$\dot{x}(t_0) \quad \text{specified} , \tag{2.65}$$

$$\lambda(t_f) = \phi_x^T . \tag{2.66}$$

This is a TPBVP. Note that (2.62) and (2.63) are coupled since $u(t)$ depends on $\lambda(t)$ through (2.64) and the coefficients in (2.63) depend on $x(t)$ and $u(t)$.

An Integral of the TPBVP

If the problem does not depend explicitly on time t, an integral of the TPBVP (2.62) to (2.66) exists that is called the *first integral* in the classical calculus of variations. It is derived by considering the time derivative of $H(x, \lambda, u, t)$:

$$\dot{H} = H_t + H_x \dot{x} + H_\lambda \dot{\lambda} + H_u \dot{u} ,$$
$$= L_t + \lambda^T f_t + \dot{\lambda}^T f + H_x \dot{x} + H_u \dot{u} ,$$
$$= L_t + \lambda^T f_t + H_u \dot{u} + (H_x + \dot{\lambda}^T) f .$$

Using (2.58) this becomes

$$\dot{H} = L_t + \lambda^T f_t + H_u \dot{u} .$$

However, from (2.61), $H_u = 0$ if $u(t)$ is a stationary control history. If, in addition, $L = L(x, u)$ and $f = f(x, u)$, then $L_t + \lambda^T f_t = 0$ and $\dot{H} = 0$, or

$$H = \text{constant for } t_0 \leq t \leq t_f . \tag{2.67}$$

Second-Order Sufficient Conditions

From Section 1.4, it follows that sufficient conditions for J to be a local minimum are: (a) $H_u = 0$ over the whole path, and (b) δJ is positive for arbitrary $\delta u(t)$ to second order, while holding $\dot{x} = f$, i.e.,

$$\delta J = \frac{1}{2}\left[\delta x^T \phi_{xx}\delta x\right]_{t=t_f} + \frac{1}{2}\int_{t_0}^{t_f}\left[\delta x^T \quad \delta u^T\right]\begin{bmatrix} H_{xx} & H_{xu} \\ H_{ux} & H_{uu}\end{bmatrix}\begin{bmatrix}\delta x \\ \delta u\end{bmatrix}dt \geq 0 , \quad (2.68)$$

where $\delta(\dot{x} - f) = 0$, or

$$\delta\dot{x} = f_x\delta x + f_u\delta u . \tag{2.69}$$

Clearly (2.68) will be positive for nonzero $\delta u(t)$ if

$$\phi_{xx} > 0 , \tag{2.70}$$

and, over the whole path,

$$\begin{bmatrix} H_{xx} & H_{xu} \\ H_{ux} & H_{uu}\end{bmatrix} > 0 . \tag{2.71}$$

By this we mean that these symmetric matrices have positive eigenvalues.

The conditions (2.70) and (2.71) are not necessary. *Second-order necessary conditions* are discussed in Chapter 8 using the concepts of neighboring optimal paths and perturbation feedback.

Example 2.3.1—Hamilton's Principle in Mechanics

The motion of a conservative mechanical system for $t_0 \leq t \leq t_f$ is such that the action integral

$$A \triangleq \int_{t_0}^{t_f} L(u, q, t)dt , \tag{2.72}$$

has a stationary value (Ref. Lz), where

$$L \triangleq T(u, q, t) - V(q, t) = \text{ the Lagrangian },$$
$$T = \text{ kinetic energy },$$
$$V = \text{ potential energy },$$
$$q = \text{ generalized coordinate vector },$$
$$u = \dot{q} = \text{ generalized velocity vector },$$
$$q(t_0) = q_0 .$$

The Hamiltonian is then

$$H = L + \lambda^T u . \tag{2.73}$$

In mechanics the Hamiltonian is defined as $H = -L + p^T u$, so that to minimize L one must maximize H; $p \equiv \lambda$ is called the generalized momentum vector.

The EL equations are then

$$\dot{\lambda}^T = -H_q \equiv -L_q ,$$

$$0 = H_u \equiv L_u + \lambda^T .$$

Eliminating λ^T between these last two relations, we have

$$\frac{d}{dt} (L_{\dot{q}}) - L_q = 0 , \tag{2.74}$$

which are *Lagrange's equations of motion for a conservative system.*

If L is not an explicit function of time, an integral of the motion is $H = $ constant, or

$$H = L - L_u u = T - V - T_u u = \text{ constant} .$$

Now T is a homogeneous quadratic function in u, so that

$$T_u u = 2T .$$

Hence, we have

$$-H = T + V = \text{ constant} , \tag{2.75}$$

i.e., the sum of the kinetic and potential energies is constant during the motion.

Example 2.3.2—A Variational Principle for Nonconservative Mechanical Systems

The motion of a nonconservative mechanical system for $t_0 \leq t \leq t_f$ is such that

$$\delta \int_{t_0}^{t_f} T(u, q)dt + \int_{t_0}^{t_f} Q^T(u, q)\delta q \, dt = 0 , \tag{2.76}$$

where

$$\dot{q} = u , \quad q(t_0) = q_0 , \tag{2.77}$$

$T(q, u)$ is the kinetic energy, and $Q(u, q)$ is the generalized force vector (Ref. Wh). $Q(u, q)$ is defined by the fact that the differential of *virtual work* done on the system by the generalized forces at time t is

$$Q^T(u, q)dq . \tag{2.78}$$

The second term in (2.76) is thus the time integral of the *virtual work*. It is *not* the variation of the work, since the work is a line integral along the path in the phase space:

$$W = \oint_{q(t_0)}^{q(t_f)} Q^T(u, q)dq ,$$

$$= \int_{t_0}^{t_f} Q^T(u, q)u\, dt ,$$

which prevents us from defining a "mechanics Hamiltonian" for nonconservative systems.

However, we can adjoin the constraint (2.77) to (2.76) with a Lagrange multiplier vector as follows:

$$\int_{t_0}^{t_f} \left[T_u \delta u + T_q \delta q + Q^T \delta q + \lambda^T (\delta u - \delta \dot{q}) \right] dt = 0 . \tag{2.79}$$

Integrating the last term by parts, we have

$$-\lambda(t_f)\delta q(t_f) + \int_{t_0}^{t_f} \left[\left(T_u + \lambda^T \right) \delta u + \left(T_q + Q^T + \dot{\lambda}^T \right) \delta q \right] dt = 0 . \tag{2.80}$$

We choose $\lambda(t)$ to make the coefficient of δq vanish, i.e.,

$$\dot{\lambda}^T = -T_q - Q^T , \quad \lambda(t_f) = 0 . \tag{2.81}$$

The integral in (2.80) can then vanish for arbitrary $\delta u(t)$ only if

$$\lambda^T = -T_u . \tag{2.82}$$

Eliminating λ between (2.81) and (2.82) gives

$$\frac{d}{dt}\left(T_{\dot{q}} \right) - T_q = Q^T , \tag{2.83}$$

which are *Lagrange's equations of motion for a nonconservative mechanical system.*

Example 2.3.3—Newton's Min Drag Nose Shape

This was one of the first problems solved in the calculus of variations (about 1686) by Isaac Newton. He apparently invented the calculus of variations to solve it without knowing the methods developed by Jakob Bernoulli. At that time, he presented only the solution to the problem, and not his method. For a given radius $r(0) = a$ and a given length ℓ (see Figure 2.8), the problem is to find $\theta(x) =$ angle between the velocity direction and the local tangent to the nose, as a function of distance x, to minimize the drag coefficient C_D of an axially symmetric nose

$$C_D \triangleq \frac{D}{q\pi a^2} = -2 \int_{x=0}^{\ell} C_p(\theta)r\, dr , \tag{2.84}$$

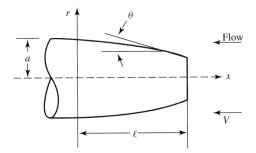

FIGURE 2.8   Nomenclature for Analyzing Minimum Drag Nose Shapes (Newton)

where (x, ℓ) are in units of a, and

$$\frac{dr}{dx} = -\tan\theta \; , \tag{2.85}$$

$$C_p = \begin{cases} 2\sin^2\theta \; , & \theta \geq 0 \\ 0 \; , & \theta \leq 0 \; , \end{cases} \tag{2.86}$$

and

$$q = \text{dynamic pressure} = \rho V^2/2 \; ,$$
$$V = \text{velocity of oncoming gas} \; ,$$
$$\rho = \text{density of oncoming gas} \; .$$

The expression for the pressure coefficient $C_p(\theta)$ was derived by Newton, who assumed that when a particle of gas hits the nose, it loses its normal component of velocity and then 'slides' along the nose. Nearly three centuries later, wind tunnel experiments showed that this expression is a good approximation at *hypersonic velocities,* where a shock wave turns the flow abruptly and lies very close to the surface of the nose.

We shall let

$$u \stackrel{\Delta}{=} \tan\theta \tag{2.87}$$

be the control variable, and allow for the possibility of a blunt tip by writing (2.84) in the form

$$C_D = 2[r(\ell)]^2 + 4\int_0^\ell \frac{ru^3}{1+u^2}dx \; . \tag{2.88}$$

The Hamiltonian for the system is

$$H = \frac{4\,ru^3}{1+u^2} + \lambda(-u) \; . \tag{2.89}$$

The EL equations are

$$\frac{d\lambda}{dx} = -H_r \equiv -\frac{4u^3}{1+u^2} \, , \tag{2.90}$$

$$0 = H_u \equiv \frac{4\,ru^2(3+u^2)}{(1+u^2)^2} - \lambda \, , \tag{2.91}$$

with boundary conditions

$$r(0) = 1 \, , \tag{2.92}$$

$$\lambda(\ell) = \phi_r(\ell) = 4\,r(\ell) \, . \tag{2.93}$$

Equations (2.90) to (2.93) must be solved with

$$\frac{dr}{dx} = -u \, . \tag{2.94}$$

Since neither the integrand in (2.88) nor the right-hand side of (2.94) are functions of the independent variable x, this second-order system has a first integral, namely $H = $ constant. Substituting λ from (2.91) into (2.89) gives

$$H = -\frac{8ru^3}{(1+u^2)^2} = \text{constant} \, . \tag{2.95}$$

Using (2.93) in (2.91) at $x = \ell$ yields

$$r(\ell)\left[1 - \frac{u^2(3+u^2)}{(1+u^2)^2}\right]_{x=\ell} = 0 \, ,$$

which is satisfied by either $r(\ell) = 0$ or

$$u(\ell) = 1 \, . \tag{2.96}$$

Using (2.96) in (2.95) gives the constant value of H:

$$H = -2r(\ell) \, , \tag{2.97}$$

and (2.97) and (2.95) give the radius r of the nose shape in terms of the slope u:

$$\frac{r}{r(\ell)} = \frac{(1+u^2)^2}{4u^3} \, . \tag{2.98}$$

This suggests integrating (2.94) in the form

$$-\int_\ell^x dx = \int_1^u \frac{dr}{du}\frac{du}{u}$$

which yields

$$\frac{\ell - x}{r(\ell)} = \int_1^u \frac{d}{du}\left(\frac{(1+u^2)^2}{4u^3}\right)\frac{du}{u} \, . \tag{2.99}$$

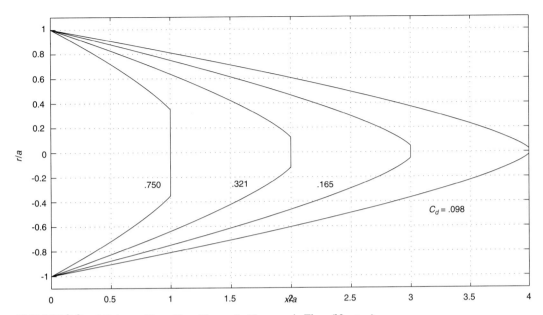

FIGURE 2.9   Minimum Drag Nose Shapes in Hypersonic Flow (Newton)

This integral can be evaluated in terms of tabulated functions:

$$\frac{\ell - x}{r(\ell)} = \frac{1}{4}\left(\frac{3}{4u^4} + \frac{1}{u^2} - \frac{7}{4} + \log u\right) . \tag{2.100}$$

Equations (2.98) and (2.100) are parametric equations for the optimum nose shape; choice of the slope u_0 at $x = 0$ determines the fineness ratio a/ℓ. Figure 2.9 shows optimum nose shapes for several values of a/ℓ.

The minimum drag can be found by integrating (2.88) using u as the independent variable:

$$C_D = \frac{u_0^2}{(1 + u_0^2)^4}\left(3 + 10u_0^2 + 17u_0^4 + 2u_0^6 + 4u_0^4 \log \frac{1}{u_0}\right) . \tag{2.101}$$

It is surprising that the hypersonic optimum nose shape has a blunt tip. However, for small fineness ratios, the blunt tip is hardly perceptible. In fact, as $a/\ell \to 0$, it can be shown that

$$\frac{r}{a} \to \left(\frac{\ell - x}{\ell}\right)^{3/4} , \tag{2.102}$$

$$C_D \to \frac{27}{16}\left(\frac{a}{\ell}\right)^2 . \tag{2.103}$$

Problems

2.3.1 Velocity Direction Programming (VDP) for Max Range with Gravity: This is the continuous version of Problem 2.1.1 and Example 2.1.1 where it was solved as the limiting case as $N \to \infty$. The equations of motion (EOM) are

$$\dot{V} = g \sin \gamma \ , \quad \dot{x} = V \cos \gamma \ , \quad \dot{y} = V \sin \gamma \ .$$

(a) For $V(0) = x(0) = 0$ and a given final time t_f, show that the path that maximizes $x(t_f)$ is the cycloid

$$V = \frac{2}{\pi} \cos \gamma \ , \quad x = \frac{2}{\pi^2}\left(\frac{\pi}{2} - \gamma - \frac{\sin 2\gamma}{2}\right) , \quad y = \frac{2}{\pi^2} \cos^2 \gamma \ , \quad \gamma = \frac{\pi}{2}(1 - t) \ ,$$

where V is in units of gt_f, (x, y) are in units of gt_f^2, and t is in units of t_f.

(b) Plot the optimal path y vs. x and compare your result with Figure 2.2.

2.3.2 VDP for Max Range with $u_c = Vy/h$; a Zermelo Problem: This is the continuous version of Problem 2.1.2 and Figure 2.4.

(a) For $x(0) = y(0) = 0$ and given t_f, show that the $\theta(t)$ function that maximizes $x(t_f)$ is the *linear tangent law*:

$$\tan \theta(t) = t_f - t \ ,$$

where t, t_f are in units of h/V.

(b) Using θ as the independent variable, show that the parametric equations of the optimal path are

$$x = \tfrac{1}{2}[\Lambda(\theta_0) - \Lambda(\theta) + \tan \theta \sec \theta - \tan \theta_0 \sec \theta_0] + t \sec \theta_0 \ ,$$

$$+ \sec \theta_0(\tan \theta - \tan \theta_0) \ ,$$

$$y = \sec \theta_0 - \sec \theta, \quad t = \tan \theta_0 - \tan \theta \ ,$$

where $\theta_0 = \tan^{-1} t_f$, (x, y) are in units of h, and $\Lambda(\theta) \overset{\Delta}{=}$ the *Lambda function* (cf. Dwight 640):

$$\Lambda(\theta) \overset{\Delta}{=} \sinh^{-1}(\tan \theta) \equiv \log(\tan \theta + \sec \theta) \equiv \log \tan(\theta/2 + \pi/4) \ .$$

(c) Plot the optimal path for $t_f = 2$ and compare your result with Figure 2.4.

2.3.3 VDP for Min Distance to a Meridian; a Geodesic Problem: This is the continuous version of Problem 2.1.3. We wish to find the heading angle β (positive counterclockwise from east) as a function of the longitude ϕ to minimize the distance traveled in going along the surface of a unit sphere from a point with (latitude, longitude) $= (\theta_0, \phi_0)$ to the meridian where $\phi = \phi_f$ (see Figure 2.5). The distance traveled is

$$J = \int_0^{\phi_f} \sec \beta \cos \theta d\phi \ ,$$

where $d\theta/d\phi = \tan \beta \cos \theta$, $\theta(0) = \theta_0$.

(a) Show that the min-distance path is $\tan \theta = \tan \theta_0 \cos(\phi_f - \phi)/\cos \phi_f$, and the optimum heading program is $\cos \beta = \sec \theta / \sec \theta_f$, where θ_f is given by $\tan \theta_f = \tan \theta_0 \sec \phi_f$. This is a *great circle* that ends perpendicular to the meridian ($\beta_f = 0$).

(b) Plot the min-distance path from $\phi_0 = 0$, $\theta_0 = 40$ deg, to the meridian $\phi_f = 50$ deg and compare the result with Figure 2.5.

2.3.4 VDP for Max Range with $V = V_0(1 + y/h)$; a Fermat Problem: This is the continuous version of Problem 2.1.4. In the (x, y) plane, the velocity magnitude of a particle varies linearly with y:

$$V = V(y) = V_0\left(1 + \frac{y}{h}\right).$$

By choosing the direction of the velocity $\theta(t)$, the particle can be steered along different paths:

$$\dot{x} = V(y)\cos\theta , \quad \dot{y} = V(y)\sin\theta .$$

(a) Starting from $x = y = 0$ and given the time t_f, show that paths with maximum $x(t_f)$ are given by the parametric equations

$$\frac{x}{h} = \frac{\sin\theta_0 - \sin\theta}{\cos\theta_0} , \quad \frac{y}{h} = -1 + \frac{\cos\theta}{\cos\theta_0} , \quad \frac{V_0 t}{h} = \Lambda(\theta_0) - \Lambda(\theta) ,$$

where $\Lambda(\theta)$ was defined in Problem 2.3.2(b).

(b) Show that the paths are segments of a circle of radius $h\sec\theta_0$ whose center is at $y = -h$, $x = h\tan\theta_0$ ending where $\theta = 0$.

(c) Plot the optimal path for normalized $t_f = 2$ and compare it with Figure 2.6.

2.3.5 VDP for Max Range with Gravity and Thrust: Consider a particle acted upon by a constant thrust specific force a along the path, in addition to the gravitational force, so that the equations of motion are (y positive down)

$$\dot{V} = a + g\sin\gamma , \quad \dot{x} = V\cos\gamma , \quad \dot{y} = V\sin\gamma .$$

(a) Show that the path that produces maximum $x(t_f)$ from zero initial state is given parametrically in terms of γ (which goes from $\pi/2$ to 0) as

$$\frac{V}{V_f} = \frac{\cos\gamma}{1 + a\sin\gamma} ,$$

$$\frac{x}{V_f^2} = \frac{1}{1 - a^2}\left[\alpha\left(\frac{\pi}{2}\right) - \alpha(\gamma)\right] - \frac{(a + \sin\gamma)\cos\gamma}{2(1 - a^2)(1 + a\sin\gamma)^2} ,$$

$$\frac{y}{V_f^2} = \frac{\cos^2\gamma + 2a\sin\gamma(1 - \sin\gamma)}{2(1 + a)^2(1 + a\sin\gamma)^2} ,$$

$$\frac{t}{V_f} = \frac{2}{1 - a^2}\left[\alpha\left(\frac{\pi}{2}\right) - \alpha(\gamma)\right] - \frac{a\cos\gamma}{(1 - a^2)(1 + a\sin\gamma)} ,$$

where a is in units of g, t is in units of t_f, V is in units of gt_f, (x, y) are in units of gt_f^2, and

$$\alpha(\gamma) \triangleq \frac{1}{\sqrt{1 - a^2}}\tan^{-1}\left[\frac{a + \tan(\gamma/2)}{\sqrt{1 - a^2}}\right] , \quad a < 1 ,$$

$$\triangleq \frac{-1}{\sqrt{a^2 - 1}}\tanh^{-1}\left[\frac{a + \tan(\gamma/2)}{\sqrt{a^2 - 1}}\right] , \quad a > 1 .$$

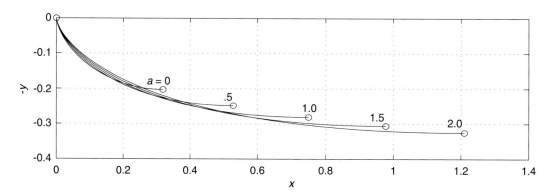

FIGURE 2.10   VDP for Max Range with Gravity and Thrust

V_f is determined by putting $t = 1$, $\gamma = 0$ in the equation for t. If $a = 1$, show that

$$x = \frac{3}{4}\cot^3\left(\frac{\gamma}{2}\right) , \quad y = \frac{9}{32}\frac{(1 - \sin\gamma)(1 + 3\sin\gamma)}{(1 + \sin\gamma)^3} , \quad t = \frac{1}{2}\cot\left(\frac{\gamma}{2}\right) + \frac{1}{6}\cot^3\left(\frac{\gamma}{2}\right) .$$

(b) Plot the optimal paths y vs. x for $a = [0 .5\ 1.0\ 1.5\ 2.0]$ and compare your results with Figure 2.10.

2.3.6 VDP for Max Range with Gravity, Thrust, and Drag: In Problem 2.3.5, add a specific drag force proportional to the square of the velocity, so that the equations of motion are (y positive up):

$$\dot{V} = a - g\sin\gamma - \frac{V^2}{\ell} , \quad \dot{x} = V\cos\gamma , \quad \dot{y} = V\sin\gamma ,$$

If we measure t in units of $\sqrt{\ell/g}$, V in $\sqrt{g\ell}$, (x, y) in ℓ, and a in g, then we may put $g = \ell = 1$ in the equations above.

A possible application is a submarine that wishes to get as far away horizontally as possible from its initial position in a given time. The value of ℓ for a submarine is on the order of 1500 ft.

(a) Show that the γ that maximizes $\dot{x}$ in steady state ($\dot{V} = 0$) is

$$\gamma_s = -\sin^{-1}\left[\sqrt{\left(\frac{a}{3}\right)^2 + \frac{1}{3}} - \frac{a}{3}\right] ,$$

and the corresponding steady velocity is

$$V_s = \sqrt{\frac{2a}{3} + \sqrt{\left(\frac{a}{3}\right)^2 + \frac{1}{3}}} .$$

Thus a fast (but nonoptimal) path is to dive vertically ($\gamma = -\pi/2$) until V_s is reached then switch to $\gamma = \gamma_s$. Calculate this path for $a = .05$ and compare your results with the dashed curve in Figure 2.11.

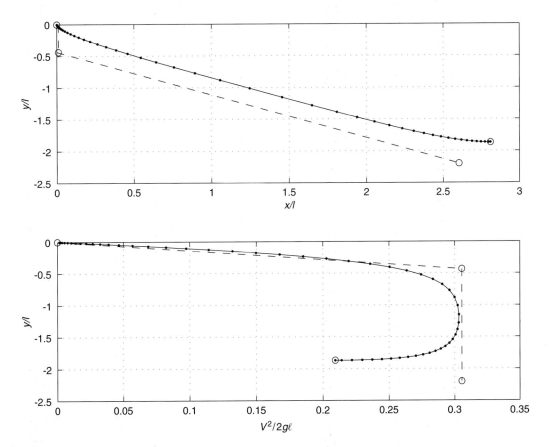

FIGURE 2.11   VDP for Max Range with Gravity, Thrust, and Drag; $a = .05$, $t_f = 5$

(b) Using $H_\gamma = 0$ and $\lambda_v(t_f) = 0$ show that $\gamma(t_f) = 0$.

(c) Using $H_\gamma = 0$ and the first integral (H=constant), show that

$$V_f = V \sec \gamma - V(a - V^2) \tan \gamma \ ,$$

where $V_f = V(t_f)$.

(d) Differentiate the first integral in (c) to show that

$$V\dot\gamma = \cos \gamma [1 - (a - 3V^2) \sin \gamma)] \ .$$

Since $V(0) = 0$ and $\dot\gamma(0)$ is finite, it follows that $\gamma(0) = -\pi/2$.

(e) Use the result in (c) with ODE23 in MATLAB to find the max range path starting from zero initial state for $t_f = 5$, $a = .05$ and compare your solution with the one shown in Figure 2.11. *Hint*: for $t \ll t_f$, show that $V \approx (1 + a)t$, $\gamma \approx -\pi/2 + kV$, where $k = $ constant; then choose k to yield $\gamma(t_f) = 0$ starting from $t = .03$.

2.3.7 Min Distance on a Sphere Using a Penalty Function: Find the heading angle β as a function of the longitude ϕ to minimize the distance traveled in going along the surface of a sphere between a point with (latitude, longitude) $= (\theta_0, 0)$ and a point with (latitude, longitude) $= (\theta_f, \phi_f)$. The distance traveled is

$$\int_0^{\phi_f} \sec \beta \cos \theta d\phi \ ,$$

where

$$\frac{d\theta}{d\phi} = \tan \beta \cos \theta \ , \quad \theta(0) = \theta_0 \ , \quad \theta(\phi_f) = \theta_f \ .$$

To handle the terminal constraint we use a *quadratic penalty* on the deviation of the final latitude from the desired final latitude, so that the performance index becomes

$$J = \frac{1}{2} S_f [\theta(\phi_f) - \theta_f]^2 + \int_0^{\phi_f} \sec \beta \cos \theta d\phi \ ,$$

(a) Show that the min distance path and optimal heading β are given by

$$\tan \theta = \tan \theta_m \sin(\phi - \alpha) \ , \quad \cos \beta = \frac{\cos \theta_m}{\cos \theta} \ ,$$

where

$$\tan \theta_0 = \tan \theta_m \sin(-\alpha) \ , \quad \tan \theta(\phi_f) = \tan \theta_m \sin(\phi_f - \alpha) \ ,$$

$$S_f [\theta(\phi_f) - \theta_f] = -\sqrt{1 - \frac{\cos^2 \theta_m}{\cos^2 \theta(\phi_f)}} \ .$$

These three equations determine θ_m, α, and $\theta(\phi_f)$. As $S_f \to \infty$, $\theta(\phi_f) \to \theta_f$ leaving two equations for determining θ_m and α (see Problem 3.3.3).

(b) Plot the min distance path, θ vs. ϕ, between Tokyo (latitude 35.7 deg north, longitude 139.7 deg east) and New York (latitude 40.7 deg north, longitude 73.8 deg west) using $S_f \to \infty$. Compare your results with Figure 2.12.

2.4 Numerical Solution with Gradient Methods

Using the *Mayer formulation* (see Section 2.2) we first describe an iterative gradient algorithm FOP0 (for Function OPtimization with 0 terminal constraints) which is the continuous version of DOP0. As in Section 2.2, the MATLAB command FMINU can also be used, which does not require analytical gradients. In Section 2.6, we describe an NR algorithm FOP0N and shooting algorithms FOP0F and FOP0B that yield exact solutions for the optimal control and state histories.

A gradient algorithm FOP0 is outlined below for solving continuous dynamic optimization problems without terminal constraints. It is identical to DOP0 described in Section 2.2, except that *differential equations are integrated* forward and backward instead of sequencing difference equations.

- Choose integration step $\Delta T = t_f / N$, where $N =$ even integer.
- Enter data, including a guess of $u(t)$ at the $N + 1$ points $t - t_0 = 0, \cdots, N \Delta T$.
- (*) Forward integration. Compute and store $x(t)$ at $t - t_0 = \Delta T, \cdots, N \Delta T$.

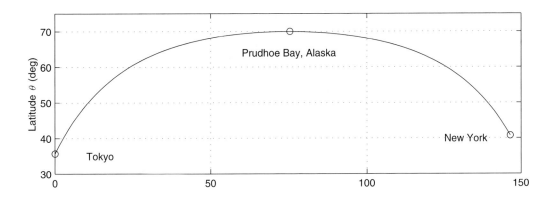

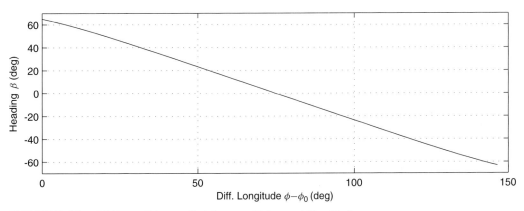

FIGURE 2.12   Minimum Distance Path between Tokyo and New York

- Evaluate $\phi[x(t_f)]$ and $\lambda^T(t_f) = \phi_x(t_f)$.
- Backward integration. Compute and store impulse response function $H_u(t)$. For $t - t_0 = N\Delta T, \cdots, 0$:

$$H_u = \lambda^T f_u \ , \tag{2.104}$$

$$\dot{\lambda}^T = -\lambda^T f_x \ . \tag{2.105}$$

- Compute $\delta u(t)$ and δu_{avg}. Choose $k < 0$ if maximizing or $k > 0$ if minimizing. Then, for $t - t_0 = 0, \cdots, N\Delta T$:

$$\delta u(t) = -k H_u^T(t) \ , \tag{2.106}$$

$$\delta u_{avg} = \sqrt{\frac{1}{t_f} \int_0^{t_f} \delta u^T(t) \delta u(t) dt} \ . \tag{2.107}$$

- If $|\delta u_{avg}| < tol$ then stop.

- Compute new $u(t)$:

$$u(t) \leftarrow u(t) + \delta u(t) . \qquad (2.108)$$

- Go to $(*)$.

FOP0—A MATLAB Code for Continuous Dynamic Optimization without Constraints

A MATLAB code for implementing the FOP0 algorithm is listed in Table 2.2. It is identical to the DOP0 code of Section 2.2 except that integration of differential equations replaces sequencing of difference equations. It uses the MATLAB integration code ODE23 that requires the use of a subroutine (FOP0-F) for the forward integration and another (FOP0-B) for the backward integration. The user must create a function file 'name.m' that contains the data for the problem to be solved.

The control histories $u(t)$ and 'name' are passed on to the ODE23 subroutine that calculates the derivative given t and s. In the subroutine input after t, s, insert the word "flag" and then tu, $name$, where tu is a table whose first column is the time history and whose $(i + 1)^{st}$ column is the ith control history. Since ODE23 uses variable time steps, the time steps in the backward integration are, in general, different from those in the forward integration. Since these integrations are interdependent some variables must be interpolated from one integration to the next; this is done using the MATLAB interpolation command TABLE1 (this command will be replaced by INTERP1 in versions of MATLAB after Version 5.1). The author wishes to thank Paul M. Montgomery for showing him how to use the TABLE1 command (1995) for this purpose.

The initial estimate of the optimal control $u(t)$ is in the form of a table $tu \stackrel{\Delta}{=} [t \ \ u]$, where t is a column vector of times (not necessarily evenly spaced) and the ith column of u contains the corresponding values of the ith element of the control vector.

Example 2.4.1—Min Drag Nose Shape Using FOP0

This is the same problem solved analytically in Example 2.3.3. The data file is listed below for the case $\ell/a = 4$. Note the independent variable is distance x instead of time, and the state vector is $[d \ \ r]^T$, where $4d$ is the cumulative drag and r is the radius.

```
function [f1,f2]=noshp(u,s,t,flg)
% t --> x = distance; state s=[d r]'; u=- tan(theta);
% lengths in r(0); drag in q*pi*(r(0))^2; q=dyn. pressure.
d=s(1); r=s(2);
if flg==1,                              % f1 = f
  f1=[4*r*u^3/(1+u^2); -u];
elseif flg==2,                          % f1=phi, f2=phis
  f1=2*r^2+d;
  f2=[1 4*r];
elseif flg==3,                          % f1=fs, f2=fu
  f1=zeros(2,2); f1(1,2)=4*u^3/(1+u^2);
  f2=[4*r*u^2*(3+u^2)/(1+u^2)^2; -1];
end
```

TABLE 2.2   A MATLAB Code for FOP0

```
function [tu,ts,la0]=fop0(name,tu,tf,s0,k,told,tols,mxit)
% Function OPtim. w. 0 terminal constraints; name must be in single
% quotes; function file 'name' computes sdot=f(s,u) for flg=1,
% (phi,phis) for flg=2, fs for flg=3, fu for flg=4; inputs: tu where
% tu=[t(N1),1) u(N1,nc)]= initial estimate of u(t); s0(ns,1)=initial
% state, tf=final time, k=step size parameter, told=reltol for ODE23;
% stops when norm(dua)< tols or no. iterations > mxit; outputs:
% optimal tu and ts where ts=[t(N,1) s(N,ns)]), la0=initial adjoint
% vector.
%
% BASIC version AEB (1984); MATLAB versions Sun H. Hur (1990),
% Fred A. Wiesinger (1991), Paul M. Montgomery (1995), AEB (1997)
%
if nargin<8, mxit=10; end; it=0; dua=1;
disp('     Iter.        phi        dua');
options=odeset('reltol',told); [dum,nc1]=size(tu); nc=nc1-1;
dum=zeros(nc,1);
while norm(dua)>tols,
  [t,s]=ode23('fop0_f',[0 tf],s0,options,tu,name); ts=[t s];
  Ns=length(t); [phi,phis]=feval(name,dum,s(Ns,:)',tf,2);
  [tb,la]=ode23('fop0_b',[tf 0],phis',options,tu,ts,name);
  N=length(tb); la0=la(N,:); u=zeros(N,nc); du=u;
  for i=1:N,
    s1=table1(ts,tb(i)); u1=table1(tu,tb(i));
    [fs,fu]=feval(name,u1,s1,tb(i),3); Hu=la(i,:)*fu;
    du(i,:)=-k*Hu; u(N-i+1,:)=u1+du(i,:);
  end
  for j=1:nc, dua(j)=norm(du(:,j))/sqrt(N); end;
  if mxit==0, disp([it phi dua]); break; end
  disp([it phi dua]);
  t1=zeros(N,1); for i=1:N, t1(i)=tb(N-i+1); end; tu=[t1 u];
  if it>=mxit, break, end;
  it=it+1;
end

function yp=fop0_f(t,s,flag,tu,name)
% Subroutine for FOP0 - fwd integ. sdot=f(s,u).
%
u=table1(tu,t); yp=feval(name,u,s,t,1);

function yp=fop0_b(t,y,flag,tu,ts,name)
% Subroutine for FOP0; bkwd. integ. ladot=-fs'*la.
%
la=y; u=table1(tu,t); s=table1(ts,t);
fs=feval(name,u,s,t,3); yp=-fs'*la;
```

The MATLAB script file below provides an initial guess of the nose slope $u(t)$ at a finite number of points and uses FOP0 to solve the problem.

```
% Script e02_4_1.m; min drag nose shape using FOP0;
%
u=[.1826 .1850 .1875 .1902 .1932 .1964 .1998 .2037 .2078 ...
    .2127 .2178 .2237 .2305 .2386 .2483 .2598 .2765 .2942 ...
    .3373 .3816 .6767]';
N=length(u)-1; tf=4; t=tf*[0:1/N:1]'; tu=[t u]; s0=[0 1]';
name='noshp'; k=.12; told=1e-5; tols=3e-4; mxit=10;
[tu,ts,la0]=fop0(name,tu,tf,s0,k,told,tols,mxit);
t1=ts(:,1); r=ts(:,3); t2=tu(:,1); u=tu(:,2);
```

Use of the MATLAB Code FMINU

The FMINU command in MATLAB can also be used to solve continuous dynamic optimization problems. The user must provide a MATLAB function file that computes $f(u)$ given u, which, in this case, means integrating (2.105) forward *with equal time steps* to calculate ϕ (which must be called 'f' when using FMINU).

Example 2.4.2—Min Drag Nose Shape Using FMINU

The MATLAB function file below calculates the drag given a slope function $u(t)$. The code ODEU is a code like ODE23 except that it uses *equal time steps*.

```
function f=noshp_f(u,tf)
% t --> x=distance; s=[d r]'; u=-tan(theta); lengths in a=r(0); drag in
% q*pi*a^2 where q = dynamic pressure;
%
s0=[0 1]'; N=length(u); [t,s]=odeu('noshp',u,s0,tf); f=2*s(2,N)^2+ s(1,N);
```

The MATLAB script file below provides an initial guess of the nose slope $u(t)$ at a finite number of points and some options for FMINU.

```
% Script e02_4_2.m; min drag nose-shape pb. using FMINU; t --> x=distance;
% s=[d r]'; u=- tan(theta); lengths in a=r(0); drag in q*pi*a^2 where q=
% dynamic pressure.
%
% Converged u(t);
u=[.1826 .1850 .1875 .1902 .1932 .1964 .1998 .2037 .2078 .2127 .2178 ...
    .2237 .2305 .2386 .2483 .2598 .2765 .2942 .3373 .3816 .6767];
tf=4; optn(1)=1; optn(14)=500; u=fminu('noshp_f',u,optn,[],tf);
```

The results are identical to those obtained with FOP0 but required many more iterations since the gradients are computed by finite differencing.

Problems

2.4.1 to 2.4.7:   Do Problems 2.3.1 to 2.3.7 numerically using FOP0 or FMINU.

2.5 Direct Solution Methods for Discrete Systems

First-order gradient algorithms like DOP0 find the optimal solution with sufficient accuracy for most practical problems but often require many iteration steps. After a few steps the control may be close enough to optimal to use direct methods that converge very quickly. We discuss two direct methods in this section; both start with the results from a nearly converged first-order gradient solution. The first method, which is a quasi–NR method, is similar to DOP0 except that we use FSOLVE to perturb $u(i)$ until $H_u(i)$ is "zero" to the desired accuracy. The second method, called "shooting," sequences the coupled EL equations either forward (or backward), computing the optimal control from $H_u = 0$ at each step; an estimate of the initial adjoint vector $\lambda(0)$ [or the final state $s(N)$] is obtained from DOP0 and perturbed using FSOLVE until the final (or initial) boundary conditions are satisfied to the desired accuracy.

The NR method is *robust* since it sequences the plant equations forward and the adjoint equations backward. The shooting method is *not robust since it sequences the plant and adjoint equations together,* which is an inherently unstable process; it works well for systems with no energy dissipation like spacecraft flight paths, but not for long aircraft flight paths where there is dissipation of energy through the aerodynamic drag (the unstable computation overflows or underflows the computer).

An NR Algorithm DOP0N

The NR algorithm given here (for scalar u) requires a good initial guess of the the control history $u(t)$. However it does not require analytical second derivatives of the input data. Another type of NR algorithm, discussed in Chapter 8, requires analytical second derivatives and calculates the second variation using a matrix Riccati equation.

A MATLAB implementation of a NR algorithm called DOP0N is listed in Table 2.3; it starts with a nearly converged solution from DOP0 and uses the same data subroutine as DOP0.

Example 2.5.1—DVDP for Max Range Using DOP0N

A script for solving Example 2.1.1 using DOP0N is listed below. It uses the same subroutine DVDP0 used with DOP0.

```
% Script e02_5_1.m; DVDP for max range with gravity using N;
%
N=5; u=(pi/2)*[1:-1/N:1/N]; name='dvdp0'; s0=[0 0]'; tf=1;
optn(1)=1; optn(14)=500; u=fsolve('dop0n',u,optn,[],name,s0,tf);
[f,s,la0]=dop0n(u,name,s0,tf); v=s(1,:); x=s(2,:);
```

TABLE 2.3   A MATLAB Code for DOP0N

```
function [f,s,la0]=dop0n(u,name,s0,tf)
% Discrete OPtim. w. 0 term. constr. using a Newton-raphson algorithm and
% FSOLVE; function file 'name' should be in the MATLAB path & calculate f
% for flg=1, (phi,phis) for flg=2, and (fs,fu) for flg=3; u=initial guess
% of (scalar) control sequence; s0=initial state; tf=final time.
%
% Forward sequence, storing s(:,i):
N=length(u); ns=length(s0); dt=tf/N; s(:,1)=s0;
for i=1:N, t=(i-1)*dt; s(:,i+1)=feval(name,u(i),s(:,i),dt,t,1); end
%
% Performance index and boundary condition phis:
[phi,phis]=feval(name,0,s(:,N+1),0,0,2); la=phis';
%
% Backward sequence, storing Hu(i):
for i=N:-1:1, t=(i-1)*dt;
 [fs,fu]=feval(name,u(i),s(:,i),dt,t,3); Hu(i)=la'*fu; la=fs'*la;
end; la0=la; f=Hu;
```

An edited diary of the solution is listed below.

```
% Edited diary of solution of Example 2.5.1;
e02_5_1
f-COUNT       RESID     STEP-SIZE      GRAD/SD  LINE-SEARCH
    6    0.00323107           1     -0.00646
    -          -              -          -
   38 1.64777e-027            1      5.82e-022  int_step
Optimization Terminated Successfully
v=       0  0.1975  0.3757  0.5172  0.6080  0.6392
x=       0  0.0031  0.0291  0.0923  0.1925  0.3157
u= 1.4137  1.0996  0.7854  0.4712  0.1571
```

A Forward Shooting Algorithm DOP0F

In this algorithm the coupled EL equations are sequenced forward using an estimate of $\lambda(0)$ from DOP0 (or even better, from DOP0N); the optimal control $u(i)$ is determined at each step. Then the initial adjoint vector $\lambda(0)$ is perturbed using FSOLVE until the final boundary conditions are satisfied to the desired accuracy. This is a sensitive procedure since the EL equations are unstable for sequencing either forward or backward. A poor guess of $\lambda(0)$ results in computer overflow or underflow before the final step is reached, i.e., the computation stops and the computer gives an error message.

The discrete EL equations were given in (2.12) through (2.16). Using the Mayer formulation $L \equiv 0$ (scc Section 2.2), they may be written as

$$s(i + 1) = f[s(i), u(i)] , \qquad (2.109)$$

$$\lambda(i) = f_s^T[s(i), u(i)]\lambda(i + 1) , \qquad (2.110)$$

$$0 = H_u \equiv f_u^T[s(i), u(i)]\lambda(i + 1) . \qquad (2.111)$$

Equations (2.110) and (2.111) determine $\lambda(i + 1)$ and $u(i)$ given $s(i)$, $\lambda(i)$. Once $u(i)$ is determined, $s(i + 1)$ may be found from (2.109). Explicit solutions of (2.110) and (2.111) for $\lambda(i + 1)$ and $u(i)$ are sometimes possible but, in general, they are nonlinear implicit equations and must be solved using an iterative Newton procedure [i.e., linearized about an initial guess of $\lambda(i + 1)$ and $u(i)$ and then the perturbed linear equations solved]. This provides an improved estimate, and the procedure is iterated until the equations are satisfied to the desired precision. An algorithm for forward shooting is outlined below.

- Use DOP0 or DOP0N to obtain a good estimate of $\lambda(0)$.
- Sequence the coupled EL equations forward from $i = 0$ to $i = N$.
- At each step use the MATLAB code FSOLVE to calculate optimal $u(i)$ and $\lambda(i + 1)$ from (2.110) and (2.111). This requires the same data subroutine used for DOP0 and DOP0N.
- Calculate the error in the final boundary condition $e_f \overset{\Delta}{=} \lambda(N) - \phi_s$, which again uses the same data subroutine used for DOP0 and DOP0N.
- Use the MATLAB code FSOLVE to perturb $\lambda(0)$ until $e_f(N) = 0$ to the desired accuracy.

A MATLAB code that implements this algorithm is listed in Table 2.4. It uses FSOLVE to solve the implicit equations (2.110) and (2.111) in the subroutine DOPFU, which is listed below DOP0F.

Example 2.5.2—DVDP for Max Range Using DOP0F

This is the same problem solved in Examples 2.2.1 and 2.5.1. The script below solves the problem using DOP0F and the same subroutine DVDP0 used with DOP0.

```
% Script e02_5_2.m;; DVDP for max range with gravity using DOP0F;
%
N=5; s0=[0 0]'; tf=1; N1=N+1; la0=[.6392 1]'; u0=1.4137;
name='dvdp0'; optn(1)=1; optn(14)=100;
la0=fsolve('dop0f',la0,optn,[],name,u0,s0,tf,N);
[f,s,u,la]=dop0f(la0,name,u0,s0,tf,N); v=s(1,:); x=s(2,:);
```

The results agree exactly with those of Example 2.5.1 as expected.

A Backward Shooting Algorithm DOP0B

In this algorithm the coupled EL equations are sequenced *backward* using an estimate of $s(N)$ from DOP0 or DOP0N; the optimal control $u(i)$ is determined at each step using

TABLE 2.4   A MATLAB Code for DOP0F

```
function [f,s,u,la]=dop0f(la0,name,u0,s0,tf,N)
% Disc. OPtim. w. 0 term. constr., Fwd shooting; name must be single quotes;
% function file 'name' must be in the MATLAB path, giving f for flg=1,
% (phi,phis) for flg=2, and (fs,fu) for flg=3; (u0,la0)=estimate of initial
% (control, adjoint vector); s0=initial state; tf=final time; N=number of
% steps; f=la-phis'= error in final boundary condition;
%
dt=tf/N; ns=length(s0); s=zeros(ns,N+1); la=s;
u=zeros(1,N); s(:,1)=s0; la(:,1)=la0;
for i=1:N, t=(i-1)*dt;
 if i==1, u1=u0; else u1=u(i-1); end;
 l1=la(:,i); z0=[u1; l1];
 z1=fsolve('dopfu',z0,[],[],name,s(:,i),la(:,i),dt,t);
 u(i)=z1(1); la(:,i+1)=z1([2:ns+1]);
 s(:,i+1)=feval(name,u(i),s(:,i),dt,t,1);
end
[phi,phis]=feval(name,u(N),s(:,N+1),dt,tf,2); f=la(:,N+1)-phis';

function f=dopfu(z0,name,s,la,dt,t)
% Subroutine for DOP0F; finds u(i), lambda(i+1) given
% s=s(:,i), la=lambda(i), and initial guess z0.
%
[ns,dum]=size(s); u1=z0(1); l1=z0([2:ns+1]);
[fs,fu]=feval(name,u1,s,dt,t,3); f=[fs'*l1-la; fu'*l1];
```

a Newton procedure. Then $s(N)$ is perturbed using FSOLVE until the initial boundary conditions are satisfied to the desired accuracy. This again is a sensitive procedure since the EL equations are unstable for sequencing backward as well as forward.

Equations (2.109) and (2.111) determine $s(i)$ and $u(i)$ given $s(i + 1)$, $\lambda(i + 1)$. Then $\lambda(i)$ is determined from (2.110). Again, for simple problems, explicit solutions of (2.109) and (2.111) are sometimes possible but, in general, they are nonlinear implicit equations and must be solved using the procedure described above in the forward shooting algorithm. A MATLAB code for backward shooting is listed in Table 2.5. It uses a subroutine DOPBU to solve the implicit equations (2.109) and (2.111) using FSOLVE, which is listed below DOP0B.

Example 2.5.3—DVDP for Max Range with Gravity Using DOP0B

This is the same problem solved in Examples 2.5.1 and 2.5.2 and uses the same subroutine. The results agree exactly with the results in Examples 2.5.1 and 2.5.2 as expected. The script below solves the problem using DOP0B.

TABLE 2.5   A MATLAB Code for DOP0B

```
function [f,s,u,la]=dop0b(sf,name,uf,s0,tf,N)
% Disc. OPtim. w. 0 term. constr., Bkwd shooting; name must be single
    quotes;
% function file 'name' must be in the MATLAB path, giving f for flg=1,
% (phi,phis) for flg=2, and (fs,fu) for flg=3; (uf,sf)=estim. final
    (control,
% state); s0=desired initial state; N=no. steps; f=error in init. state.
%
ns=length(sf); dt=tf/N; [phi,phis]=feval(name,uf,sf,dt,1,2);
s(:,N+1)=sf; la(:,N+1)=phis';
%
% Backward sequencing of Euler-Lagrange eqns:
for i=N:-1:1, t=(i-1)*dt;
  if i==N, u1=uf; else u1=u(i+1); end
  s1=s(:,i+1); z0=[u1; s1];
  z1=fsolve('dopbu',z0,[],[],name,s(:,i+1),la(:,i+1),dt,t);
  u(i)=z1(1); s(:,i)=z1([2:ns+1]);
  [fs,fu]=feval(name,u(i),s(:,i),dt,t,3); la(:,i)=fs'*la(:,i+1);
end
f=s(:,1)-s0;

function f=dopbu(z0,name,s,la,dt,t)
% Subroutine for DOP0B; finds u(i), s(i) given s=s(:,i+1),
% la=lambda(:,i+1), & initial guess z0.
%
ns=length(s); u1=z0(1); s1=z0([2:ns+1]); fi=feval(name,u1,s1,dt,t,1);
[fs,fu]=feval(name,u1,s1,dt,t,3); f=[s-fi; fu'*la];
```

```
% Script e02_5_3.m; DVDP for max xf w. gravity using DOP0B;
%
N=5; sf=[.6392 .3157]'; uf=0; s0=[0 0]'; tf=1; name='dvdp0';
optn(1)=1; optn(14)=50;
sf=fsolve('dop0b',sf,optn,[],name,uf,s0,tf,N);
[f,s,u,la]=dop0b(sf,name,uf,s0,tf,N); v=s(1,:); x=s(2,:);
```

Problems

2.5.1 to 2.5.7:   Do Problems 2.2.1 to 2.2.7 using DOP0N and DOP0F or DOP0B. Note that Problem 2.5.6 is impossible using single shooting for $t_f >$ about 4; the optimal path is mostly at steady state with transition segments of time duration $\Delta t \approx .25$ at both ends (see Problem 5.2.4 for an analytic example).

TABLE 2.6   A MATLAB Code for FOP0N

```
function [f,s,la0]=fop0n(p,name,s0,tf)
% Function OPtimization with 0 term. constraints using a NR algor. with
% 'fsolve'; plant eqns fwd., then adjoint eqns bkwd; a NR code since FSOLVE
% perturbs p=u; name must be in single quotes; func. file 'name' computes
% sdot=f(s,u) for flg=1, (phi,phis) for flg=2, and (fs,fu) for flg=3;
% inputs: u = guess of control history (1 by N+1); tf = spec. final time;
% s0(ns,1)=initial state; outputs: f=Hu; s = optimal state histories;
% la0=initial adjoint vector;
%
ns=length(s0); N=length(p)-1; s=zeros(ns,N+1); Hu=zeros(1,N+1); u=p([1:N+1]);
[t,s]=odeu(name,u,s0,tf); [Hu,phi,la0]=odeh(name,u,s,tf); f=Hu;
```

2.6 Direct Solution Methods for Continuous Systems

As in the previous section on discrete systems, we present an NR code and forward and backward shooting algorithms. The main difference is that we integrate differential equations instead of sequencing difference equations. In the MATLAB codes below we use ODEU (which uses equal time steps) for the NR code and ODE23 (which uses variable time steps) for the shooting codes. The comments made about the discrete codes regarding sensitivity and robustness also apply here, namely, the NR algorithm is robust while the shooting algorithms are not.

An NR Algorithm FOP0N

A MATLAB code FOP0N that implements a continuous NR algorithm for continuous dynamic systems is listed in Table 2.6. The forward and backward differential equation subroutines ODEU and ODEH are equal time-step RK codes; they are included on the disk that accompanies this book.

A MATLAB script for an example using FOP0N is listed below. It uses the same data subroutine 'noshp' used by FOP0.

Example 2.6.1—Min Drag Nose Shape Using FOP0N

```
% Script e02_6_1.m; min drag nose shape using FOP0N;
%
% Converged initial guess from FOP0:
u0=[.1820 .1844 .1869 .1896 .1926 .1958 .1992 .2030 .2072 .2118 .2170 ...
    .2229 .2296 .2375 .2470 .2587 .2737 .2944 .3259 .3854 .8795];
s0=[0 1]'; tf=4; name='noshp'; optn(1)=1; optn(14)=500;
u=fsolve('fop0n',u0,optn,[],name,s0,tf); [t,s]=fop0n(u,name,s0,tf);
```

The results of running this script are shown below in an edited MATLAB diary.

TABLE 2.7   A MATLAB Code for FOP0F

```
function [f,t,y]=fop0f(p,name,s0,tf)
% Fcn. OPtim. w. 0 term. constr. using Fwd shooting; p=la0' from a nearly
% converged solution of FOP0; s0(1,ns)=initial state vector, tf=final
% time; name must be in single quotes; 2 function files 'namee' & 'name'
% must be on the MATLAB path, 'namee' ==> EL equations & 'name' ==>
% (phi,phis); use FSOLVE to iterate p until f=0 to desired accuracy.
%
ns=length(s0); la0=p([1:ns])'; y0=[s0; la0]; tol=1e-5;
options=odeset('reltol',tol);
[t,y]=ode23([name,'e'],[0 tf],y0,options); N=length(t);
sf=y(N,[1:ns])'; [phi,phis]=feval(name,0,sf,tf,2);
f=y(N,[ns+1:2*ns])'-phis';
```

```
% Diary e02_6_1.dia
e02_6_1
f-COUNT        RESID    STEP-SIZE        GRAD/SD  LINE-SEARCH
  22   0.0000147179            1    -0.0000294

   -       - -                 -           -
 168   1.80728e-021            1     4.77e-019  int_step
Optimization Terminated Successfully
```

FOP0F—A MATLAB Forward Shooting Code for Continuous Dynamic Optimization without Terminal Constraints

A MATLAB code FOP0F that implements a continuous forward shooting algorithm is listed in Table 2.7. It uses the MATLAB command ODE23, which is an accurate variable stepsize code for solving ordinary differential equations. The EL equations must be in subroutine 'namee', and (ϕ, ϕ_s) must be in subroutine 'name' (the same subroutine used for FOP0 and FOP0N). The reason for a separate subroutine for the EL equations is to take advantage of cases where it is possible to find *explicit solutions $u = u(x, \lambda)$ of the optimality condition* $H_u(x, \lambda, u) = 0$. It is possible to solve the optimality condition at every integration step using FSOLVE, but this is not only much slower but also less reliable.

Example 2.6.2—Min Drag Nose Shape Using FOP0F

As an example we do the same problem done in Example 2.6.1. It is possible here to find the optimal $u(t)$ explicity in terms of the state and adjoint variables (see 'noshpe.m' below).

```
% Script e02_6_2.m; fwd. shooting solution of min drag
% nose shape;
%
tf=4; s0=[0 1]'; name='noshp'; la0=[1 .3785]; p0=la0; optn(1)=1;
```

TABLE 2.8   A MATLAB Code for FOP0B

```
function [f,t,y]=fop0b(sf,name,s0,tf)
% Fcn. OPtim. w. 0 term. constr. using Bkwd. shooting; sf is an estimate
% from a converged solution of FOP0; s0=initial state vector; tf=final
% time; name must be in single quotes; function files 'namee' & 'name'
% must be on the MATLAB path; 'namee' ==> EL eqns, 'name' ==> (phi,phis)
% for flg=2; use FSOLVE to iterate p until f=0 to desired accuracy;
%
ns=length(s0); [phi,phis]=feval(name,0,sf,tf,2); yf=[sf; phis'];
tol=1e-5; options=odeset('reltol',tol);
[t,y]=ode23([name,'e'],[tf 0],yf,options); N=length(t);
f=y(N,[1:ns])'-s0;
```

```
optn(14)=500; p=fsolve('fop0f',p0,optn,[],name,s0,tf);
[f,t,y]=fop0f(p,name,s0,tf); d=y(:,1); r=y(:,2); N=length(t);
```

```
function yp=noshpe(t,y)
% Subroutine for Example 2.6.2; EL eqns. for shooting soln. of
% min drag nose shape; t --> x, state = [dD/dx r]'; u = tan(theta);
% (r,l) in units of "a" = r(0), drag in q*pi*a^2, q = dynamic pressure;
%
r=y(2); lr=y(4); A=4*r-lr; B=12*r-2*lr; C=-lr;
u=sqrt((-B+sqrt(B^2-4*A*C))/(2*A));
yp=[4*r*u^3/(1+u^2); -u; 0; -4*u^3/(1+u^2)];
```

An edited diary of the solution is shown below.

```
e02_6_2
f-COUNT      RESID      STEP-SIZE      GRAD/SD    LINE-SEARCH
      3 4.57559e-008          1      -9.15e-008
      9 8.85595e-009       1.44       4.03e-008    int_step
     14 9.36308e-022          1       5.76e-015    int_step
Optimization Terminated Successfully
```

FOP0B—A MATLAB Backward Shooting Code for Continuous Dynamic Optimization with 0 Terminal Constraints

A MATLAB code FOP0B that implements a continuous backward shooting algorithm is listed in Table 2.8. It uses the same subroutines 'namee' and 'name' used by FOP0F.

Example 2.6.3—Min Drag Nose Shape Using FOP0B

As an example we do the same problem done in Examples 2.6.1 and 2.6.2.

```
% Script e02_6_3.m; min drag nose shape using FOP0B;
%
tf=4; s0=[0 1]'; name='noshp'; sf=[.0971 .0228]'; optn(1)=1; optn(14)=500;
sf=fsolve('fop0b',sf,optn,[],name,s0,tf);
[f,t,y]=fop0b(sf,name,s0,tf); d=y(:,1); r=y(:,2); N=length(t);
```

The results are identical to those from the forward shooting solution, as they should be.

Problems

2.6.1 to 2.6.7:   Do Problems 2.4.1 to 2.4.7 using FOP0N and FOP0F or FOP0B. Note that Problem 2.6.6 is impossible using single shooting for $t_f >$ about 4; the optimal path is mostly at steady state with transition segments of time duration $\Delta t \approx .25$ at both ends (see Problem 5.2.4 for an analytic example).

2.7 Chapter Summary

Discrete Dynamic Optimization with No Terminal Constraints

Choose the sequence $u(i)$, $i = 0, \ldots, N - 1$ to minimize

$$J = \phi[x(N)] + \sum_{0}^{N-1} L[x(i), u(i), i] ,$$

subject to

$$x(i + 1) = f[x(i), u(i), i] , \quad x(0) = x_o ,$$

where N is specified.

Necessary Conditions for a Stationary Path
Let

$$H(i) \stackrel{\Delta}{=} L[x(i), u(i), i] + \lambda^T(i + 1)f[x(i), u(i), i] .$$

Then

$$x(i + 1) = f[x(i), u(i), i] ,$$
$$\lambda(i) = H_x^T(i) \equiv L_x^T(i) + f_x^T(i)\lambda(i + 1) , \quad \lambda(N) = \phi_x^T ,$$

where $u(i)$ is determined from

$$H_u(i) \equiv L_u(i) + \lambda^T(i + 1)f_u(i) = 0 .$$

This forms a TPBVP for $x(i)$ and $\lambda(i)$.

Continuous Dynamic Optimization without Terminal Constraints

Choose the function $u(t)$, $t_0 \leq t \leq t_f$ to minimize

$$J = \phi[x(t_f)] + \int_{t_0}^{t_f} L[x(t), u(t), t]dt \ ,$$

subject to

$$\dot{x} = f[x(t), u(t), t] \ , \quad x(t_0) = x_o \ ,$$

where t_f is specified.

Necessary Conditions for a Stationary Path

Let

$$H(t) \overset{\triangle}{=} L[x(t), u(t), t] + \lambda^T(t) f[x(t), u(t), t] \ .$$

Then

$$\dot{x} = f[x(t), u(t), t] \ ,$$
$$\dot{\lambda} = -H_x^T(t) \equiv -L_x^T(t) - f_x^T(t)\lambda(t) \ , \quad \lambda(t_f) = \phi_x^T \ ,$$

where $u(t)$ is determined from

$$H_u(t) \equiv L_u(t) + \lambda^T(t) f_u(t) = 0 \ .$$

This forms a TPBVP for $x(t)$ and $\lambda(t)$.

Gradient Algorithms DOP0 and FOP0

1. Guess $u(i)$ or $u(t)$ at N points.
2. Find $x(i)$ or $x(t)$ by forward sequencing or integration.
3. Evaluate ϕ, ϕ_x.
4. Find $H_u(i)$ or $H_u(t)$ by backward sequencing or integration.
5. $\delta u(i) = -k H_u^T(i)$ or $\delta u(t) = -k H_u^T(t)$, where $k > 0$.
6. If $\delta u_{avg} < \epsilon$ then stop.
7. $u(i) = u(i) + \delta u(i)$ or $u(t) = u(t) + \delta u(t)$.
8. Go to (2).

NR Algorithms DOP0N and FOP0N

These algorithms are very similar to DOP0 and FOP0 except that FSOLVE is used to make $H_u(i) = 0$ to the desired accuracy. A good initial guess of $u(i)$ and $\lambda(0)$ is required from DOP0 or FOP0.

Forward Shooting Algorithms DOP0F and FOP0F

1. Estimate the initial adjoint vector $\lambda(0)$ from a DOP0 or FOP0 solution.
2. Find $s(i)$, $\lambda(i)$ or $s(t)$, $\lambda(t)$ by forward sequencing or integration of the EL equations, determining $u(i)$ or $u(t)$ at each step by setting $H_u(i) = 0$ or $H_u(t) = 0$.
3. Evaluate the error $e_f \overset{\Delta}{=} \lambda(N) - \phi_s$ or $e_f \overset{\Delta}{=} \lambda(t_f) - \phi_s$.
4. Use FSOLVE to vary the elements of $\lambda(0)$ to make $e_f = 0$.
5. This 'single shooting' algorithm will FAIL for problems with energy dissipation and t_f large compared to characteristic times (see Problems 2.5.6, 2.6.6, and 5.2.4).

Backward Shooting Algorithms DOP0B and FOP0B

1. Estimate the final state $s(N)$ or $s(t_f)$ from a DOP0 or FOP0 solution.
2. Evaluate $\lambda(N) = \phi_s$ or $\lambda(t_f) = \phi_s$.
3. Find $s(i)$, $\lambda(i)$ or $s(t)$, $\lambda(t)$ by backward sequencing or integration of the EL equations, determining $u(i)$ or $u(t)$ at each step by setting $H_u(i) = 0$ or $H_u(t) = 0$.
4. Evaluate the error $e_0 \overset{\Delta}{=} s(0) - s_0$.
5. Use FSOLVE to vary the elements of $s(N)$ or $s(t_f)$ to make $e_0 = 0$.
6. This 'single shooting' algorithm will FAIL for problems with energy dissipation and t_f large compared to characteristic times (see Problems 2.5.6, 2.6.6, and 5.2.4).

Appendix A Alternative Derivation of the Discrete EL Equations

In Subsection 2.1.1, consider differential changes in J and in the difference equation constraints:

$$dJ = \phi_x dx(N) + \sum_{i=0}^{N-1} [L_x(i)dx(i) + L_u(i)du(i)] \ , \tag{2.112}$$

$$dx(i+1) = f_x(i)dx(i) + f_u(i)du(i) \ , \ i = 0, \ldots, N-1 \ , \tag{2.113}$$

$$dx(0) = 0 \ . \tag{2.114}$$

For a stationary path, $dJ = 0$ for arbitrary $du(i)$ subject to the constraints. The sequence $du(i)$ completely determines the sequence $dx(i)$ from (2.113) and (2.114). Hence, for $dJ = 0$ to be consistent with the constraints, (2.112) must be linearly dependent on (2.113) and (2.114), i.e.,

$$dJ = \sum_{i=0}^{N-1} \lambda^T(i+1)[dx(i+1) - f_x(i)dx(i) - f_u du(i)] + \lambda^T(0)dx(0) \ . \tag{2.115}$$

Changing indices of summation on the first term in (2.115) gives

$$dJ = \lambda^T(N)dx(N) + \sum_{i=0}^{N-1} \left\{ [\lambda^T(i) - \lambda^T(i+1)f_x(i)]dx(i) - \lambda^T(i+1)f_u(i)du(i) \right\} \ . \tag{2.116}$$

Equating coefficients of $dx(i)$ and $du(i)$ in (2.112) and (2.116) yields the discrete EL equations for $i = 0, \ldots, N - 1$:

$$\lambda^T(N) = \phi_x \ , \tag{2.117}$$

$$\lambda^T(i) = \lambda^T(i+1) f_x(i) + L_x(i) \ , \tag{2.118}$$

$$0 = \lambda^T(i+1) f_u(i) + L_u(i) \ . \tag{2.119}$$

Appendix B   Alternative Derivation of the Continuous EL Equations

In Subsection 2.3.1, consider a variation of J and of the differential equation constraints:

$$\delta J = \phi_x \delta x(t_f) + \int_{t_0}^{t_f} (L_x \delta x + L_u \delta u) \, dt \ , \tag{2.120}$$

$$\dot{\delta x} = f_x \delta x + f_u \delta u \ , \ 0 \le t \le t_f \ , \ \delta x(t_0) = 0 \ . \tag{2.121}$$

For a stationary path, $\delta J = 0$ for arbitrary $\delta u(t)$ subject to the constraints. The function $\delta u(t)$ completely determines the function $\delta x(t)$ from (2.121). Hence, for $\delta J = 0$ to be consistent with the constraints, (2.120) must be linearly dependent on (2.121), i.e.,

$$\delta J = \int_{t_0}^{t_f} \lambda^T (\dot{\delta x} - f_x \delta x - f_u \delta u) dt \ . \tag{2.122}$$

Integrating the $\lambda^T \dot{\delta x}$ term in (2.122) by parts and using (2.121) gives

$$\delta J = \lambda^T(t_f) \delta x(t_f) + \int_{t_0}^{t_f} \left[(-\dot{\lambda}^T - \lambda^T f_x) \delta x - \lambda^T f_u \delta u \right] dt \ . \tag{2.123}$$

Equating coefficients of δx and δu in (2.120) and (2.123) yields the continuous EL equations for $0 \le t \le t_f$:

$$\lambda^T(t_f) = \phi_x \ , \tag{2.124}$$

$$\dot{\lambda}^T = -\lambda^T f_x - L_x \ , \tag{2.125}$$

$$0 = \lambda^T f_u + L_u \ . \tag{2.126}$$

Dynamic Optimization with Terminal Constraints

3.1 Discrete Dynamic Systems

In many dynamic optimization problems there are *terminal constraints,* i.e., some functions of the terminal states are specified:

$$\psi[x(N)] = 0 \ ,$$

where ψ has dimension $q \leq n$.

Solution of such problems requires an extension of the theory and the algorithms developed in Sections 2.1 and 2.2. Using the Mayer form (see Section 2.2), the problem is to choose $u(i)$ to minimize

$$J = \phi[x(N)] \ , \tag{3.1}$$

subject to

$$x(i+1) = f[x(i), u(i), i] \ , \ i = 0, \cdots, N-1 \ , \tag{3.2}$$

$$x(0) = x_0 \ , \quad \psi[x(N)] = 0 \ . \tag{3.3}$$

Necessary Conditions for a Stationary Solution

Adjoin the constraints to the performance index with Lagrange multiplier vectors v and $\lambda(i)$:

$$\bar{J} = \phi + v^T \psi + \sum_{i=0}^{N-1} \lambda^T(i+1) \left\{ f[x(i), u(i), i] - x(i+1) \right\} + \lambda^T(0)[x_0 - x(0)] \ . \tag{3.4}$$

93

The procedure is the same as in Section 2.1 except that $L = 0$ and ϕ is replaced by

$$\Phi \stackrel{\Delta}{=} \phi + v^T \psi \ . \tag{3.5}$$

Hence

$$H(i) \stackrel{\Delta}{=} \lambda^T(i+1)f[x(i), u(i), i] \ , \tag{3.6}$$

and

$$\lambda^T(i) = H_x(i) \equiv \lambda^T(i+1)f_x \ , \tag{3.7}$$

$$\lambda^T(N) = \Phi_x \equiv \phi_x + v^T\psi_x \ , \tag{3.8}$$

$$0 = H_u(i) \equiv \lambda^T(i+1)f_u \ , \quad i = 0, \cdots, N-1 \ . \tag{3.9}$$

A TPBVP for a Stationary Solution

The optimality condition (3.9) together with the state equations (3.2), the boundary conditions (3.3), the adjoint equations (3.7), and the adjoint boundary conditions (3.8), constitute a TPBVP for determining a stationary solution $x(i)$, $\lambda(i)$, $u(i)$. The q constant Lagrange multipliers v must be chosen so as to satisfy the q terminal constraints in (3.3).

Example 3.1.1—Discrete Thrust Direction Programming (DTDP)

A constant specific thrust force a is applied to a body in a direction that makes an angle θ with respect to an inertial direction (the x-axis in Figure 3.1). The EOM are

$$\dot{u} = a\cos\theta \ , \quad \dot{v} = a\sin\theta \ , \quad \dot{y} = v \ , \quad y(0) = 0 \ . \tag{3.10}$$

If θ is held constant for time intervals of length ΔT, it is straightforward to show that

$$u(i+1) = u(i) + a\Delta T \cos\theta(i) \ , \quad v(i+1) = v(i) + a\Delta T \sin\theta(i) \ ,$$
$$y(i+1) = y(i) + \Delta T v(i) + \tfrac{1}{2}a(\Delta T)^2 \sin\theta(i) \ . \tag{3.11}$$

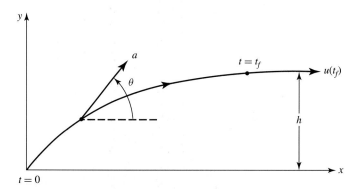

FIGURE 3.1   Nomenclature for Thrust-Direction Programming (TDP)

The *orbit injection* problem is to find the $\theta(i)$ sequence that maximizes $u(N)$ with $y(N) = y_f$, $v(N) = 0$, for zero initial conditions and given t_f ($\Delta T = t_f/N$), i.e.,

$$\phi = u(N), \quad \psi = \begin{bmatrix} y(N) - y_f \\ v(N) \end{bmatrix}. \tag{3.12}$$

If we measure time in units of t_f, (u, v) in units of at_f, (x, y, y_f) in units of at_f^2, then we can put $a = 1$, $\Delta T = 1/N$ in the equations above. The $H(i)$ sequence is then

$$H(i) = \lambda_u(i+1)u(i+1) + \lambda_v(i+1)v(i+1) + \lambda_y(i+1)y(i+1), \tag{3.13}$$

and the augmented performance index is

$$\Phi = u(N) + v_y[y(N) - y_f] + v_v v(N). \tag{3.14}$$

The discrete EL equations are then

$$\lambda_u(i) = H_u(i) = \lambda_u(i+1), \quad \lambda_v(i) = H_v(i) = \lambda_v(i+1) + \frac{1}{N}\lambda_y(i+1),$$
$$\lambda_y(i) = H_y(i) = \lambda_y(i+1), \tag{3.15}$$

and the optimality condition $H_\theta(i) = 0$ yields

$$0 = -\lambda_u(i+1)\sin\theta(i) + \lambda_v(i+1)\cos\theta(i) + \frac{1}{2N}\lambda_y(i+1)\cos\theta(i), \tag{3.16}$$

with terminal conditions

$$\lambda_u(N) = \Phi_u = 1, \quad \lambda_v(N) = \Phi_v = v_v, \quad \lambda_y(N) = \Phi_y = v_y. \tag{3.17}$$

The adjoint equations are easily solved in this case:

$$\lambda_u(i) = 1, \quad \lambda_y(i) = v_y, \quad \lambda_v(i) = v_v + v_y\left(1 - \frac{i}{N}\right), \tag{3.18}$$

so that the optimality condition is the *linear tangent law*

$$\tan\theta(i) = v_v + v_y \frac{N - i - 1/2}{N}. \tag{3.19}$$

The constants v_v and v_y must be determined to satisfy $y(N) = y_f$ and $v(N) = 0$.

If we assume that $v(i)$ is symmetric about the midpoint, then $\theta(i)$ must be antisymmetric, which implies that $\theta(N-1) = -\theta(0)$; this, in turn, implies that

$$v_v = -\frac{v_y}{2}, \tag{3.20}$$

so that

$$\tan\theta(i) = \tan\theta(0)\left(1 - \frac{2i}{N-1}\right), \tag{3.21}$$

where $\tan\theta(0) \equiv (1 - 1/N)v_y/2$. The choice of $\theta(0)$ determines $y(N) = y_f$ and vice-versa.

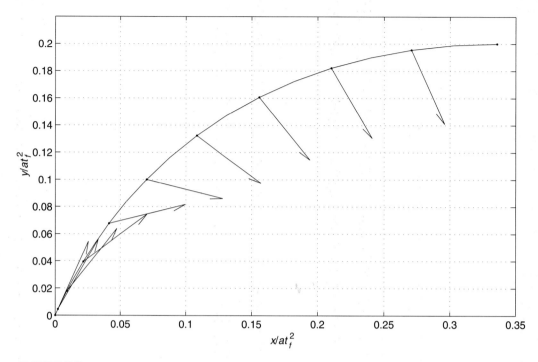

FIGURE 3.2   DTDP for Max u_f with $v_f = 0$ and $y_f = .2$; x vs. y

For the case $N = 10$, $\theta(0) = 65.13$ deg, yields $y_f = .20$. Figure 3.2 shows x vs. y, while Figure 3.3 shows (θ, u, v) vs. t.

Problems

3.1.1 DVDP for Max Range with Gravity and Specified y_f: This is the same as 2.1.1 except that we add the constraint that y_f is specified.

(a) Derive the necessary conditions for a stationary solution for an arbitrary y_f.
(b) A solution in tabulated functions of the necessary conditions in (a) does not seem to be possible, but the necessary conditions can be solved numerically for any feasible value of y_f using the MATLAB code FSOLVE. Do this for the case $y_f = .1$, and plot the optimal path.

3.1.2 DVDP for Max Range with $u_c = Vy/h$ and Specified y_f: This is the same as Problem 2.1.2 except that we add the constraint that y_f is specified.

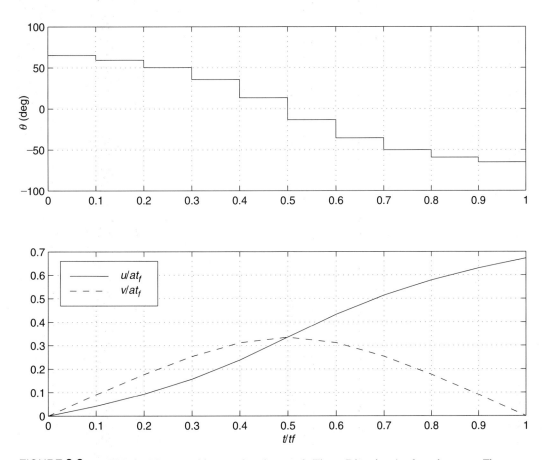

FIGURE 3.3 DTDP for Max u_f with $v_f = 0$ and $y_f = .2$; Thrust Direction Angle and u, v vs. Time

(a) Show that the optimality criterion may be written as

$$\tan \theta(i) = \tan \theta(0) - i\Delta T \ ,$$

where $\theta(0)$ is chosen so that $y(N) = h$.

(b) Find $\theta(0)$ for the case $Vt_f/h = 6$, $y(N) = 0$, and $N = 20$. Plot y/h vs. x/h with arrows showing the heading of the ship at each point where it changes.
Answer: $\theta(0) = 1.233$ rad.

3.1.3 DVDP for Min Distance Path between Two Points on a Sphere:

This is the same as Problem 2.1.3 except that we add the constraint that the final latitude is specified:

$$\theta(N) = \theta_f \ .$$

(a) Show that the min distance path is given by solving the discrete step TPBVP of Problem 2.1.3 with

$$\lambda(N) = \nu \ ,$$

where ν is chosen to give $\theta(N) = \theta_f$.

(b) Find the min distance path from Tokyo to New York and plot it on a latitude/longitude grid. The (latitude, longitude) of Tokyo and New York are (35.7 deg N, 139.7 deg E) and (40.7 deg N, 73.8 deg W) respectively. Note that the northernmost point on the path is quite close to Prudhoe Bay, Alaska. Compare the min distance found with the exact distance using spherical trigonometry and Figure 2.12.

3.1.4 DVDP for Max Range with $V = V_o(1 + y/h)$ and Specified y_f: This is the same as Problem 2.1.4 except that we add the constraint that y_f is specified.

(a) Show that the optimality criterion may be written as

$$\tan \theta(i) = [\nu + N - i - 1/2]\frac{V\Delta T}{h} \ ,$$

where

$$\Delta T = \frac{t_f}{N} \ , i = 0, \cdots, N - 1 \ ,$$

and ν is a constant to be chosen so that $y(N) = y_f$.
(b) Plot y/h vs. x/h and θ vs. Vt/h for the case $Vt_f/h = 2$, $y_f/h = 0$, and $N = 11$.

3.1.5 DTDP for Max x_f with $y_f = 0$:   A rocket is launched with velocity u_0 parallel to the x-axis and v_0 parallel to the y-axis. With constant thrust specific force a, we wish to find the DTDP $\theta(i)$ for max $x(N)$ with $y(N) = y_f$. The discretized EOM are (3.14) to (3.16) and $\phi(N) = x(N)$, $\psi(N) = y(N) - y_f$.

(a) Show that optimal $\theta(i)$ and max $x(N)$ are

$$\theta(i) = \sin^{-1}(y_f - 2v_0) = \text{ constant} , \quad x(N) = u_0 + \cos\theta/2 \ ,$$

where x is in units of at_f^2 and u_0, v_0 are in units of at_f.
(b) Calculate and plot the optimal path for $N = 20$, $u_0 = v_0 = 1/\sqrt{2}$, $y_f = 0$.

3.1.6 DTDP for Max u_f to $v_f = 0$ and Specified y_f: Verify the solution of Example 3.1.1 shown in Figs. 3.2 and 3.3.

3.1.7 DTDP for Max u_f to $v_f = 0$ and Specified y_f, x_f: This is an extension of Example 3.1.1 to include specification of x_f, a rendezvous problem instead of an injection problem.

(a) Show that the optimal thrust direction angle $\theta(i)$ is given by the *bilinear tangent law*:

$$\tan \theta(i) = \frac{\nu_v + \nu_y[1 - \Delta T(i + 1/2)]}{1 + \nu_x[1 - \Delta T(i + 1/2)]} \ .$$

The three parameters ν_v, ν_y, ν_x may be determined to satisfy the three final conditions on v, y, x. Thus one can solve this problem by sequencing the four difference equations forward and interpolating these parameters to satisfy the final conditions. A code like MATLAB's FSOLVE will do this interpolation for you.
(b) Find the vector ν for the case $v_f = 0$, $y_f = .2$, $x_f = .15$, and plot the optimal path.

3.1.8 DTDP for Max x_f with Gravity and $y_f = 0$:   The same as Problem 3.1.5 except that gravity acts down the y-axis and $y(N) = y_f$ where $y_f \neq 0$. For the plot take $g/a = 1/3$.

3.1.9 DTDP for Max u_f with Gravity to $v_f = 0$ and Specified y_f: This is the same as Example 3.1.1 except that we add a constant gravitational force per unit mass g so that

$$v(i+1) = v(i) + \Delta T[a \sin \theta(i) - g] ,$$

$$y(i+1) = y(i) + \Delta T v(i) + .5(\Delta T)^2[a \sin \theta(i) - g] .$$

The EL equations are unchanged, and the linear tangent law still applies. The optimal path will, of course, be different.

Find and plot the path for max u_f for the case $g/a = 1/3$, $v_f = 0$, $y_f = .2at_f^2$ starting from zero initial state. Compare your result with Figure 3.4.

3.1.10 DTDP for Max u_f with Gravity to $v_f = 0$ and Specified y_f, x_f: This is the same as Problem 3.1.7 except that we add a constant gravitational force per unit mass g as in Problem 3.1.9. The EL equations are unchanged, and the bilinear tangent law still applies. The optimal path will, of course, be different.

Find and plot the path for max u_f for the case $g/a = 1/3$, $v_f = 0$, $y_f = .2$, $x_f = .15$ starting from zero initial state.

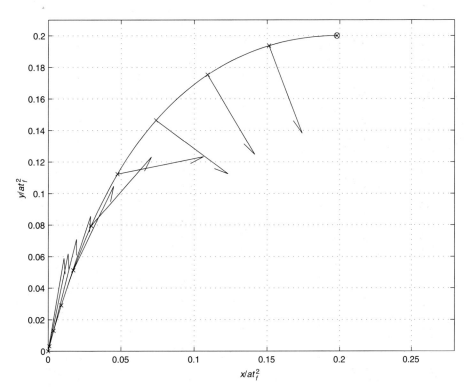

FIGURE 3.4   DVDP for Max u_f with Gravity and $v_f = 0$, $y_f = .2at_f^2$, $g/a = 1/3$

3.2 Numerical Solution with Gradient Methods

We first describe *a gradient algorithm* that we call DOPC (for Discrete OPtimization with Constraints); as with DOP0, this code requires guessing a step-size parameter k. Then we show how to use the MATLAB command CONSTR, so that one does *not* have to guess k. Later, in Section 3.5, we describe a shooting algorithm that converges quickly, provided a good initial guess is available from DOPC.

The system equations (3.2) are sequenced forward with a guessed $u(i)$ sequence, starting with the specified initial conditions $x(0)$. It is, of course, unlikely that the terminal constraints (3.4) will be satisfied. The adjoint equations are then sequenced backward to determine a gradient sequence; this sequence is used to make small changes in the control sequence $du(i)$ that cause the next solution to be closer to the desired terminal conditions and to decrease the magnitude of the gradient. This process is repeated until the terminal constraint error and the gradient sequence are negligibly small. Details are described below.

Derivation of the DOPC Algorithm

Differential changes in the performance index and the terminal constraints are given by

$$d\phi = \sum_{i=0}^{N-1} H_u^\phi(i) du(i) \ , \tag{3.22}$$

$$d\psi_j = \sum_{i=0}^{N-1} H_u^{\psi_j}(i) du(i) \ , \quad j = 1, \cdots, q \ , \tag{3.23}$$

where the *performance index pulse response sequence* $H_u^\phi(i)$ is computed by a backward sequencing of the adjoint equations. For $i = N - 1, \cdots, 0$:

$$\lambda^\phi(i) = f_x^T \lambda^\phi(i+1) \ , \quad \lambda^\phi(N) = \phi_x^T \ , \tag{3.24}$$

$$\left[H_u^\phi(i) \right]^T = f_u^T \lambda^\phi(i+1) \ . \tag{3.25}$$

At the same time, the *q terminal constraint pulse response sequences* $H_u^{\psi_j}(i)$, $j = 1, \cdots, q$ are computed:

$$\lambda^{\psi_j}(i) = f_x^T \lambda^{\psi_j}(i+1) \ , \quad \lambda^{\psi_j}(N) = [\psi_x^{(j)}]^T \ , \tag{3.26}$$

$$\left[H_u^{\psi_j}(i) \right]^T = f_u^T \lambda^{\psi_j}(i+1) \ . \tag{3.27}$$

From here onward the algorithm is *exactly the same* as the parameter optimization algorithm POP in Section 1.3 if we replace L by ϕ, f by ψ, λ by ν, and y by u, which is treated as an $N * n_c \times 1$ column vector:

$$u \triangleq \begin{bmatrix} u(:, 1) \\ \vdots \\ u(:, n_c) \end{bmatrix} ,$$

where $n_c = $ the number of controls, $N = $ the number of steps, and

$$u(:, i) \triangleq \begin{bmatrix} u(1, i) \\ \vdots \\ u(N, i) \end{bmatrix} ,$$

i.e., a column vector of all the N steps of the ith control (MATLAB notation).

In this notation (3.22) and (3.23) can be written, for finite increments, as

$$\Delta\phi \approx H_u^\phi \Delta u , \quad \Delta\psi \approx H_u^\psi \Delta u ,$$

where $H_u^\phi = $ a row vector with $M \triangleq N \times n_c$ elements, and $H_u^\psi = $ a matrix with q rows and M columns.

Making these replacements we have

$$v = -Q^{-1}g , \tag{3.28}$$

where

$$g = H_u^\psi [H_u^\phi]^T , \quad Q = H_u^\psi [H_u^\psi]^T , \tag{3.29}$$

The $q \times q$ matrix Q is the *terminal constraint controllability matrix* and it must be non-singular; it corresponds to the $f_y f_y^T$ matrix in POP (Section 1.3).

Next, we put $\Delta\psi = -\eta\psi$. Then from the POP algorithm we have

$$\Delta u = -k[H_u^\phi + v^T H_u^\psi]^T - \eta[H_u^\psi]^T Q^{-1}\psi , \tag{3.30}$$

where ψ is the calculated value for the nominal path and $0 < \eta \leq 1$. Then $u + \Delta u$ is an improved control sequence that will reduce the terminal constraint errors and the performance index.

The algorithm is iterated until the norm of Δu is less than some specified small value. In the first few iterations, it may be necessary to concentrate on meeting the terminal constraints, i.e., to find a feasible path. To do this put $k = 0$ and, if necessary, use $\eta < 1$ to avoid large values of $|\Delta u|$. After a feasible path is found ($\psi \approx 0$), a well-chosen value of k will improve the performance index ϕ while maintaining $\psi \approx 0$.

This algorithm is summarized below.

Summary of DOPC

- Guess $u(i)$ for $i = 0$ to $N - 1$, where u is $n_c \times 1$, i.e., there are n_c controls.
- (∗) Forward sequence the system equations from $x(0)$ using $u(i)$ and store $x(i) =$ where x is $n \times 1$, i.e., there are n states.
- Evaluate ϕ and ψ, where ψ is $q \times 1$, i.e., there are q terminal constraints, and set

$$\lambda^\phi(N) = \phi_x^T , \quad \lambda^\psi(N) = \psi_x^T ,$$

  where λ^ϕ is $n \times 1$ and λ^ψ is $n \times q$.
- Backward sequence and store the response sequences $H_u^\phi(i)$ and $H_u^\psi(i)$, where H_u^ϕ is $1 \times n_c$ and H_u^ψ is $q \times n_c$. For $i = N - 1, \cdots, 0$:

$$\lambda^\phi(i) = f_x^T(i)\lambda^\phi(i + 1) , \quad \left[H_u^\phi(i)\right]^T = f_u^T(i)\lambda^\phi(i + 1) ,$$

$$\lambda^\psi(i) = f_x^T(i)\lambda^\psi(i + 1) , \quad \left[H_u^\psi(i)\right]^T = f_u^T(i)\lambda^\psi(i + 1) ,$$

- Determine v, which is $q \times 1$:

$$v = -Q^{-1}g ,$$

  where

$$g = H_u^\psi[H_u^\phi]^T , \quad Q = H_u^\psi[H_u^\psi]^T ,$$

  so g is $q \times 1$ and Q is the symmetric $q \times q$ controllability matrix.
- Compute Δu. Choose $k < 0$ if maximizing, $k > 0$ if minimizing, and $0 < \eta \le 1$:

$$\Delta u = -k[H_u^\phi + v^T H_u^\psi]^T - \eta[H_u^\psi]^T Q^{-1}\psi .$$

- Compute $\Delta u_{avg} \triangleq$ norm of Δu divided by $\sqrt{N}$ for each control.
- If max $\Delta u_{avg} < tol$ then end.
- Compute new u:

$$u \leftarrow u + \Delta u .$$

- Go to (∗) .

A MATLAB Code for DOPC

A MATLAB code for implementing the DOPC algorithm with multiple controls is listed in Table 3.1. The data for the problem being solved are put into a subroutine 'name'.

Example 3.2.1—Discrete Thrust Direction Programming

This is the same problem solved analytically in Example 3.1.1 and shown in Figs. 3.2 and 3.3. The data subroutine DTDPC is listed below.

```
function [f1,f2]=dtdpc(th,s,dt,t,flg)
% Subroutine for Ex. 3.2.1; DTDP for max uf to vf=0 and
% spec. yf; t in units of tf; (u,v) in a*tf; (x,y) in a*tf^2;
% s = [u,v,y,x]' = state vector; th = theta = control;
%
```

TABLE 3.1   A MATLAB Code for DOPC

```
function [u,s,nu,la0]=dopc(name,u,s0,tf,k,tol,mxit,eta)
% Disc. OPtim. w. term. Constr., tf given; u is nc by N; name must be
% in single quotes; function file 'name' should be in the MATLAB path;
% it computes f for flg=1, Phi=(phi,psi) and Phis=(phis,psis) for flg
% 2, and (fs,fu) for flg=3; u(nc,N)=initial guess of control sequence;
% s0(ns,1)=initial state; tf=final time; k=step-size parameter; u
% should be scaled so that one unit of each element is of approximate-
% ly the same significance; stops when both dua and norm(psi)<tol or
% no. of iterations >= mxit; 0 < eta <= 1 where d(psi)=-eta*psi is
% desired change in term. conditions on next iteration; BASIC version
% 1984; scalar ctrl MATLAB version 1994; vector control version
% Note Hu is a 3-D array, nt1 by nc by N; 9/12/96; rev. 1/11/98.
%
if nargin<8, eta=1; end; if nargin<7, mxit=10; end
ns=length(s0); [nc,N]=size(u); N1=N+1; dum=zeros(nc,1);
Phi=feval(name,dum,s0,1,0,2); nt1=length(Phi); nt=nt1-1;
s=zeros(ns,N1); la=zeros(ns,nt1); Hu=zeros(nt1,nc,N); dt=tf/N;
nu=zeros(nt,1); dua=1; pmag=1; it=0; s(:,1)=s0; n2=[2:nt1];
disp('      Iter.     phi     norm(psi)      dua')
while max([norm(dua) pmag])> tol,
 % Forward sequencing and store state histories s:
  for i=1:N,
    s(:,i+1)=feval(name,u(:,i),s(:,i),dt,(i-1)*dt,1);
  end
 % Performance index, terminal constraints & gradients:
  [Phi,Phis]=feval(name,dum,s(:,N1),dt,N*dt,2);
  phi=Phi(1); psi=Phi(n2); pmag=norm(psi)/sqrt(nt);
  if mxit==0, disp([it phi pmag]); break; end
  la=Phis';
 % Backward sequencing, storing Hu(:,i):
  for i=N:-1:1
   [fs,fu]=feval(name,u(:,i),s(:,i),dt,(i-1)*dt,3);
   Hu(:,:,i)=la'*fu; la=fs'*la;
  end
 % nu and la0:
  ga=zeros(nt1,1); Qa=zeros(nt1); n1=[1:nt1]; dua=zeros(nc,1);
  for i=1:N
   ga=ga+Hu(n1,:,i)*Hu(1,:,i)'; Qa=Qa+Hu(n1,:,i)*Hu(n1,:,i)';
  end; g=ga(n2); Q=Qa(n2,n2); nu=-Q\g; la0=la*[1; nu];
 % New u(:,i):
  for i=1:N, Huphi=Hu(1,:,i); Hupsi=Hu(n2,:,i);
   du(:,i)=-k*(Huphi'+Hupsi'*nu)-eta*Hupsi'*(Q\psi);
   dua=dua+norm(du(:,i))/sqrt(N);
  end
  u=u+du; disp([it phi norm(psi) dua'])
  it=it+1; if it>=mxit, break, end
end
```

```
u=s(1); v=s(2); y=s(3); x=s(4); co=cos(th); si=sin(th);
if flg==1,                                        % f1 = f
 f1=s+dt*[co; si; v+dt*si/2; u+dt*co/2];
elseif flg==2,           % f1 = Phi = [phi;psi]; f2 = Phix
 f1=[u; v; y-.2];
 f2=[eye(3) zeros(3,1)];
elseif flg==3,                            % f1 = fx; f2 = fu
 f1=[1 0 0 0; 0 1 0 0; 0 dt 1 0; dt 0 0 1];
 f2=dt*[-si; co; dt*co/2; -dt*si/2];
 f3=zeros(16,4);      % f3=fxx, f4=fxu, f5=fuu
 f4=zeros(4);
 f5=dt*[-co; -si; -dt*si/2; -dt*co];
end
```

An edited diary showing the computation of the solution is listed below.

```
th0=[1:-2/9:-1]; x0=[0 0 0 0]'; tf=1; k=-7; tol =.00005;
[th,s]=dopc('dtdpc',th0,x0,tf,k,tol);
     Iter.      phi    norm(psi)     dua
         0     0.8082    0.0379     0.2033
    1.0000     0.6888    0.0039     0.0477

       -         -         -          -
    7.0000     0.6716    0.0000     0.0001
    8.0000     0.6716    0.0000     0.0000
th =
    1.1366    1.0332    0.8753    0.6233    0.2352   -0.2352
   -0.6233   -0.8753   -1.0332   -1.1366
```

Use of the MATLAB Code CONSTR

The CONSTR command (for CONSTRained optimization) in the MATLAB Optimization Toolbox finds the minimum of a function 'f' with respect to a parameter vector (or matrix) 'u', with constraints $g(u) = 0$. All that is required is a MATLAB function file that computes f and g given u, which, in this case, means sequencing (3.2) forward to calculate ϕ and ψ, which must be called 'f' and 'g' when using CONSTR.

Example 3.2.2—DTDP Using the MATLAB Code CONSTR

Here we solve the same problem solved in Example 3.2.1, using CONSTR instead of DOPC. The function file to calculate the performance index 'f' and the constraint vector 'g' is listed below.

```
function [f,g,u,v,y,x]=dtdp_fg(th,N,yf)
% Subroutine for e03_2_2; DTDP for max uf to vf=0 and spec. yf;
% t in tf, (u,v) in a*tf; (x,y) in a*tf^2; s = [u,v,y,x]'= state
```

```
% vector; th = control; x is included for trajectory plot;
%
u(1)=0; v(1)=0; y(1)=0; x(1)=0; dt=1/N;
for i=1:N ,
 u(i+1)=u(i)+dt*cos(th(i)); v(i+1)=v(i)+dt*sin(th(i));
 y(i+1)=y(i)+dt*v(i)+dt^2*sin(th(i))/2;
 x(i+1)=x(i)+dt*u(i)+dt^2*cos(th(i))/2;
end
f=-u(N+1); g=[v(N+1) y(N+1)-yf]';
```

The script file listed below generates an initial guess of $u(i)$ and supplies the options required by CONSTR.

```
% Script e03_2_2.m; DTDP for max uf to vf=0 and spec. yf using
% CONSTR; t in units of tf,(u,v) in a*tf, (x,y) in a*tf^2; s=
% [u,v,y,x]'; th=control; x is included for trajectory plot.
%
N=40; yf=.2; un=ones(1,N); th=(pi/3)*[1:-2/N:-1+2/N]; t=[0:N]/N;
optn(1)=1; optn(4)=2e-4; optn(13)=2; optn(14)=500;
vub=(pi/2)*un; vlb=-vub; th=constr('dtdp_fg',th,optn,vlb,vub,[],N,yf);
[f,g,u,v,y,x]=dtdp_fg(th,N,yf);
```

An edited diary of the solution is given below.

```
% Diary DTDP with zero gravity, using MATLAB CONSTR command
e03_4_2
f-COUNT    FUNCTION        MAX{g}         STEP  Procedures
   11     -0.823968     0.0866025          1
    -          -            -                        -
  345     -0.671605 5.67872e-010           1     mod Hess(2)
Optimization Converged Successfully
th = 1.1365    1.0334    0.8751    0.6234    0.2350   -0.2352
     -0.6233   -0.8751   -1.0331   -1.1367
```

Problems

Problems 3.2.1 to 3.2.10:   Do Problems 3.1.1 to 3.1.10 using DOPC or CONSTR.

3.2.11 DVDP for Max Range with Gravity, Thrust, and Specified y_f: A body is acted upon by a constant thrust specific force a along the path, and a constant gravitational specific force g, so that the equations of motion EOM are (cf. Problem 2.2.5)

$$\dot{V} = a - g \sin \gamma , \quad \dot{y} = V \sin \gamma , \quad \dot{x} = V \cos \gamma ,$$

with zero initial conditions. We wish to find $\gamma(t)$ to maximize $x(t_f)$ with specified $y(t_f)$.

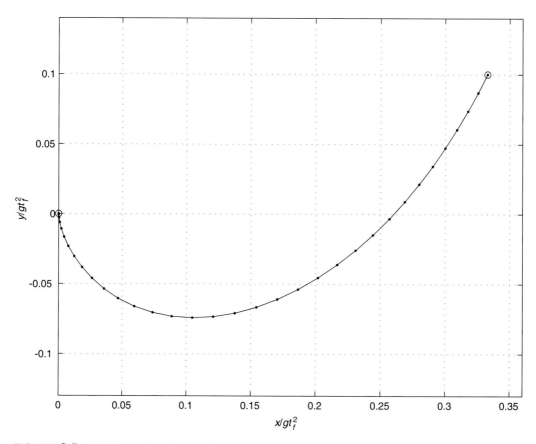

FIGURE 3.5   DVDP for Max Range with Gravity, Thrust ($a/g = .5$, $y_f = .1$)

(a) Discretize the EOM.
(b) For $a/g = .5$, use a computer code such as DOPC or CONSTR to find the path that
maximizes $x(t_1)$ for $y(t_1) = .1$. Compare your result with Figure 3.5.

3.2.12 DVDP for Max Range with Gravity, Thrust, Drag, and Specified y_f: This is
the same as Problem 3.2.11 except we add a drag force proportional to V^2 so the EOM are

$$\dot{V} = a - g \sin \gamma - \frac{V^2}{\ell} , \quad \dot{y} = V \sin \gamma , \quad \dot{x} = V \cos \gamma ,$$

We wish to find $\gamma(t)$ to maximize $x(t_f)$ with specified $y(t_f)$.

(a) Discretize the equations of motion (cf. Problem 2.2.6). Use (x, y) in units of the length
ℓ, V in $\sqrt{g\ell}$, t in $\sqrt{\ell/g}$, and a in g.
(b) For $a = .05$, $t_f = 5$ use DOPC or CONSTR to find the path that maximizes $x(t_f)$ with
$y(t_f) = -.3$. Compare your results with Figure 3.19, which shows the solution to the
corresponding continuous problem.

3.3 Continuous Dynamic Systems

This is the continuous equivalent of Section 3.1. Using the Mayer formulation, the problem is to choose $u(t)$ to minimize

$$J = \phi[x(t_f)] \, , \tag{3.31}$$

subject to

$$\dot{x} = f[x(t), u(t), t] \, , \tag{3.32}$$

$$x(t_0) = x_0 \, , \quad \psi[x(t_f)] = 0 \, , \tag{3.33}$$

where t_f is specified.

Necessary Conditions for a Stationary Solution

Adjoin the constraints to the performance index with Lagrange multipliers v and $\lambda(t)$:

$$\bar{J} = \phi + v^T \psi + \int_0^{t_f} \lambda^T(t) \{ f[x(t), u(t), t] - \dot{x} \} \, dt \, . \tag{3.34}$$

The procedure is the same as in Section 2.3.1 except that $L = 0$ and ϕ is replaced by

$$\Phi = \phi + v^T \psi \, . \tag{3.35}$$

Hence

$$H(t) = \lambda^T(t) f[x(t), u(t), t] \, , \tag{3.36}$$

and

$$\dot{\lambda}^T = -H_x \equiv -\lambda^T f_x \, , \tag{3.37}$$

$$\lambda^T(t_f) = \Phi_x \equiv \phi_x + v^T \psi_x \, , \tag{3.38}$$

$$0 = H_u \equiv \lambda^T f_u \, , \ 0 \le t \le t_f \, . \tag{3.39}$$

A TPBVP for a Stationary Solution

The optimality condition (3.39) together with the state equations (3.32), the boundary conditions (3.33), the adjoint equations (3.37), and the boundary conditions (3.38), constitute a TPBVP for determining a stationary solution $x(t)$, $\lambda(t)$, $u(t)$. The q Lagrange multipliers v must be chosen so as to satisfy the q terminal constraints in (3.33).

Example 3.3.1—Thrust Direction Programming (TDP) for Max u_f with (v_f, y_f) Specified

This is the continuous version of Example 3.1.1 (see Figure 3.1). Measuring t in t_f, (u, v) in at_f, and (x, y) in at_f^2, the problem is to find $\theta(t)$ to maximize $u(1)$ subject to

$$\dot{u} = \cos\theta \, , \quad \dot{v} = \sin\theta \, , \quad \dot{y} = v \, ,$$

where initial conditions are all zero and $v(1) = 0$, $y(1) = y_f$. The Hamiltonian function $H(t)$ is then

$$H(t) = \lambda_u(t) \cos \theta(t) + \lambda_v(t) \sin \theta(t) + \lambda_y(t) v(t) .$$

The augmented performance index is

$$\Phi = u(1) + \nu_y[y(1) - y_f] + \nu_v v(1) .$$

The augmented equations are then

$$\dot{\lambda}_u = -H_u = 0 , \quad \dot{\lambda}_v = -H_v = -\lambda_y , \quad \dot{\lambda}_y = -H_y = 0 ,$$

and the optimality condition is

$$0 = H_\theta = -\lambda_u \sin \theta + \lambda_v \cos \theta .$$

The terminal conditions are

$$\lambda_u(1) = \Phi_u = 1 , \quad \lambda_v(1) = \Phi_v = \nu_v , \quad \lambda_y(1) = \Phi_y = \nu_y .$$

The adjoint equations are easily integrated in this case:

$$\lambda_u(t) = 1 , \quad \lambda_y(t) = \nu_y , \quad \lambda_v(t) = \nu_v + \nu_y(1 - t) ,$$

so that

$$\tan \theta(t) = \nu_v + \nu_y(1 - t) .$$

The constants ν_v and ν_y must be determined to satisfy $y(1) = y_f$ and $v(1) = 0$.

The state histories can be found in terms of tabulated functions by using θ as the independent variable instead of time t. Differentiating the optimality condition gives:

$$\dot{\theta} = -\nu_y \cos^2 \theta .$$

Dividing $\dot{v}$ by $\dot{\theta}$ gives

$$\frac{dv}{d\theta} = \frac{-\sin \theta}{\nu_y \cos^2 \theta} \quad \Rightarrow \quad v = \frac{\sec \theta(0) - \sec \theta}{\nu_y} ,$$

where we have used $v(0) = 0$. Using $v(1) = 0$ we see that

$$\theta(1) = -\theta(0) \quad \Rightarrow \quad \nu_v = -\tan \theta(0) , \quad \nu_y = 2 \tan \theta(0) ,$$

which gives

$$\tan \theta = \tan \theta(0)(1 - 2t) \quad \Rightarrow \quad t = \frac{\tan \theta(0) - \tan \theta}{2 \tan \theta(0)} .$$

Dividing $\dot{u}$, $\dot{y}$, $\dot{x} = u$ by $\dot{\theta}$ and integrating, we obtain:

$$u = \frac{\Lambda[\theta(0)] - \Lambda(\theta)}{2 \tan \theta(0)} , \quad v = \frac{\sec \theta(0) - \sec \theta}{2 \tan \theta(0)} , \quad \Lambda(z) \triangleq \sinh^{-1} \tan z .$$

$$y = \frac{t \sec \theta(0) - v \tan \theta - u}{4 \tan \theta(0)} , \quad x = \frac{v - u \tan \theta}{2 \tan \theta(0)} .$$

$\theta(0)$ is determined by $y(1) = y_f$:

$$4y_f = \frac{1}{\sin \theta(0)} - \frac{\Lambda[\theta(0)]}{\tan^2 \theta(0)} .$$

For $y_f = .2$ we used FSOLVE in MATLAB to determine that $\theta(0) = 65.13$ deg (cf. Problem 3.3.6). Not surprisingly, the resulting path was very close to the one obtained with DTDP in Figures 3.2 and 3.3.

Problems

3.3.1 VDP for Max Range with Gravity and y_f Specified: This is a continuous version of Problem 3.1.1. A bead slides on a wire without friction in a gravitational field. Find the angle of the wire $\gamma(t)$ to maximize the horizontal range in a given time t_f ending with $y(t_f) = y_f$.

The EOM are

$$\dot{V} = g \sin \gamma , \quad \dot{x} = V \cos \gamma , \quad \dot{y} = V \sin \gamma .$$

(a) For $V(0) = x(0) = y(0) = 0$ and a given final time t_f, show that the path that maximizes $x(t_f)$ with the terminal constraint $y(t_f) = y_f$, is given by the cycloid

$$\alpha x = t - \frac{\sin \alpha t}{\alpha} , \quad \alpha^2 y = 1 - \cos \alpha t , \quad 2\gamma = \pi - \alpha t , \quad \alpha V = 2 \sin \left(\frac{\alpha t}{2} \right) ,$$

where α is determined from the implicit equation $\alpha^2 y_f = 1 - \cos \alpha$, and t is in units of t_f, V in gt_f, and (x, y) in gt_f^2.

(b) Using (a) plot the *reachable space* in time t_f by selecting several values of α and plotting the corresponding optimal paths. Compare your results with Figure 3.6.

3.3.2 VDP for Max Range with $u_c = Vy/h$ and y_f Specified: This is a continuous version of Problem 3.1.2. The normalized system equations are

$$\dot{x} = \cos \theta + y , \quad \dot{y} = \sin \theta ,$$

where $y(1) = y_f$, (x, y) are in units of h, and (t, t_f) are in units of h/V.

(a) Using θ as the independent variable, show that the optimal path is given by

$$y = \sec \theta_o - \sec \theta , \quad t = \tan \theta_o - \tan \theta ,$$

$$x = \tfrac{1}{2} [\sinh^{-1}(\tan \theta_o) - \sinh^{-1}(\tan \theta) + \tan \theta \sec \theta - \tan \theta_o \sec \theta_o] + t \sec \theta_o ,$$

where $\Lambda(z) \overset{\Delta}{=} \sinh^{-1}(\tan z)$. The constants θ_0, θ_f are determined by the implicit equations

$$t_f = \tan \theta_0 - \tan \theta_f , \quad y_f = \sec \theta_0 - \sec \theta_f .$$

(b) Plot the optimal path for the case $t_f = 1$, $y_f = 0$. Compare your result with the corresponding path in Figure 7.4.

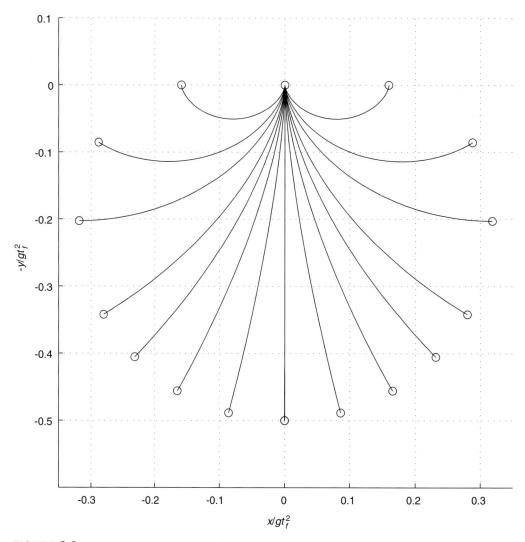

FIGURE 3.6   VDP with Gravity; Reachable Space in Time t_f

3.3.3 VDP for Min Distance Between Two Points on a Sphere: This is a continuous version of Problem 3.1.3. We wish to find the heading angle β as a function of the longitude ϕ to minimize the distance traveled in going along the surface of a sphere from a point with (latitude, longitude) = $(\theta_0, \ 0)$ to a point with (latitude, longitude) = $(\theta_f, \ \phi_f)$. The distance traveled is

$$J = \int_0^{\phi_f} \sec\beta \cos\theta \, d\phi \ ,$$

where

$$\frac{d\theta}{d\phi} = \tan \beta \cos \theta \ , \quad \theta(0) = \theta_o \ , \quad \theta(\phi_f) = \theta_f \ .$$

(a) Show that the min distance path is generated by

$$\cos \beta = \frac{\cos \theta_m}{\cos \theta} \ ,$$

where $\theta_m = $ max latitude on the great circle path through the two points.

(b) Show that the min distance path (a great circle) is given by

$$\tan \theta = \tan \theta_m [\cos(\phi_m - \phi)] \ ,$$

where ϕ_m, θ_m are given by

$$\tan \phi_m = \frac{\tan \theta_0 \cos \phi_f - \tan \theta_f \cos \phi_0}{\tan \theta_f \sin \phi_0 - \tan \theta_0 \sin \phi_f} \ , \quad \tan \theta_m = \frac{\tan \theta_0}{\cos(\phi_m - \phi_0)} \ .$$

(c) Find and plot the min distance path from Tokyo to New York (see Problem 2.3.7).

3.3.4 VDP for Max Range with $V = V_o(1 + y/h)$ and y_f Specified: This is the same as Problem 2.3.4 (which see) except that y_f is specified.

(a) Starting from $x = y = 0$, show that path with max $x(t_f)$ and specified t_f and y_f is given by the relations in Problem 2.3.4(a) where θ_0 and θ_f are determined by the implicit equations

$$t_f = \Lambda(\theta_0) - \Lambda(\theta_f) \ , \quad y_f = \frac{\cos \theta_f}{\cos \theta_0} - 1 \ .$$

(b) The dual problem is to minimize t_f with x_f and y_f specified. How would you determine t_f given (x_f, y_f) using the analytical solution in 2.3.4(a)?

(c) Plot a few members of the one-parameter family of optimal paths starting from the origin (parameter θ_0). Determine and plot the "wavefronts", i.e., the contours of constant t_f. Compare your plot with Figure 3.7.

3.3.5 TDP for Max x_f with Specified y_f: This is the continuous version of Problem 3.1.5.

(a) Show that the EOM are

$$\dot{u} = \cos \theta \ , \quad \dot{v} = \sin \theta \ , \quad \dot{x} = u \ , \quad \dot{y} = v \ ,$$

with $u(0) = u_0$, $v(0) = v_0$, $x(0) = y(0) = 0$, $y(1) = y_f$, where (u, v) are in units of $a t_f$, (x, y) are in units of $a t_f^2$, t is in units of t_f.

(b) Show that the optimal θ is *constant,* given by

$$\theta = \sin^{-1}[2(y_f - v_0)] \ ,$$

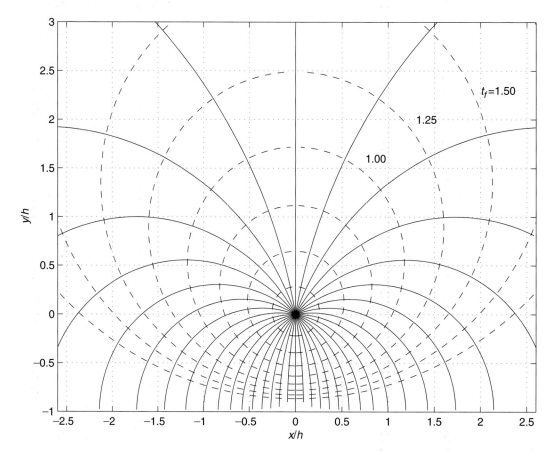

FIGURE 3.7   VDP with $V = V_o(1 + y/h)$; Reachable Space in a Given Time (Fermat Problem)

and the max range $x(1)$ is given by

$$x(1) = u_0 + \cos\theta/2 .$$

(c) Calculate and plot the optimal path for $u_0 = v_0 = .5/\sqrt{2}, \ y_f = .2$.

3.3.6 TDP for Max u_f to $v_f = 0$ and Specified y_f: Use the implicit analytic solution of Example 3.3.1 and a code like FSOLVE in MATLAB to determine $\theta(0)$ for $y_f = .2$ and plot the resulting path. Compare your results with Figures 3.2 and 3.3 (which show the discrete version of the problem).

3.3.7 TDP for Max u_f to $v_f = 0$ and Specified y_f, x_f: This is an extension of Example 3.3.1 to include specification of x_f, making it an orbital rendezvous problem instead of an orbit injection problem. Add the state x by including the differential equation $\dot{x} = u$.

(a) Show that the optimal $\theta(t)$ is given by the *bilinear tangent law*:

$$\tan \theta = \frac{v_v + v_y(t_f - t)}{1 + v_x(t_f - t)} \ ,$$

where (v_v, v_y, v_x) are constants to be determined in order to satisfy the three boundary conditions $v(1) = 0$, $y(1) = y_f$, $x(1) = x_f$.

(b) Show that the bilinear tangent law in (a) can also be written as a *rotated linear tangent law* (rotated through an angle α):

$$\tan(\theta - \alpha) = \tan(\theta_0 - \alpha) - mt \ ,$$

where (θ_0, α, m) are constants.

(c) Show that the problem can be solved in terms of tabulated functions by using $\bar{\theta} \triangleq \theta - \alpha$ as the independent variable instead of time. Derive or verify the following solution (Ref. BR1):

$$\begin{bmatrix} u \\ v \end{bmatrix} = \frac{1}{m} M \begin{bmatrix} f_u \\ f_v \end{bmatrix} , \quad \begin{bmatrix} x \\ y \end{bmatrix} = \frac{1}{m^2} M \begin{bmatrix} \bar{x} \\ \bar{y} \end{bmatrix} ,$$

where

$$f_u \triangleq f(\bar{\theta}_0) - f(\bar{\theta}) , \quad f_v \triangleq \sec \bar{\theta}_0 - \sec \bar{\theta} ,$$
$$f_x \triangleq f_v - f_u \tan \bar{\theta} , \quad f_y \triangleq \tfrac{1}{2}(t \sec \bar{\theta}_0 - f_v \tan \bar{\theta} - f_u) ,$$

and

$$\bar{\theta} \triangleq \theta - \alpha , \quad \bar{\theta}_0 \triangleq \theta_0 - \alpha , \quad f(z) \triangleq \log[\tan(z) + \sec(z)] \equiv \sinh^{-1}(\tan z) ,$$
$$t = \frac{1}{m}(\tan \bar{\theta}_0 - \tan \bar{\theta}) , \quad M \triangleq \begin{bmatrix} \cos \alpha & -\sin \alpha \\ \sin \alpha & \cos \alpha \end{bmatrix} ,$$

where t is in units of t_f, (u, v) are in units of at_f, and (x, y) are in units of at_f^2.

(d) Show that m can be expressed in terms of $\bar{\theta}_0$ and $\bar{\theta}_f$ as follows:

$$m = \tan \bar{\theta}_0 - \tan \bar{\theta}_f ,$$

and that

$$v(1) = 0 \ \Rightarrow \ f_u \sin \alpha + f_v \cos \alpha = 0 \Rightarrow \tan \alpha = -\frac{f_v}{f_u} \ ,$$

$$\begin{bmatrix} x(1) \\ y(1) \end{bmatrix} = \begin{bmatrix} \cos \alpha & -\sin \alpha \\ \sin \alpha & \cos \alpha \end{bmatrix} \begin{bmatrix} f_x \\ f_y \end{bmatrix} ,$$

where f_u, f_v, f_x, f_y are evaluated with $\bar{\theta} = \bar{\theta}_f$.

(e) Use the implicit analytic solution of (d) with a code like FSOLVE in MATLAB to determine $(\bar{\theta}_0,\ \bar{\theta}_f,\ \alpha,\ m)$ for the case $y(1) = .2,\ x(1) = .2$ and plot the optimal flight path. Compare your results with Figures 3.8 and 3.9.

3.3.8 TDP for Max $x(t_f)$ with Gravity and $y(t_f) = 0$:   The same as Problem 3.3.5 except that gravity acts down the y-axis and $y(1) = y_f$, where $y_f \neq 0$. For the plot take $g/a = 1/3$.

3.3.9 TDP for Max u_f with Gravity to $v_f = 0$ and y_f Specified: This is the continuous version of Problem 3.1.9; it may also be considered an extension of Problem 3.3.6 by adding gravity. The EOM are

$$\dot{u} = a\cos\theta\ ,\quad \dot{v} = a\sin\theta - g\ ,\quad \dot{y} = v\ ,$$

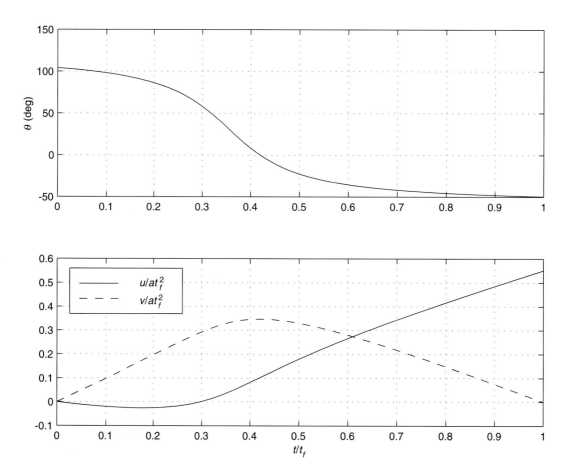

FIGURE 3.8   TDP for Max u_f with $t_f = 1,\ v_f = 0,\ y_f = .2,\ x_f = .2$; x vs. y

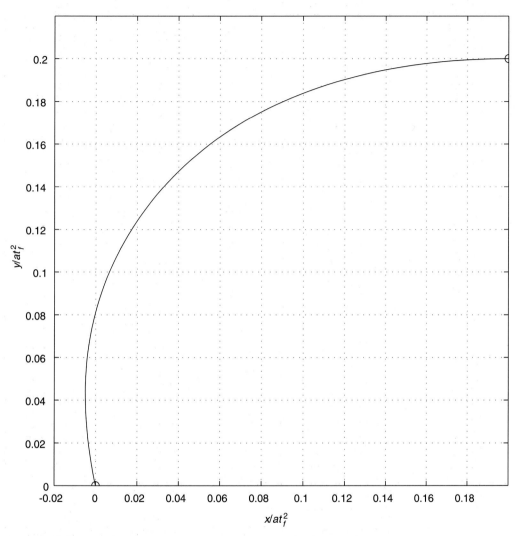

FIGURE 3.9   TDP for Max u_f with $t_f = 1$, $v_f = 0$, $y_f = .2$, $x_f = .2$; θ, u, v vs. t

with zero initial conditions and $v(t_f) = 0$, $y(t_f) = y_f$. We wish to find $\theta(t)$ to maximize $u(t_f)$, where t_f is specified.

(a) If we approximate g as constant, show that the EL equations are unchanged and the linear tangent law still applies:

$$\tan \theta = \tan \theta_0 - mt \; ,$$

where θ_0 and m are constants.

(b) The analytical solution for constant g follows easily from Example 3.3.1 by simply subtracting gt from v and $gt^2/2$ from y, noting that θ is no longer symmetric about the midpoint. Using t in t_f, (u, v) in at_f, (x, y) in at_f^2, and g in a, show that

$$t = \frac{1}{m}(\tan\theta_0 - \tan\theta) , \quad u = \frac{1}{m}[\Lambda(\theta_0) - \Lambda(\theta)] , \quad v = \frac{1}{m}(\sec\theta_0 - \sec\theta) - gt ,$$

$$x = \frac{1}{m}[(v + gt) - u\tan\theta] , \quad y = \frac{1}{2m}[t\sec\theta_0 - (v + gt)\tan\theta - u] - \tfrac{1}{2}gt^2 ,$$

where

$$\Lambda(z) \triangleq \log[\tan(z) + \sec(z)] \equiv \sinh^{-1}(\tan z) ,$$

(c) Show that $t(\theta_f) = 1$, $v(\theta_f) = 0$, $y(\theta_f) = y_f$ determine θ_0, θ_f and m from

$$\tan\theta_f = \frac{cg - \sqrt{c^2 - 1 + g^2}}{1 - g^2} , \quad m = \tan\theta_0 - \tan\theta_f ,$$

$$2m^2 y_f = m(\sec\theta_0 - g\tan\theta_f) - \Lambda(\theta_0) + \Lambda(\theta_f) - gm^2 ,$$

where $c \triangleq \sec\theta_0 - g\tan\theta_0$.

(d) For $g/a = 1/3$, use FSOLVE in MATLAB to determine θ_0 to give $y(t_f) = .2$.

3.3.10 TDP for Max u_f with Gravity to $v_f = 0$ and Specified y_f, x_f: This is the continuous version of Problem 3.1.10; it may also be considered an extension of Problem 3.3.7 by adding gravity, or an extension of Problem 3.3.9 by adding the constraint on x_f.

(a) If we approximate g as constant, show that the EL equations are unchanged from those in Problem 3.3.8 and the "rotated linear tangent law" still applies:

$$\tan(\theta - \alpha) = \tan(\theta_0 - \alpha) - mt ,$$

where θ_0, α, and m are constants.

(b) The analytical solution for constant g follows easily from Problem 3.3.7 by simply subtracting gt from v and $gt^2/2$ from y. Using t in t_f, (u, v) in at_f, (x, y) in at_f^2, and g in a, show that

$$\begin{bmatrix} u \\ v \end{bmatrix} = T\begin{bmatrix} \bar{u} \\ \bar{v} \end{bmatrix} - \begin{bmatrix} 0 \\ gt \end{bmatrix} , \quad \begin{bmatrix} x \\ y \end{bmatrix} = T\begin{bmatrix} \bar{x} \\ \bar{y} \end{bmatrix} - \begin{bmatrix} 0 \\ gt^2/2 \end{bmatrix} ,$$

where

$$\bar{u} \triangleq \frac{1}{m}[\Lambda(\bar\theta_0) - \Lambda(\bar\theta)] , \quad \bar{v} \triangleq \frac{1}{m}(\sec\bar\theta_0 - \sec\bar\theta) ,$$

$$\bar{x} \triangleq \frac{1}{m}(\bar{v} - \bar{u}\tan\bar\theta) , \quad \bar{y} \triangleq \frac{1}{2m}(t\sec\bar\theta_0 - \bar{v}\tan\bar\theta - \bar{u}) ,$$

and

$$\bar\theta \triangleq \theta - \alpha , \quad \bar\theta_0 \triangleq \theta_0 - \alpha , \quad \Lambda(z) \triangleq \log[\tan(z) + \sec(z)] \equiv \sinh^{-1}(\tan z) ,$$

$$t = \frac{1}{m}(\tan\bar\theta_0 - \tan\bar\theta) , \quad T \triangleq \begin{bmatrix} \cos\alpha & -\sin\alpha \\ \sin\alpha & \cos\alpha \end{bmatrix} .$$

(c) Show that m and α are determined by $\bar{\theta}_0$, $\bar{\theta}_f$ as:

$$m = \tan\bar{\theta}_0 - \tan\bar{\theta}_f , \quad \alpha = \tan^{-1}\frac{y_f + g/2}{x_f} - \tan^{-1}\frac{\bar{x}}{\bar{y}} .$$

(d) For $g/a = 1/3$, use the implicit analytic solution of (b) and (c) with FSOLVE in MATLAB to determine (θ_0, θ_f) so that $y_f = .2$, $x_f = .15$ and plot the optimal flight path.

3.3.11 VDP for Max Range with Gravity and Thrust with y_f Specified:

(a) This is the continuous version of Problem 3.2.11 which is the same as Problem 2.3.5 with the added constraint that y_f is specified. Show that the adjoint equations and the optimality condition can be solved for V as a function of γ, V_f, and γ_f:

$$V \sec \gamma = \frac{V_f}{\cos \gamma_f - a \sin(\gamma - \gamma_f)} .$$

(b) Differentiating the expression in (a) with respect to time and using $\dot{V} = a - \sin\gamma$, show that

$$\dot{\gamma} = \frac{V_f \sec \gamma_f}{(V \sec \gamma)^2} .$$

3.3.12 VDP for Max Range with Gravity, Thrust, Drag, and y_f Specified:

(a) This is the continuous version of Problem 3.2.12 which is the same as Problem 2.3.6 with the added constraint that y_f is specified. Show that the adjoint equations can be solved for γ as a function of V, V_f, and γ_f

$$V \sec \gamma = V_f \sec \gamma_f + V(a - V^2)(\tan\gamma - \tan\gamma_f) .$$

(b) Differentiating the expression in (a) with respect to time and using $\dot{V} = a - \sin\gamma - V^2$, show that

$$\dot{\gamma} = \frac{V_f \sec \gamma_f}{(V \sec \gamma)^2} + 2V \cos\gamma .$$

(c) Integrate (b) and $\dot{V} = a - \sin\gamma - V^2$, $\dot{x} = V \cos\gamma$ and $\dot{y} = V \sin\gamma$ numerically and interpolate V_f, γ_f to make $y(1) = y_f$, $\gamma(1) = \gamma_f$. Use the data of Problem 3.2.12 and compare your results with Figure 3.19.

3.3.13 VDP to Enclose Max Area in a Constant Wind:

An A/C has a constant velocity V with respect to the air, and the wind velocity u is constant. Find the closed curve, as projected on the ground, that the A/C should fly to enclose the max area in a given time T. The EOM are

$$\dot{x} = V \cos\theta + u , \quad \dot{y} = V \sin\theta ,$$

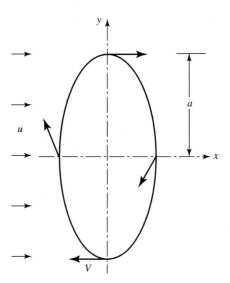

FIGURE 3.10   A/C Path in a Wind to Enclose Max Area in a Given Time

where we have chosen the x-axis to be in the direction of the wind. If the A/C flies a closed curve, the area enclosed is

$$A = \int_0^T y\dot{x}dt \ .$$

Answer: The curve is an ellipse (see Figure 3.10) with eccentricity u/V, minor axis parallel to the wind, and the max area enclosed is

$$A_{max} = \frac{V^2T^2}{4\pi}\left(1 - \frac{u^2}{V^2}\right)^{3/2} \ .$$

3.3.14 Surface of Min Area Connecting Two Circular Loops: Given two coaxial circular loops of radius a, a distance 2ℓ apart, we wish to find the surface of revolution connecting the two loops that has min area (this is the shape of a soap film stretched between the two rings). Using cylindrical coordinates (r, x) as shown in Figure 3.11, the surface area is

$$A = 2\pi \int_{-\ell}^{\ell} r\sqrt{1 + u^2}dx \ ,$$

where

$$\frac{dr}{dx} = u \ , \ r(-\ell) = a \ , \ r(\ell) = a \ .$$

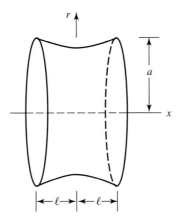

FIGURE 3.11   Surface of Min Area Connecting Equal Circular Loops

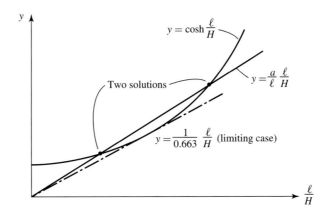

FIGURE 3.12   The Two Solutions of the Min Surface Area Problem

(a) Show that the minimizing surface is the catenary

$$r = H \cosh \frac{x}{H} \ ,$$

where H/ℓ is determined by $(a\ell)(\ell/H) = \cosh \ell/H$.

(b) Show that this equation has two solutions for $0 \le \ell/a \le .663$ and no solution for $\ell/a > .663$ (see Figure 3.12).

(c) Show that the min area is given by:

$$A_{min} = \begin{cases} 2\pi a^2[\tanh(\ell/H) + (\ell/H)\text{sech}^2(\ell/H)] , & 0 \le \ell/a \le .528 , \\ 2\pi a^2, & \ell/a \ge .528 . \end{cases}$$

where $2\pi a^2 =$ the flat area within the two circular loops.

(d) Find the surface of revolution of min area connecting two *unequal* coaxial circular loops a distance ℓ apart, where one loop has radius a and the other loop has radius b. For each given value of b/a, show that a limiting value of ℓ/a exists beyond which the min surface is two flat discs within the circular loops. See Problem 10.3.1 and Figure 10.9 for further discussion of this problem.

3.3.15 Min Surface Area for a Given Volume: In building a dwelling it is of interest to minimize surface area for a given volume, i.e., to use the least amount of material to surround a given volume. Assuming symmetry about a vertical axis and a circular floor area of specified radius a, show that the answer is to build a tent (or igloo or indoor sportsdome) whose surface is a segment of a spherical surface (see Figure 3.13).

Hint: use cylindrical coordinates (r, z) with the z-axis vertical. The problem then becomes to find $\alpha(r)$ to minimize

$$J = 2\pi \int_0^a r \sec \alpha \cdot dr ,$$

subject to

$$V = 2\pi \int_0^a rz \cdot dr , \quad \frac{dz}{dr} = \tan \alpha , \quad z(a) = 0 .$$

3.3.16 Shape of a Beam for Min Weight: We wish to find the depth profile of a beam $b(x)$ to minimize its weight, given that it is to carry a uniform load per unit length w, have

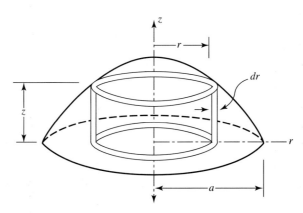

FIGURE 3.13   Nomenclature for Min Surface Area Problem

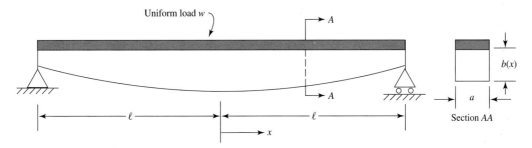

FIGURE 3.14   Nomenclature for Beam Shape Problem

constant width a, have a rectangular cross-section [a by $b(x)$], have length 2ℓ, be simply supported at both ends, and deflect a distance ϵ in the middle (see Figure 3.14).

Using the symmetry of the problem, we can formulate it for only the right half of the beam. Let $y(x)$ be the deflection of the beam under load (positive downward); then

$$\frac{dy}{dx} = s , \quad y(0) = \epsilon , \quad y(\ell) = 0 , \quad \frac{ds}{dx} = -\frac{\ell^2 - x^2}{u^3} , \quad s(0) = 0 ,$$

where $s = $ slope of the deflection curve, $u(x) = \sqrt{Ea/6w}\; b(x)$, $E = $ elastic modulus of the beam material.

The weight of the beam is proportional to the volume J, so it is proportional to

$$J = \int_0^\ell u \cdot dx .$$

Show that the optimal $b(x)$ is given by

$$b(x) = b_o \left(1 - \frac{x}{\ell} \right)^{1/2} \left(1 + \frac{x}{\ell} \right)^{1/4} , \quad 0 \le x \le \ell .$$

where b_o is a constant related to ϵ.

Note that optimal $b(x) \Rightarrow b(\ell) = 0$, which violates assumptions used in deriving the Euler beam equations. We should introduce a constraint that $b(x) \ge b_{\min} > 0$ (see Chapter 9).

3.3.17 TDP for Max Radius Orbit Transfer; Small Δr: For orbit transfer where the change in radius is small compared to the initial radius and the velocity change is small compared to the initial (circular) velocity, the EOM are well-approximated by the linearized equations below (still nonlinear in θ). The problem of transfer from one circular orbit to another, may then be stated as

$$\max_{\theta(t)} J = r(t_f)$$

subject to

$$\dot{r} = v , \quad \dot{v} = r + 2u + a\sin\theta , \quad \dot{u} = -v + a\cos\theta ,$$

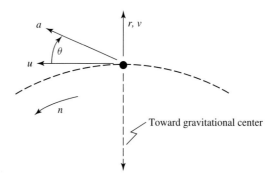

FIGURE 3.15   Nomenclature for Orbit Transfer
Problem

with BCs $r(0) = v(0) = v(t_f) = u(0) = 2u(t_f) + r(t_f) = 0$. Here $r =$ perturbation in
radius, $v =$ radial velocity, $u =$ perturbation in tangential velocity, $a =$ thrust acceleration
(constant), $t_f =$ specified final time, and $\theta =$ thrust direction. r is measured in units of
initial radius, u and v are in units of circular velocity at the initial radius, and time is in
units of $1/n$, where $n =$ circular orbit rate at the initial radius (see Figure 3.15).

Show that the optimal thrust direction program may be written as

$$\tan \theta = \frac{\sin(t - \alpha)}{2[\cos(t - \alpha) - \cos(\delta)]} ,$$

where α and δ are to be determined so that $v(t_f) = 0$ and $2u(t_f) + r(t_f) = 0$. With
$t_f = 3.35, \alpha = 1.76$, and $\delta = .40$, this $\theta(t)$ history is a fair approximation to the nonlinear
max radius orbit transfer of Example 3.4.1.

3.3.18 The Shape of a Hanging Chain: This problem was posed in 1638 by Galileo,
who conjectured (incorrectly) that the shape was a parabola. It was solved using the calculus
of variations by the Bernoulli brothers Jakob and Johannes and by Leibnitz in a series of
papers from 1690 to 1692 (Ref. Go).

For static equilibrium the potential energy of a system of particles must be a minimum,
which follows from Hamilton's principle; see Section 2.3. Given a chain of length $2L$
hanging from two points on the same level a distance 2ℓ apart, where $\ell < L$, we wish to
find the shape of the chain $y(x)$ that minimizes its potential energy in the earth's gravitational
field, i.e., find $u(x)$ to minimize

$$J = -g\sigma \int_{-\ell}^{\ell} y \, ds ,$$

subject to the constraints

$$2L = \int_{-\ell}^{\ell} ds , \quad y(-\ell) = 0 , \quad y(\ell) = 0 ,$$

where g = gravitational force per unit mass, σ = mass per unit length of chain, $ds = (1 + u^2)^{1/2}dx$, and $dy/dx = u$.

Show that the curve is the "catenary"

$$y = H \left(\cosh \frac{x}{H} - \cosh \frac{\ell}{H} \right) ,$$

where H is determined by $L/H = \sinh \ell/H$.

3.3.19 Bank Angle Program for Min Fuel Holding Path: Air traffic controllers some-times ask an arriving aircraft to 'hold', i.e., delay its arrival for a specified time. It is of interest to do this using as little fuel as possible. For constant altitude, constant velocity flight, treating the aircraft as a point mass, the EOM are

$$\dot{\theta} = u , \quad \dot{x} = \cos\theta , \quad \dot{y} = \sin\theta ,$$

where $\theta(0) = x(0) = y(0) = y(t_f) = 0$, $x(t_f) = x_f$, $\theta(t_f) = 0$ or 2π. Here θ = heading angle, $u = \tan\sigma$, and σ = the bank angle. Time is in units of V/g, (x, y, x_f) are in units of V^2/g, where V is the aircraft velocity, and g = gravitational force per unit mass. The incremental fuel used per unit time due to turning is proportional to u^2; hence, to minimize incremental fuel used, the performance index is

$$\min_{u(t)} J = \int_0^{t_f} u^2 dt .$$

(a) Show that the optimal heading angle history satisfies the nonlinear differential equation

$$\ddot{\theta} = A\sin\theta + B\cos\theta , \quad \theta(0) = 0 , \quad \dot{\theta}(0) = C .$$

where (A, B) are constants. The three constants (A, B, C) may be used to meet the three terminal conditions.

(b) If $t_f - x_f \ll t_f$ then the aircraft simply stretches the path by turning slightly left and then right (or vice-versa). Hence $|\theta| \ll 1$ and an LQ approximation to the (x, y) equations is

$$\dot{x} \approx 1 - \frac{\theta^2}{2} , \quad x(0) = 0 , \quad \dot{y} \approx \theta , \quad y(0) = 0 .$$

For this case show that the optimal path is given by

$$\theta = D\sin(\omega t) , \quad y = \frac{D}{\omega}[1 - \cos(\omega t)] , \quad x = \left(1 - \frac{D^2}{4}\right)t + \frac{D^2}{8\omega}\sin(2\omega t) ,$$

where $\omega = 2\pi/t_f$, $D = 2\sqrt{1 - x_f/t_f}$. For $t_f = 60$ and $x_f = 40$, use codes like

MATLAB's ODE23 and FSOLVE to find A, B, C in (a) to calculate the optimal path and then plot σ vs. time and the path (x vs. y).

(c) If $x_f \ll t_f$ then the A/C must make a 360 deg turn, so that $\theta_f = 2\pi$. Let $\theta = 2\pi t/t_f + \delta\theta$ and assume $|\delta\theta| \ll 1$. For this case show that the optimal path is

$$\theta = 2\pi \frac{t}{t_f} - \frac{2x_f}{t_f} \sin \omega t , \quad y = \frac{1 - \cos \omega t}{\omega} + \frac{x_f}{t_f} \frac{1 - \cos(2\omega t)}{2} ,$$

$$x = \frac{\sin \omega t}{\omega} + \frac{x_f}{t_f} \left[t - \frac{\sin(2\omega t)}{2\omega} \right] .$$

For $t_f = 60$ and $x_f = 10$, use codes like MATLAB's ODE23 and FSOLVE to find A, B, C in (a) to calculate the optimal path and then plot σ vs. time and the path (x vs. y). Compare your plot with Figure 3.16, where the solid curves are the nonlinear solution and the dashed curves are the LQ approximation.

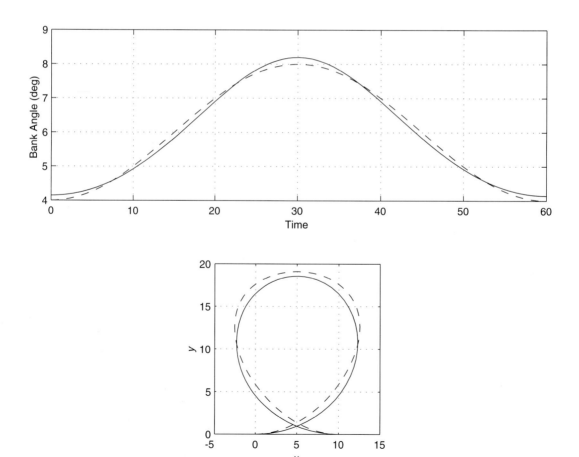

FIGURE 3.16   Min Fuel Holding Path for $t_f = 60$ and Distance-To-Go $x_f = 10$

3.4 Numerical Solution with Gradient Methods

As with discrete systems, only very simple problems can be solved analytically. Hence we now discuss algorithms for numerical solution. These algorithms involve extensive calculations that were almost impossible before digital computers. We first discuss gradient algorithms; shooting algorithms are treated in Section 3.6.

The derivation of the gradient algorithm FOPC follows the same steps as the derivation of DOPC, except that sequences are replaced with functions, sequencing is replaced by integration, and summations are replaced with integrals. Hence we just give a summary of the algorithm at the end of the chapter.

A MATLAB Code for FOPC

A MATLAB code for implementing the FOPC algorithm with multiple controls is listed in Table 3.2. Data for the problem being solved must be placed in a subroutine 'name'. See the discussion in Section 2.4 regarding interpolation using ODE23. The forward integration code FOPC-F is identical to FOP0-F listed in Table 2.2. The backward integration code FOPC-B differs from FOP0-B in that it must integrate to find g and Q. The subroutines

TABLE 3.2   A MATLAB Code for FOPC

```
function [tu,ts,nu,la0]=fopc(name,tu,tf,s0,k,told,tols,mxit,eta)
% Function OPtim. w. terminal Constraints
% Name must be in single quotes; function file 'name' computes
% f(s,u)=sdot for flg=1, (Phi,Phis) for flg=2, [fs,fu] for flg=3,
% inputs: global variables name, tu, ts, ns, np, where np-1 = no.
% terminal constraints, tu=[t(N1),1) u(N1,nc)]= init. estim. of
% u(t); s0(ns,1)=init. state, tf=final time, k=step size param,
% told=reltol for ode23; stops when norm(dua)<tols or number of
% iterations > mxit; 0<eta<=1 where d(psi)=-eta*psi is desired
% change in term. constraints on next iteration; outputs: opt.
% tu and ts=[t(N,1) s(N,ns)]), nu and la0=initial adjoint vector.
% BASIC version AEB '84; MATLAB versions Sun H. Hur '90 and
% Fred A. Wiesinger '91, AEB 11/94, 2/2/98
%
if nargin<9, eta=1; end; if nargin<8, mxit=10; end;
it=0; dua=1; pmag=1;
disp('    Iter.         phi         norm(psi)      dua');
options=odeset('reltol',told); [dum,nc1]=size(tu); nc=nc1-1;
while max([norm(dua) pmag])>tols,
  [t,s]=ode23('fopc_f',[0 tf],s0,options,tu,name); ts=[t s];
  Ns=length(t); ns=length(s0); dum=zeros(nc,1);
  [Phi,Phis]=feval(name,dum,s(Ns,:)',tf,2); np=length(Phi);
  phi=Phi(1); psi=Phi([2:np]); pmag=norm(psi)/sqrt(np-1);
  yf=[formm(Phis','c'); zeros(np*(np+1)/2,1)]; ny=length(yf);
```

(continued)

TABLE 3.2   *(continued)*

```
[tb,y]=ode23('fopc_b',[tf 0],yf,options,tu,ts,name,ns,np);
N=length(tb); Qg=forms(y(N,[np*ns+1:ny])'); g=Qg([2:np],1);
Q=Qg([2:np],[2:np]); nu=-Q\g;
la0=formm(y(N,[1:np*ns])',ns)*[1; nu]; du=zeros(N,nc); u=du;
for i=1:N,
   s1=table1(ts,tb(i)); u1=table1(tu,tb(i));
   [fs,fu]=feval(name,u1,s1,tb(i),3);
   la=formm(y(i,[1:np*ns])',ns); HuPhi=la'*fu;
   Huphi=HuPhi(1,:); Hupsi=HuPhi([2:np],:);
   du(i,:)=-k*(Huphi+nu'*Hupsi)-eta*(psi'/Q)*Hupsi;
   u(N-i+1,:)=u1+du(i,:);
end
for j=1:nc, dua(j)=norm(du(:,j))/sqrt(N); end;
if mxit==0, disp([it phi pmag dua]); break; end
disp([it phi pmag dua]);
t1=zeros(N,1); for i=1:N, t1(i)=tb(N-i+1); end; tu=[t1 u];
if it>=mxit, break, end; it=it+1;
end
function yp=fopc_b(t,y,flag,tu,ts,name,ns,np)
% Subroutine for FOPC - Bkwd integ. of adjoint eqns and
% Qg=int(HuPhi*HuPhi')dt;
%
la=formm(y([1:np*ns]),ns); u=table1(tu,t);
s=table1(ts,t); [fs,fu]=feval(name,u,s,t,3);
HuPhi=la'*fu; Qgd=-HuPhi*HuPhi';
lad=-fs'*la; yp=[formm(lad,'c'); forms(Qgd)];
```

FORMM and FORMS were developed by Sun H. Hur to convert general and symmetric matrices (respectively) into vectors in order to use the vector differential equation solver ODE23 in MATLAB.

Example 3.4.1—TDP for Max Radius Orbit Transfer in a Given Time

Given a rocket engine with constant thrust T, operating for a given length of time t_f, we wish to find the thrust-direction history, $\theta(t)$, to transfer a S/C from a given initial circular orbit to the largest possible circular orbit. The nomenclature is defined in Figure 3.17; r = radial distance of S/C from attracting center, u = radial component of velocity, v = tangential component of velocity, m = mass of S/C, $-\dot{m}$ = fuel consumption rate (constant), θ = thrust direction angle, μ = gravitational constant of attracting center. Using time in units of $\sqrt{r_o^3/\mu}$, r in units of r_o, u and v in units of $\sqrt{\mu/r_o}$, m in m_o, thrust in units of $\mu m_o/r_o^2$, the problem may be stated as: Find $\theta(t)$ to maximize $r(t_f)$ subject to

$$\dot{r} = u \ , \quad \dot{u} = \frac{v^2}{r} - \frac{1}{r^2} + a \sin\theta \ , \quad \dot{v} = -\frac{uv}{r} + a \cos\theta \ ,$$

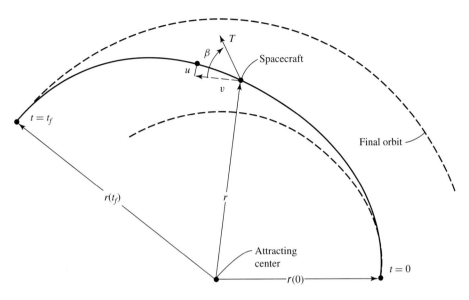

FIGURE 3.17   Max Radius Orbit Transfer in a Given Time (or Min Time for a Given Final Radius

with $r(0) = 1$, $u(0) = 0$, $v(0) = 1$, $\psi_1 = u(t_f) = 0$, $\psi_2 = v(t_f) - 1/\sqrt{r(t_f)} = 0$, and $a \triangleq T/(1 - |\dot{m}|t)$.

The Hamiltonian is, therefore,

$$H = \lambda_r u + \lambda_u \left(\frac{v^2}{r} - \frac{1}{r^2} + a \sin\theta \right) + \lambda_v \left(\frac{-uv}{r} + a \cos\theta \right) ,$$

and

$$\Phi = r(t_f) + \nu_1 u(t_f) + \nu_2 \left[v(t_f) - \sqrt{\frac{1}{r(t_f)}} \right] .$$

Thus, the necessary conditions become

$$\dot{\lambda}_r = -H_r = -\lambda_u \left(-\frac{v^2}{r^2} + \frac{2}{r^3} \right) - \lambda_v \left(\frac{uv}{r^2} \right) ,$$

$$\dot{\lambda}_u = -H_u = -\lambda_r + \lambda_v \frac{v}{r} , \quad \dot{\lambda}_v = -H_v = -\lambda_u \frac{2v}{r} + \lambda_v \frac{u}{r} ,$$

$$0 = H_\theta = (\lambda_u \cos\theta - \lambda_v \sin\theta)a \Rightarrow \tan\theta = \frac{\lambda_u}{\lambda_v} ,$$

$$\lambda_r(t_f) = \Phi_r = 1 + \frac{\nu_2}{2[r(t_f)]^{3/2}} , \quad \lambda_u(t_f) = \Phi_u = \nu_1 , \quad \lambda_v(t_f) = \Phi_v = \nu_2 .$$

These six differential equations are to be solved subject to the six boundary conditions, with the choice of ν_1 and ν_2 available to satisfy the additional two boundary conditions. The control $\theta(t)$ is determined in terms of λ_u and λ_v from the optimality condition.

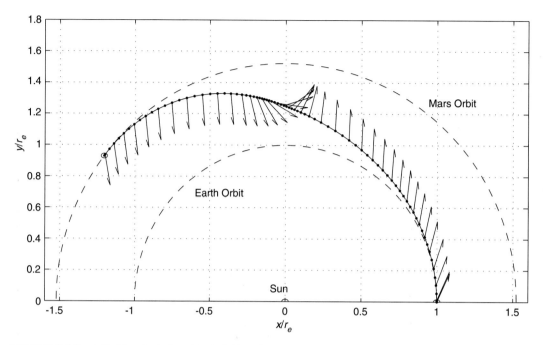

FIGURE 3.18   Min Time Path from Earth Orbit to Mars Orbit Using Low Thrust; Trip Time = 193 Days

A numerical solution of this problem for

$$T = .1405 , \quad |\dot{m}| = .07489 , \quad t_f = 3.3155 ,$$

was given in Ref. MP and is repeated below. Interpreted for a 10,000-lb S/C moving out from the Earth's orbit, the thrust would be 0.85 lb, the fuel consumption 12.9 lb/day, and the trip time 193 days. The optimal thrust direction and the resulting trajectory are shown in Figure 3.18. The radial component of thrust is outward for the first half (roughly) of the flight, and inward for the second half, with a rapid change in thrust direction angle in the middle of the flight.

The data subroutine MARC is listed below.

```
function [f1,f2]=marc(be,x,t,flg)
% th=x(4) is ignorable but included for plotting.
%
T=.1405; B=.07489; a=T/(1-B*t); r=x(1); u=x(2); v=x(3); c=cos(be); s=sin(be)
ud=v^2/r-1/r^2+a*s; vd=-u*v/r+a*c;
if flg==1
 f1=[u; ud; vd; v/r];
elseif flg==2
 f1=[r; u; v-1/sqrt(r)]; f2=[1 0 0 0; 0 1 0 0; 1/(2*(r)^(3/2)) 0 1 0];
```

```
elseif flg==3
 f1=[0 1 0 0;-(v/r)^2+2/r^3 0 2*v/r 0; u*v/r^2 -v/r -u/r 0; 0 0 0 0];
 f2=a*[0; c; -s; 0];
end
```

The MATLAB script listed below solves the problem.

```
% Script e03_4_1.m; max radius transfer in given time using FOPC;
%
clear; name='marc'; be=[.5:.25:5.5]; N=length(be)-1; tf=3.3155;
t=tf*[0:1/N:1]; tu=[t' be']; s0=[1 0 1 0]'; k=-3; told=1e-5;
tols=5e-4; mxit=10;
[tu,ts,nu,la0]=fopc(name,tu,tf,s0,k,told,tols,mxit); t1=ts(:,1);
r=ts(:,2); u=ts(:,3); v=ts(:,4); th=ts(:,5); N1=length(t1);
t2=tu(:,1); be=tu(:,2);
```

Use of the MATLAB Code CONSTR

Since the control function $u(t)$ is determined only at a finite number $(N + 1)$ of points, we can again use the MATLAB code CONSTR to solve continuous dynamic optimization problems with terminal constraints, just as in Section 3.2 for discrete dynamic optimization problems. All that is required is a MATLAB function file to calculate the performance index and the terminal constraints for an arbitrary $u(t)$, which means integrating (3.39) forward and calculating ϕ and ψ from (3.38) and (3.41), which must be called 'f' and 'g' when using CONSTR. This can be done by using the code ODEU.

Example 3.4.2—TDP for Max Radius Orbit Transfer Using CONSTR

The same max radius orbit transfer problem solved in Example 3.4.1 with the code FOPC is solved here with the MATLAB code CONSTR. Even with a good initial guess, the convergence was quite slow; FOPC needed only 69 function evaluations to converge from the poor guess.

```
% Script e03_4_2.m; TDP for max radius from earth orbit
% using CONSTR; ; s=[r,u,v,th]';
%
be0=[.4 .5 .6 .7 .8 .9 1.1 1.2 1.4 1.7 2.7 4.5 4.7 ...
      4.9 5.0 5.1 5.2 5.3 5.3 5.4 5.4];      % Initial guess
N1=length(be0); N=N1-1; tf=3.3155; t=tf*[0:1/N:1];
un=ones(1,N1); vlb=.2*un; vub=6*un; optn(1)=1; optn(2)=1e-5;
optn(3)=1e-5; optn(4)=1e-5; optn(13)=2; optn(14)=1000;
s0=[1 0 1 0]';
be=constr('marc_fg',be0,optn,vlb,vub,[],s0,tf);
[f,g,s]=marc_fg(be,s0,tf);
r=s(1,:); u=s(2,:); v=s(3,:); th=s(4,:);
```

```
function [f,g,s]=marc_fg(be,s0,tf)
% Subroutine for Example 3.4.2; TDP for max radius
% orbit in time tf, starting from earth orbit;
%th=x(4) is included for plotting; 3/28/94, 3/5/97
%
[t,s]=odeu('marc',be,s0,tf);  N=length(be);
rf=s(1,N); uf=s(2,N); vf=s(3,N);
f=-rf; g=[uf vf-1/sqrt(rf)];
```

Problems

3.4.1 to 3.4.11: Do gradient solutions of Problems 3.3.1 to 3.3.11 using FOPC or CONSTR.

3.4.12 VDP for Max Range with Gravity, Thrust, Drag and y_f Specified: Do a gradient solution of Problem 3.3.12. The normalized EOM are

$$\dot{V} = a - V^2 - \sin\gamma\ , \quad \dot{y} = V\sin\gamma\ , \quad \dot{x} = V\cos\gamma\ ,$$

where t is in units of t_f, V is in gt_f, (x, y) are in gt_f^2, and a is in g. We wish to find $\gamma(t)$ to maximize $x(t_f)$ with specified $y(t_f)$ and t_f. Use FOPC with normalized $a = .05$, $t_f = 5$, and $y_f = -.3$ to verify the max x_f path shown in Fig. 3.19. Also shown in Fig. 3.19 is a path consisting of an initial vertical dive, a steady descent, and a final vertical climb; the parameters were chosen to maximize final range.

3.4.13 to 3.4.19: Do gradient solutions of Problems 3.3.13 to 3.3.19 using FOPC or CONSTR.

3.4.20 TDP for Transfer to the Orbits of Mars and Jupiter:
(a) Verify the solution to the example for the Mars transfer orbit in Fig. 3.18.
(b) Using the same S/C data, find the max radius orbit in dimensionless time $t_f = 8.22$ and make a plot similar to Fig. 3.18. The max radius is approximately equal to the radius of the orbit of Jupiter.

3.4.21 TDP for Transfer to the Orbit of Venus: Using the S/C data of Problem 3.4.20, find the min radius orbit in dimensionless time $t_f = 2.40$ and make a plot similar to Fig. 3.18. The min radius is approximately equal to the radius of the orbit of Venus.

3.4.22 Min Altitude Loss; Glider with a Parabolic Lift-Drag Polar: This is a dynamic version of Problem 1.2.8 (min steady glide angle and descent rate). We wish to find $\alpha(t)$ to minimize the altitude loss in a given time t_f or in a given horizontal distance x_f, starting and ending with specified velocity and horizontal flight for an A/C with $\alpha_m = 1/12$, $\eta = .5$.

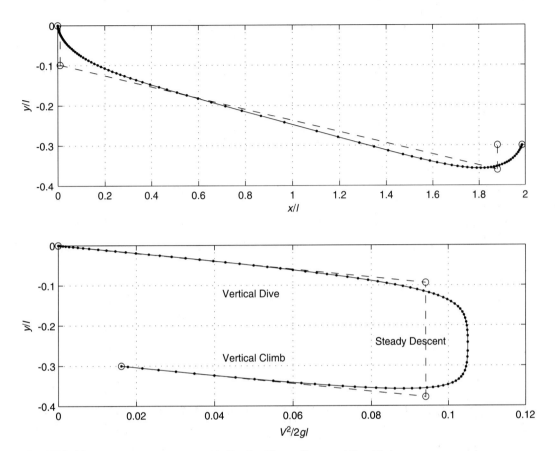

FIGURE 3.19   VDP for Max Range with Gravity, Thrust, Drag, and Specified y_f

The normalized EOM are

$$\dot{V} = T - \eta(\alpha^2 + \alpha_m^2)V^2 - \sin\gamma \ , \quad \dot{\gamma} = \alpha V - \frac{\cos\gamma}{V} \ , \quad \dot{h} = V\sin\gamma \ , \quad \dot{x} = V\cos\gamma \ .$$

where T is in units of A/C weight, (x, h) are in ℓ, V is in $\sqrt{g\ell}$, time is in $\sqrt{\ell/g}$, and

$$\ell = \frac{2m}{\rho S C_{L_\alpha}} \ , \quad \alpha_m = \sqrt{\frac{C_{D_0}}{\eta C_{L_\alpha}}} \ .$$

(a) Calculate and plot the glide path for min altitude loss in normalized time $t_f = 30$ with $T = 0$, $V_0 = V_f = 4$, $\gamma_0 = \gamma_f = 0$. The middle part of the path should be close to the min steady descent rate $\dot{h} = -.2515$ corresponding to $V = 2.614$, $\gamma = -.0964$ rad,

$\alpha = .1457$ rad. Note the phugoid oscillations in the optimal path (normalized phugoid period $= 2\pi V$).

(b) Calculate and plot the glide path for min altitude loss in normalized horizontal distance $x_f = 120$ with $T = 0$, $V_0 = V_f = 4$, $\gamma_0 = \gamma_f = 0$. To do this, change the independent variable from time to horizontal distance, so that the final value of the independent variable is specified. The middle part of the path should be close to the minEO steady descent angle $\gamma = -.0831$ rad, corresponding to $V = 3.458$, $\alpha = .1457$ rad, $\dot{h} = -.2872$. Note the phugoid oscillations in the optimal path (normalized phugoid period $= 2\pi V$).

(c) Find the energy-state approximate solutions for parts (a) and (b) and plot $V^2/2g\ell$ vs. h/ℓ for these solutions and the more accurate solutions from (a) and (b).

3.4.23 Max Altitude Climb; A/C with Parabolic Lift-Drag Polar: This is a dynamic version of Problem 1.2.12 (max steady climb rate and climb angle with given thrust for $T \ll mg$). We wish to find $\alpha(t)$ to maximize the altitude gain in a given time t_f or in a given horizontal distance x_f, starting and ending with specified velocity and horizontal flight for an A/C with $\alpha_m = 1/12$, $\eta = .5$. The equations of motion were given in Problem 3.4.22.

(a) Calculate and plot the flight path for max altitude gain in normalized time $t_f = 30$ with $T = .2$, $V_0 = V_f = 7$, $\gamma_0 = \gamma_f = 0$. The middle part of the path should be close to the max steady climb rate $\dot{h} = .4726$ corresponding to $\gamma = .1022$ rad, $V = 4.630$, $\alpha = .0466$ rad. Note the phugoid oscillations in the optimal path (normalized phugoid period $= 2\pi V$).

(b) Calculate and plot the flight path for max altitude gain in normalized horizontal distance $x_f = 120$ with $T = .2$, $V_0 = V_f = 5$, $\gamma_0 = \gamma_f = 0$. To do this, change the independent variable from time to horizontal distance, so that the final value of the independent variable is specified. The middle part of the path should be close to the max steady climb angle $\gamma = .1167$ rad, corresponding to $V = 3.464$, $\alpha = .0833$ rad, $\dot{h} = .4033$. Note the phugoid oscillations in the optimal path (normalized phugoid period $= 2\pi V$).

(c) Find the energy-state approximate solutions for parts (a) and (b) and plot $V^2/2g\ell$ vs. h/ℓ for these solutions and the more accurate solutions from (a) and (b).

3.4.24 Optimal Control Problems with Parameterized Control Histories: If the control history is approximated by a time function with open parameters (e.g., a polynomial in time with the coefficients as parameters), then the function optimization problem is replaced by a parameter optimization problem. Let p be the vector of open parameters, so that $u = u(p, t)$.

Consider the Mayer problem where we wish to choose p so as to minimize (or maximize) $J = \phi[x(t_f)]$ subject to

$$\dot{x} = f(x, u) , \quad x(0) = x_o , \quad 0 = \psi[x(t_f)] ,$$

where t_f is specified.

(a) Show that the gradients of the performance index and the constraints with respect to p may be calculated as follows:

$$\phi_p = \int_0^{t_f} H_u^\phi u_p dt \ , \quad \psi_p = \int_0^{t_f} H_u^\psi u_p dt \ ,$$

where H_u^ϕ and H_u^ψ may be determined using FOPC.

(b) Give a set of necessary conditions for a stationary solution.

(c) Give a gradient projection algorithm for determining the optimal p-vector, starting with an initial guess of p.

(d) Consider the dual of the brachistochrone problem, i.e., maximizing horizontal distance in a given time with vertical loss specified:

$$\phi = x(t_f) \ , \quad \psi = y(t_f) - y_f \ , \quad \dot{V} = g\sin u \ , \quad \dot{x} = V\cos u \ , \quad \dot{y} = V\sin u \ ,$$

and u is approximated by $u = p_1 + p_2 t$. Find ϕ_p and ψ_p for the initial guess $p_1 \neq 0$, $p_2 = 0$.

3.4.25 Max Altitude Climbing Turn for a 727 A/C: This is a dynamic version of Problem 1.3.22. There are *two controls*, angle-of-attack α and bank-angle σ. We wish to find the control histories to maximize the final altitude in a given time t_f, with terminal constraints that the A/C path be turned 90 degrees and the velocity be slightly above the stall velocity. Such a flight path might be of interest to reduce jet engine noise over an area that lies ahead of a take-off runway.

The equations of motion in normalized coordinates are

$$\dot{V} = T\cos(\alpha + \epsilon) - C_D V^2 - \sin\gamma \ , \quad V\dot{\gamma} = \left[T\sin(\alpha + \epsilon) + C_L V^2\right]\cos\sigma - \cos\gamma \ ,$$
$$V\cos\gamma\,\dot{\psi} = \left[T\sin(\alpha + \epsilon) + C_L V^2\right]\sin\sigma \ ,$$
$$\dot{h} = V\sin\gamma \ , \quad \dot{x} = V\cos\gamma\cos\psi \ , \quad \dot{y} = V\cos\gamma\sin\psi \ .$$

The normalization of variables and the expressions for $T(V)$, $C_D(\alpha)$, and $C_L(\alpha)$ are given in Problem 1.3.21. $h =$ altitude, $x =$ horizontal distance in the initial direction, and $y =$ horizontal distance perpendicular to the initial direction. (h, x, y) are in units of the characteristic length ℓ, which here is 3253 ft.

Consider a steady climbing turn at max turn rate for a climb angle of $\gamma = 6$ deg. From Problem 1.3.22 the optimal steady controls are $\alpha = 11.84$ deg and $\sigma = 29.52$ deg, the velocity is 280.4 ft/sec (normalized $V = 1.225$), and the max turn rate is 3.73 deg/sec. Choose these values of γ and σ as initial guesses of the control histories, and specify final velocity as 137.3 ft/sec (normalized $V = .60$) and the final turn angle as 90 deg. Choose the final time as $90/3.73 = 24.13$ sec (normalized time = 2.40) and the initial normalized velocity as $V = 1.00$, high enough that no part of the flight path goes below $h = 0$.

Let $s \triangleq [V \ \gamma \ \psi \ h \ x \ y]^T$ and $u \triangleq [\alpha \ \sigma \ ]^T$. The EOM above give $\dot{s} = f(s, u)$. Using FOPC find the control histories that maximize $h(t_f)$ with normalized $V(t_f) = .60$ and $\psi(t_f) = \pi/2$, with initial conditions $V = 1.00$, $\gamma = \psi = h = x = y = 0$. Compare your solution with Figs. 3.20 and 3.21. The max time is relatively insensitive to changes in the control histories so they are not determined very accurately by the gradient method; Figs. 3.20 and 3.21 were obtained using the shooting method (see Section 3.6).

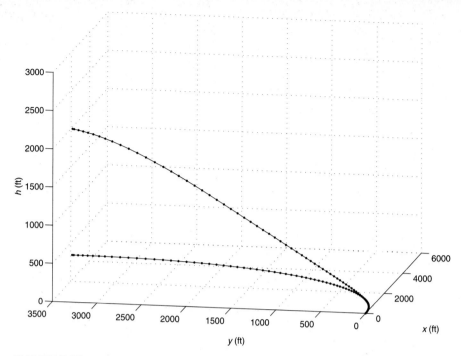

FIGURE 3.20 Max Altitude Climbing Turn for a 727 Aircraft with $V(0) = 324$ ft/sec, $t_f = 24.1$ sec, $V(t_f) = 194$ ft/sec, $\psi_f = 90$ deg—Flight Path

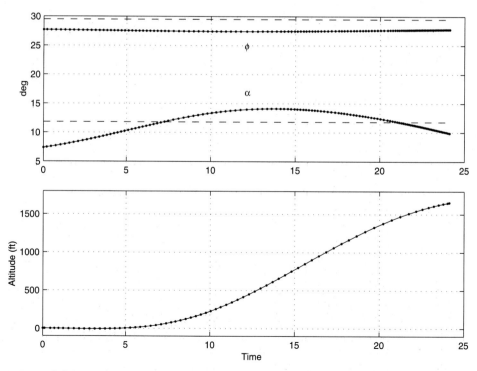

FIGURE 3.21 Max Altitude Climbing Turn for a 727 Aircraft with $V(0) = 324$ ft/sec, $t_f = 24.1$ sec, $V(t_f) = 194$ ft/sec, $\psi_f = 90$ deg—Controls and Altitude vs. Time

3.5 Direct Solution Methods for Discrete Systems

In this section we develop NR and shooting algorithms that solve discrete dynamic optimization problems with terminal constraints.

An NR Algorithm DOPCN

The differences from DOP0N (see Section 2.5) are (1) the backward sequencing of the adjoint equations has the boundary condition $\lambda^T(N) = \phi_s + v^T \psi_s$ and (2) there are terminal constraints $\psi(N) = 0$. The parameter vector used in FSOLVE is augmented to include v, i.e., $p = [u(0) \cdots u(N) \ v(1) \cdots v(q)]$ are estimates from a DOPC solution. A MATLAB implementation of DOPCN is listed in Table 3.3; it uses the same subroutines developed for DOPC, making it easy to use after finding a DOPC solution.

Example 3.5.1—DTDP for Max u_f to $v_f = 0$ and Specified y_f Using DOPCN

The initial guess for the control sequence and for v were obtained from Example 3.2.1, so that only a few iterations were required to obtain an accurate solution. The subroutine DTDPC is the same one listed in Example 3.2.1.

TABLE 3.3   A MATLAB Code for DOPCN

```
function [f,s,la0]=dopcn(p,name,s0,tf)
% Discrete OPtimization w. terminal Constraints, Newton-raphson code; plant
% eqns. sequenced forward, adjoint eqns. backward; use FSOLVE to iterate
% p=[u(i) nu] to make f=[Hu(i) psi]=0; function file 'name' computes s(i+1)
% =f[s(i),u(i)] for flg=1, (Phi,Phis) for flg=2, and (fs,fu) for flg=3
% (same file used by DOPC); inputs: u = guess of control sequence (1 by N),
% nu=guess of terminal constraint Lagrange multipliers; tf=spec. final time;
% s0=initial state (ns by 1); outputs: s=optimal state sequence, la0=initial
% lambda.
%
ns=length(s0); Phi=feval(name,0,s0,1,0,2); nt=length(Phi)-1;
N=length(p)-nt; s=zeros(ns,N+1); la=s; Hu=zeros(1,N); u=p([1:N]);
nu=p([N+1:N+nt]);
%
% Forward sequencing and store s(:,i):
dt=tf/N; s(:,1)=s0;
for i=1:N, s(:,i+1)=feval(name,u(:,i),s(:,i),dt,(i-1)*dt,1); end;
%
% Backward sequencing and store Hu(i) & psi:
[Phi,Phis]=feval(name,0,s(:,N+1),dt,tf,2); psi=Phi([2:nt+1]);
la=Phis'*[1;nu'];
for i=N:-1:1,
 [fs,fu]=feval(name,u(:,i),s(:,i),dt,(i-1)*dt,3); Hu(i)=la'*fu; la=fs'*la;
end; la0=la; f=[Hu psi'];
```

```
% Script e03_5_1.m; DTDP for max uf to vf=0 and spec. yf using DOPCN;
%
N=10; s0=[0 0 0 0]'; tf=1; t=[0:N]/N; name='dtdpc'; optn(1)=1; optn(14)=500;
th=[1.0472 .8378 .6283 .4189 .2094 0 -0.2094 -.4189 -.6283 -.8378];
nu=[-2.3963 4.7923]; p0=[th nu]; p=fsolve('dopcn',p0,optn,[], name,s0,tf);
[f,s,la0]=dopcn(p,name,s0,tf); th=p([1:N]);
u=s(1,:); v=s(2,:); x=s(4,:); y=s(3,:);
```

A Forward Shooting Algorithm DOPCF

The differences from DOP0F (see Section 2.5) are (1) the parameter vector is augmented to include ν and (2) the error vector is augmented to include $\psi = 0$. Initial estimates of $p = [\lambda(0), \ \nu]$ are obtained from a DOPC or DOPCN solution. A MATLAB implementation of DOPCF is listed in Table 3.4 (for scalar control only). It uses the subroutine DOPFU listed in Table 2.4.

Example 3.5.2—DTDP for Max u_f to $v_f = 0$ and Specified y_f Using DOPCF

The initial guesses for $\lambda(0)$ and ν were obtained from Example 3.2.1, so that only a few iterations were required to obtain the 'exact' solution. Again the subroutine 'dtdpc.m' is the same one listed in Example 3.2.1. The results agree exactly with those from Example 3.5.1.

TABLE 3.4   A MATLAB Code for DOPCF

```
function [f,s,u,la]=dopcf(p,name,u0,s0,tf,N)
% Discrete OPtimization w. terminal Constraints & Forward shooting;
% Name must be single quotes; function file 'name' must be in the
% MATLAB path, giving s(i+1) for flg=1, (Phi,Phis) for flg=2,
% (fs,fu) for flg=3; p=[la0,nu]; u0 = estimated initial control;
% s0=initial state; N = number of steps;
%
dt=tf/N; ns=length(s0); s=zeros(ns,N+1); la=s;
N1=length(p); la0=p([1:ns])'; nu=p([ns+1:N1])';
u=zeros(1,N); u(1)=u0; s(:,1)=s0; la(:,1)=la0;
for i=1:N, t=(i-1)*dt;
  if i==1, u1=u0; else u1=u(i-1); end;
  l1=la(:,i); z0=[u1; l1];
  z1=fsolve('dopfu',z0,[],[],name,s(:,i),la(:,i),dt,t);
  u(i)=z1(1); la(:,i+1)=z1([2:ns+1]);
  s(:,i+1)=feval(name,u(i),s(:,i),dt,t,1);
end
[Phi,Phis]=feval(name,u(N),s(:,N+1),dt,tf,2);
psi=Phi([2:N1-ns+1]);
f=[la(:,N+1)-Phis'*[1;nu]; psi];
```

```
% Script e03_5_2.m; DTDP for max uf to vf=0 and specified yf using DOPCF;
%
N=10; s0=[0 0 0 0]'; tf=1; t=tf*[0:N]/N; th0=1.0472;
la0=[1 2.3960 4.7923 0]; nu=[-2.3963 4.7923]; p0=[la0 nu];
name='dtdpc'; optn(1)=1; optn(2)=1e-8; optn(14)=50;
p=fsolve('dopcf',p0,optn,[],name,th0,s0,tf,N);
[f,s,th]=dopcf(p,name,th0,s0,tf,N); thh=[th th(N)];
u=s(1,:); v=s(2,:); y=s(3,:); x=s(4,:);
```

A Backward Shooting Algorithm DOPCB

The differences from DOP0B (see Section 2.5) are (1) the parameter vector is augmented to include ν and (2) the error vector is augmented to include $\psi = 0$. Initial estimates of $p = [s_f, \ \nu]$ are obtained from a DOPC or DOPCN solution. A MATLAB implementation of DOPCB is listed in Table 3.5 (for scalar control only). It uses the subroutine DOPBU listed in Table 2.5.

Example 3.5.3—DTDP for Max u_f to $v_f = 0$ and Specified y_f Using DOPCB

The initial guess for s_f and ν were obtained from Example 3.2.1, so that only a few iterations were required to obtain the 'exact' solution. Again the subroutine 'dtdpc.m' is the

TABLE 3.5   A MATLAB Code for DOPCB

```
function [f,s,u,la]=dopcb(p,name,uf,s0,N)
% Discrete OPtimization w. terminal Constraints using Bkwd shooting; name
% must be single quotes; function file 'name' must be in the MATLAB path,
% giving f for flg=1, (Phi,Phis) for flg=2, (fs,fu,fus,fuu) for flg=3;
% p=[sf,nu]; [sf,uf] = estim. final (state, control); s0=desired initial
% state; N = number of steps; f=s(:,1)-s0 = error in initial state;
%
ns=length(s0); dt=tf/N; sf=p([1:ns])'; s=zeros(ns,N+1);
la=s; u=zeros(1,N);
[Phi,Phis]=feval(name,uf,sf,dt,tf,2); nt1=length(Phi);
psi=Phi([2:nt1]); nu=p([ns+1:ns+nt1-1])';
s(:,N+1)=sf; la(:,N+1)=Phis'*[1;nu];
%
% Backward sequencing of Euler-Lagrange eqns:
for i=N:-1:1, t=(i-1)*dt;
  if i==N, u1=uf; else u1=u(i+1); end
  s1=s(:,i+1); z0=[u1; s1];
  z1=fsolve('dopbu',z0,[],[],name,s(:,i+1),la(:,i+1),dt,t);
  u(i)=z1(1); s(:,i)=z1([2:ns+1]);
  [fs,fu]=feval(name,u(i),s(:,i),dt,t,3); la(:,i)=fs'*la(:,i+1);
end
f=[s(:,1)-s0; psi];
```

same one listed in Example 3.2.1. The results agree exactly with those in Examples 3.5.1 and 3.5.2.

```
% Script e03_5_3.m; DTDP for max uf to vf=0 and spec. yf using DOPCB;
%
N=10; s0=[0 0 0 0]'; tf=1; t=[0:N]/N; thf=-.8378;
sf=[.6716 0 .2000 .3359]; nu=[-2.3963 4.7923]; p0=[sf nu];
name='dtdpc'; optn(1)=1; optn(2)=1e-8; optn(14)=50;
p=fsolve('dopcb',p0,optn,[],name,thf,s0,tf,N);
[f,s,th]=dopcb(p,name,thf,s0,tf,N); thh=[th th(N)];
u=s(1,:); v=s(2,:); y=s(3,:); x=s(4,:);
```

Problems

3.5.1 to 3.5.12:   Do Problems 3.2.1 to 3.2.12 using DOPCN and either DOPCB or DOPCF. Note that Problem 3.5.12 is impossible using single shooting for $t_f >$ about 4; the optimal path is nearly steady state with transition segments of time duration $\Delta t \approx .25$ at both ends (see Problem 5.2.4 for an analytic example).

3.6   Direct Solution Methods for Continuous Systems

In this section we develop NR and shooting algorithms that solve continuous dynamic optimization problems with terminal constraints.

An NR Algorithm FOPCN

FOPCN differs from FOPC in the boundary conditions for the backward integration of the adjoint equations; FOPCN has the boundary condition $\lambda^T(t_f) = \phi_s + v^T \psi_s$. Also FOPCN uses *equal time steps* whereas FOPC uses variable time steps. FSOLVE is used with the parameter vector $p = [u(0) \cdots u(N) \, v(1) \cdots v(q)]$; an initial estimate of $u(t)$ can be interpolated from a FOPC solution.

A MATLAB implementation of FOPCN is listed in Table 3.6 (for one control only) ; it uses the same subroutines developed for FOPC, making it easy to use after finding a FOPC solution. The ODEU and ODEHNU subroutines are equal time-step RK codes (on the disk that accompanies this book).

Example 3.6.1—TDP for Max u_f with $v_f = 0$ and Specified y_f Using FOPCN

This is the same problem solved in Example 3.4.1. A MATLAB script is listed below that solves the problem. The initial guesses for $\beta(t)$ and v were interpolated from a solution of Example 3.4.1 (the 'initial guesses' listed below are the converged solutions). The solution is exact within the limitations of the fixed-step RK integration schemes used in ODEU and ODEHNU and is very close to the solution from FOPC.

TABLE 3.6   A MATLAB Code for FOPCN

```
function [f,s,u,la0]=fopcn(p,name,s0,tf)
% Function OPtimization w. terminal Constraints using a Newton-raphson
% algorithm; plant eqns are integrated forward, adjoint eqns backward;
% used with FSOLVE to perturb p=[u nu] to make f=[Hu psi']=0; s0=initial
% state (ns by 1); tf=final time; uses same subroutine 'name' used by
% FOPC; u=guess of control history (1 by N+1); nu=guess of terminal
% constraint Lagrange multipliers (1 by nt); s=optimal state histories;
% la0=initial adjoint vector;
%
ns=length(s0); Phi=feval(name,0,s0,1,2); nt=length(Phi)-1;
N1=length(p)-nt; s=zeros(ns,N1); Hu=zeros(1,N1); u=p([1:N1]);
nu=p([N1+1:N1+nt]); [t,s]=odeu(name,u,s0,tf);
[Hu,phi,la0,psi]=odehnu(name,u,s,tf,nu); f=[Hu psi'];
```

```
% Script e03_6_1.m; max radius transfer in given time using FOPCN;
%
be=[.4310   .4663   .5045   .5457   .5902   .6380   .6893   .7441   .8026...
     .8648   .9306 1.0002 1.0735 1.1507 1.2324 1.3196 1.4151 1.5252...
    1.6675 1.9044 2.5826 3.9427 4.4209 4.6081 4.7222 4.8071 4.8768...
    4.9373 4.9913 5.0407 5.0864 5.1292 5.1695 5.2077 5.2442 5.2792...
    5.3129 5.3456 5.3773 5.4083 5.4386];
s0=[1 0 1 0]'; tf=3.3155; name='marc'; nu=[-1.4229 1.2638];
    p0=[be nu];
p=fsolve('fopcn',p0,optn,[],name,s0,tf);
[f,s,be,la0]=fopcn(p,name,s0,tf); r=s(1,:); u=s(2,:); v=s(3,:);
    th=s(4,:);
```

A Forward Shooting Algorithm FOPCF

The EL equations are integrated forward to determine ψ and the final boundary condition errors $e_f \triangleq \lambda(t_f) - \phi_x^T - \psi_x^T \nu$. In integrating the EL equations the control is determined from the optimality condition $H_u = 0$ at each integration step; in general, this must be done iteratively using $u \leftarrow u - H_{uu}^{-1} H_u$. This either maximizes or minimizes H depending on whether H_{uu} is negative or positive definite. In some problems, as in Example 3.6.3 below, an explicit solution for u can be obtained from $H_u = 0$ as a function of λ and x.

In general, the initial guess of $\lambda(t_0)$, ν does not satisfy the terminal constraints $e_f = 0$, $\psi = 0$. Hence we use FSOLVE to vary the components of $\lambda(t_0)$ and ν until $e_f = 0$, $\psi = 0$. An algorithm for forward shooting with terminal constraints is summarized below.

- Guess the Lagrange multipliers $\lambda(t_0)$ and ν.
- Integrate the EL equations forward and store $e_f = \lambda(t_f) - \phi_x^T - \psi_x^T \nu$ and ψ.

- Use FSOLVE to vary the elements of $p \triangleq [\lambda(t_0), \nu]$ until $[e_f, \psi]$ are zero to the desired accuracy.

FOPCF—A MATLAB Forward Shooting Code for Continuous Dynamic Optimization with Terminal Constraints

A MATLAB code that can be used to implement the algorithm above for the continuous case is listed in Table 3.7.

TABLE 3.7   A MATLAB Code for FOPCF

```
function [f,t,y]=fopcf(p,name,s0,tf,nc)
% Fcn. OPtim. w. terminal Constr. using Fwd. shooting; p=[la0 nu] from FOPC
% solution; s0(1,ns)= initial state vector; tf=spec. final time; name must
% be in single quotes; two function files 'namee' and 'name' must be on the
% MATLAB path, 'namee' containing the EL equations, 'name' containing (Phi,-
% Phis) for flg=2; f=laf'-phis-nu'*psis,psi'; use FSOLVE to iterate p until
% f=0 to computer accuracy; nc=no. of controls (default=1).
%
if nargin<5, nc=1; end; dum=zeros(nc,1); ns=length(s0); la0=p([1:ns])';
n1=length(p); nu=p([ns+1:n1])'; y0=[s0; la0]; tol=1e-3;
 options=odeset('reltol',tol); [t,y]=ode23([name,'e'],[0 tf],y0,options);
N=length(t); sf=y(N,[1:ns])'; laf=y(N,[ns+1:2*ns])';
[Phi,Phis]=feval(name,dum,sf,tf,2); ef=laf-Phis'*[1;nu];
 psi=Phi([2:n1-ns+1]); f=[ef' psi'];
```

Example 3.6.2—TDP for Max Radius Orbit Transfer

This is the same problem solved in Examples 3.4.1 and 3.6.1, now using FOPCF. The first subroutine is listed below; the second subroutine MARC was listed in Example 3.4.1.

```
function yp=marce(t,y)
% E-L eqns. for min tf to Mars; y=[r u v th lr lu lv lt]';
%
T=.1405; mdot=.07489; r=y(1); u=y(2); v=y(3); lr=y(5); lu=y(6);
      lv=y(7);
lt=y(8); a=T/(1-mdot*t); la=[lr lu lv lt]'; d=sqrt(lu^2+lv^2);
f=[u; v^2/r-1/r^2+a*lu/d; -u*v/r+a*lv/d; v/r];
fs=[0 1 0 0; -(v/r)^2+2/r^3 0 2*v/r 0; u*v/r^2 -v/r -u/r 0;
     -v/r^2 0 1/r 0];
yp=[f; -fs'*la];
```

A script for solving the problem is listed below.

```
% Script e03_6_2.m; max radius orbit transfer using FOPCF; guess of la0 and
% nu from FOPC;
%
la0=[1.869 .928 2.008 0]'; nu=[-1.426 1.276]'; p0=[la0' nu']; s0=[1 0 1 0]';
 tf=3.3155; name='marc'; optn(1)=1; optn(14)=500;
p=fsolve('fopcf',p0,optn,[],name,s0,tf);
[f,t,y]=fopcf(p,name,s0,tf); be=atan2(y(:,6),y(:,7)); [N,ns]=size(y);
for i=1:N, if be(i)<0; be(i)=be(i)+2*pi; end; end;
```

The slight differences between these solutions and the FOPC solution cannot be seen in Figs. 3.22 to 3.24. Note that $\lambda_v < 0$ for a brief period near the middle of the path; this gives the rather unexpected result that the thrust vector has a *backward* component for a short time in the middle of this min time path; the thrust direction angle changes very rapidly during this time. Apparently it is worthwhile to first build up the centrifugal force to accelerate outward, and then to decrease it using this period of backward thrust.

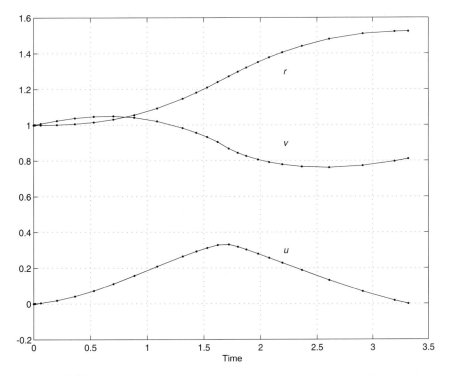

FIGURE 3.22   State Histories for Max Radius Orbit Transfer from Shooting Solution

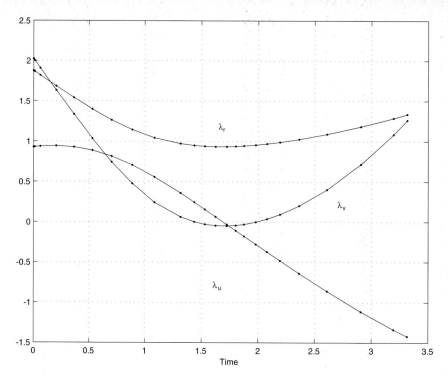

FIGURE 3.23   Adjoint Histories for Max Radius Orbit Transfer from Shooting Solution

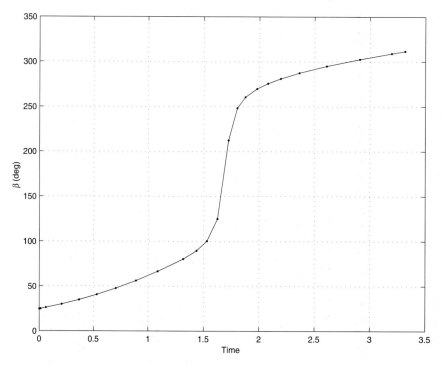

FIGURE 3.24   Thrust Direction Angle History for Max Radius Orbit Transfer from Shooting Solution

A Backward Shooting Algorithm FOPCB

The EL equations are integrated backward to determine the initial state errors. The comments above about the determining $u(t)$ apply also to backward shooting. In general, the initial guess of $s(t_f)$, ν does not yield the desired initial states or desired ψ. Hence we use FSOLVE to vary the components of $s(t_f)$ and ν until $\psi = 0$ and the initial states are correct to the desired accuracy. An algorithm for backward shooting with terminal constraints is summarized below.

- Guess $p \overset{\Delta}{=} [s(t_f),\ \nu]$.
- Integrate the EL equations backward and store $f \overset{\Delta}{=} [s(t_0) - s_0,\ \psi]$.
- Use FSOLVE to vary the elements of p until $f = 0$ to the desired accuracy.

FOPCB—A MATLAB Backward Shooting Code for Continuous Dynamic Optimization with Terminal Constraints

A MATLAB code that implements the algorithm above for the continuous case is listed in Table 3.8.

TABLE 3.8   A MATLAB Code for FOPCB

```
function [f,t,y]=fopcb(p,name,s0,tf,nc)
% Function OPtimization w. terminal Constraints using Bkwd. shooting;
% p=[sf' nu'] is an estimate from a converged soln. of FOPC; s0=
% desired initial state vector; tf=spec. final time; f=[init. state
% error, psi]; name must be in single quotes; function files 'namee'
% & 'name' must be on the MATLAB path, 'namee' containing the E-L
% eqns, 'name' containing (Phi,Phis); use FSOLVE to iterate p until
% f=0 to desired accuracy; nc=no. of controls (default=1).
%
if nargin<5, nc=1; end; dum=zeros(nc,1); ns=length(s0); sf=p([1:ns])';
[Phi,Phis]=feval(name,dum,sf,tf,2); nt1=length(Phi);
nu=p([ns+1:ns+nt1-1])'; psi=Phi([2:nt1]);
yf=[sf; Phis'*[1;nu]]; tol=1e-5; options=odeset('reltol',tol);
[t,y]=ode23([name,'e'],[tf 0],yf,options); N=length(t);
f=[y(N,[1:ns])'-s0; psi];
```

Example 3.6.3—TDP for Max Radius Orbit Transfer Using FOPCB

This is the same problem solved in Example 3.4.1, now using FOPCB. A script that solves the problem using FOPCB is listed below. The subroutine MARCE was listed in Example 3.6.2 and MARC in Example 3.4.2.

```
% Script e03_6_3.m; max radius orbit transfer using FOPCB; guess of
% sf and nu from FOPC;
%
sf=[1.5237 0 .8101 0]; nu=[-1.426 1.276]; p0=[sf nu];
s0=[1 0 1 0]'; tf=3.3155; name='marc'; optn(1)=1; optn(14)=500;
p=fsolve('fopcb',p0,optn,[],name,s0,tf); [f,t,y]=fopcb(p,name,s0,tf);
be=atan2(y(:,6),y(:,7)); N=length(t); t=tf*ones(N,1)-t;
for i=1:N, if be(i)<0; be(i)=be(i)+2*pi; end; end;
```

Problems

3.6.1 to 3.6.25:   Solve Problems 3.4.1 to 3.4.25 using FOPCN, FOPCB, and FOPCF. To
do this you will first have to do the problems with FOPC to obtain good initial guesses.
Note that Problem 3.6.12 is impossible using single shooting for $t_f >$ about 4; the optimal
path is nearly steady state with transition segments of time duration $\Delta t \approx .25$ at both ends
(see Problem 5.2.4 for an analytic example). FOPCN is for one control only, so it cannot
be used for Problem 3.6.25.

3.7 Chapter Summary

Discrete Dynamic Optimization with Terminal Constraints

Choose the sequence $u(i)$, $i = 0, \ldots, N - 1$ to minimize

$$J = \phi[x(N)] + \sum_{i=0}^{N-1} L[x(i), u(i), i] \,,$$

where N is specified and

$$x(i + 1) = f[x(i), u(i), i] \,, \quad x(0) = x_o \,, \quad \psi[x(N)] = 0 \,.$$

Necessary Conditions for a Stationary Path

Define

$$H(i) \stackrel{\Delta}{=} L[x(i), u(i), i] + \lambda^T(i + 1) f[x(i), u(i), i] \,, \quad \Phi \stackrel{\Delta}{=} \phi[x(N)] + v^T \psi[x(N)] \,.$$

Then

$$x(i + 1) = f[x(i), u(i), i] \,,$$
$$\lambda(i) = H_x^T(i) \equiv L_x^T(i) + f_x^T(i)\lambda(i + 1) \,, \quad \lambda^T(N) = \Phi_x \equiv \phi_x + v^T \psi_x \,,$$

where $u(i)$ is determined from

$$H_u(i) \equiv L_u(i) + \lambda^T(i+1) f_u(i) = 0 \ ,$$

This forms a TPBVP for $x(i)$ and $\lambda(i)$, where we must also find ν to satisfy the terminal constraints $\psi = 0$.

Continuous Dynamic Optimization with Terminal Constraints

Choose the function $u(t)$, $t_0 \leq t \leq t_f$ to minimize

$$J = \phi[x(t_f)] + \int_{t_0}^{t_f} L[x(t), u(t), t] dt \ ,$$

where t_f is specified and

$$\dot{x} = f[x(t), u(t), t] \ , \quad x(t_0) = x_o \ , \quad \psi[x(t_f)] = 0 \ .$$

Necessary Conditions for a Stationary Path

Define

$$H(t) \overset{\Delta}{=} L[x(t), u(t), t] + \lambda^T(t) f[x(t), u(t), t] \ , \quad \Phi \overset{\Delta}{=} \phi[x(t_f)] + \nu^T \psi[x(t_f)] \ .$$

Then

$$\dot{x} = f[x(t), u(t), t] \ ,$$
$$\dot{\lambda} = -H_x^T(i) \equiv -L_x^T(i) - f_x^T(t)\lambda(t) \ , \quad \lambda^T(t_f) = \Phi_x \equiv \phi_x + \nu^T \psi_x \ ,$$

where $u(t)$ is determined from

$$H_u(t) \equiv L_u(t) + \lambda^T(t) f_u(t) = 0 \ .$$

This forms a TPBVP for $x(t)$ and $\lambda(t)$, where we must also find ν to satisfy the terminal constraints $\psi = 0$.

First Integral

If $H_t = 0$, an integral of the TPBVP is $H = $ constant.

Gradient Algorithms DOPC and FOPC

1. Guess $u(i)$ or $u(t)$ at N points.
2. Find $x(i)$ or $x(t)$ by forward sequencing or integration.
3. Evaluate ϕ, ψ, ϕ_x, ψ_x.
4. Find $H_u^\phi(i)$ and $H_u^\psi(i)$ or $H_u^\phi(t)$ and $H_u^\psi(t)$ by backward sequencing or integration.

5. Determine $\nu = -Q^{-1}g$, where Q, g are quadratic summations or integrals of $H_u^\phi(i)$, $H_u^\psi(i)$ or $H_u^\phi(t)$, $H_u^\psi(t)$.
6. $\delta u(i) = -k[H_u^\phi(i) + \nu^T H_u^\psi(i)]^T - \eta[H_u^\psi(i)]^T Q^{-1}\psi$ or $t \leftarrow i$, where $k > 0$, $0 < \eta \leq 1$.
7. If $(|\delta u(i)| < \epsilon$ or $(|\delta u(t)| < \epsilon$ then stop.
8. $u(i) = u(i) + \delta u(i)$ or $u(t) = u(t) + \delta u(t)$.
9. Go to (2).

NR Algorithms DOPCN and FOPCN

These algorithms are very similar to DOPC and FOPC except that FSOLVE is used to make $H_u(i) = 0$ and $\psi = 0$ to the desired accuracy. Good initial guesses of $u(i)$ and ν are required from DOPC or FOPC.

Forward Shooting Algorithms DOPCF and FOPCF

1. Estimate ν and $\lambda(0)$ from a DOPC or FOPC solution.
2. Find $s(i)$, $\lambda(i)$ or $s(t)$, $\lambda(t)$ by forward sequencing or integration of the EL equations, determining $u(i)$ or $u(t)$ at each step by setting $H_u(i) = 0$ or $H_u(t) = 0$.
3. Evaluate the error $e_f \overset{\Delta}{=} [\lambda(N) - \phi_s + \psi_s \nu, \ \psi]$ or $e_f \overset{\Delta}{=} [\lambda(t_f) - \phi_s + \psi_s \nu, \ \nu]$.
4. Use FSOLVE to vary the elements of $\lambda(0)$ and ν to make $e_f = 0$ to the desired accuracy.
5. This 'single shooting' algorithm will FAIL for problems with energy dissipation and t_f large compared to characteristic times (see Problems 3.5.12, 3.6.12, and 5.2.4).

Backward Shooting Algorithms DOPCB and FOPCB

1. Estimate ν and the final state $s(N)$ or $s(t_f)$ from a DOPC or FOPC solution.
2. Evaluate ψ and $\lambda(N) = \phi_s - \psi_s \nu$ or $\lambda(t_f) = \phi_s - \psi_s \nu$.
3. Find $s(i)$, $\lambda(i)$ or $s(t)$, $\lambda(t)$ by backward sequencing or integration of the EL equations, determining $u(i)$ or $u(t)$ at each step by setting $H_u(i) = 0$ or $H_u(t) = 0$.
4. Evaluate the error $e_0 \overset{\Delta}{=} [s(0) - s_0, \ \psi]$.
5. Use FSOLVE to vary the elements of ν and $s(N)$ or $s(t_f)$ to make $e_0 = 0$ to the desired accuracy.

Appendix A   Alternative Derivation of the Discrete EL Equations

In Subsection 3.1.1, consider differential changes in J and in the constraints

$$dJ = \phi_x dx(N) , \tag{3.40}$$

$$0 = dx(i+1) - f_x(i)dx(i) - f_u du(i), \quad i = 0, \ldots, N-1, \tag{3.41}$$

$$0 = dx(0) , \tag{3.42}$$

$$0 = \psi_x dx(N) . \tag{3.43}$$

For a stationary path, $dJ = 0$ for arbitrary $dx(i)$ and $du(i)$ subject to the constraints. For $dJ = 0$ to be consistent with the constraints, (3.40) must be linearly dependent on (3.41) to (3.43), i.e.,

$$dJ = \sum_{i=0}^{N-1} \lambda^T(i+1)[dx(i+1) - f_x(i)dx(i) - f_u(i)du(i)] + \lambda^T(0)dx(0) - \nu^T \psi_x dx(N) .$$

$$(3.44)$$

Changing indices of summation on the first term in (3.44) gives

$$dJ = [\lambda^T(N) - \nu^T \psi_x]dx(N) + \sum_{i=0}^{N-1} \{ [\lambda^T(i) - \lambda^T(i+1)f_x(i)]dx(i)$$

$$- \lambda^T(i+1)f_u(i)du(i) \} .$$

$$(3.45)$$

Equating coefficients of $dx(i)$ and $du(i)$ in (3.41) and (3.46) yields the discrete EL equations

$$\lambda^T(N) = \phi_x + \nu^T \psi_x ,$$

$$(3.46)$$

$$\lambda^T(i) = \lambda^T(i+1)f_x(i) ,$$

$$(3.47)$$

$$0 = \lambda^T(i+1)f_u(i) ,$$

$$(3.48)$$

for $i = 0, \ldots, N-1$.

Appendix B Alternative Derivation of the Continuous EL Equations

In Subsection 3.3.1, consider a variation of J and of the constraints

$$\delta J = \phi_x \delta x(t_f) ,$$

$$(3.49)$$

$$0 = \dot{\delta} x - f_x \delta x - f_u \delta u , \quad 0 \le t \le t_f ,$$

$$(3.50)$$

$$0 = \delta x(t_0) ,$$

$$(3.51)$$

$$0 = \psi_x \delta x(t_f) .$$

$$(3.52)$$

For a stationary path, $\delta J = 0$ for arbitrary $\delta x(t)$ and $\delta u(t)$ subject to the constraints. For $\delta J = 0$ to be consistent with the constraints, (3.49) must be linearly dependent on (3.51) and (3.52), i.e.,

$$\delta J = \int_{t_0}^{t_f} \lambda^T(\delta \dot{x} - f_x \delta x - f_u \delta u)dt - \nu^T \psi_x \delta x(t_f) .$$

$$(3.53)$$

Integrating the $\lambda^T \delta \dot{x}$ term in (3.53) by parts and using (3.51) gives

$$\delta J = [\lambda^T(t_f) - \nu^T \psi_x]\delta x(t_f) + \int_{t_0}^{t_f} \left[(-\dot{\lambda}^T - \lambda^T f_x)\delta x - \lambda^T f_u \delta u \right] dt .$$

$$(3.54)$$

Equating coefficients of δx and δu in (3.52) and (3.54) yields the continuous EL equations for $0 \le t \le t_f$:

$$\lambda^T(t_f) = \phi_x + \nu^T \psi_x \, , \tag{3.55}$$

$$\dot{\lambda}^T = -\lambda^T f_x \, , \tag{3.56}$$

$$0 = \lambda^T f_u \, . \tag{3.57}$$

CHAPTER 4

Dynamic Optimization with Open Final Time

4.1 Introduction

This chapter extends the previous chapter to allow open final time. An important special case is the *min time problem,* which must have at least one terminal condition, since otherwise there is no indication of when to stop. Hence, another way to solve min time problems is to guess the min time t_f, and maximize the key terminal condition; by doing this with several values of t_f, the min time can be interpolated to satisfy the key terminal condition. This maximization problem is said to be the *dual problem* to the min time problem. We have already discussed several problems that are obvious duals of min time problems, e.g., (1) the VDP problem for max range in a given time is dual to the VDP problem for min time with given range, and (2) the TDP problem for orbit transfer with max radius in a given time is dual to the TDP problem for orbit transfer in min time with given radius.

4.2 Discrete Dynamic Systems

In many dynamic optimization problems, we wish to minimize the final time to reach some specified terminal condition, or to minimize some function of the terminal states with the final time open. Such problems require a slight extension of the theory and algorithms developed thus far. This extension is simply to regard the final time as a *parameter* to be optimized in addition to the control *functions,* so that the problem becomes a combined *calculus and calculus of variations problem.*

Necessary Conditions for a Stationary Solution

We wish to find the control vector sequence $u(i)$, $i = 0, \ldots, N-1$ *and the final time t_f to minimize*

$$J = \phi[x(N), t_f] \,, \tag{4.1}$$

where the number of steps N is specified and

$$t_f = N\Delta \tag{4.2}$$

subject to the constraints

$$x(i+1) = f[x(i), u(i), i, \Delta] \,, \quad x(0) = x_0 \,, \tag{4.3}$$

$$0 = \psi[x(N), t_f] \,, \tag{4.4}$$

for $i = 0, \cdots, N-1$. Δ is the time step. For *min time problems, $\phi = N\Delta$.*

Necessary conditions for a stationary solution can be obtained by an extension of Section 3.1.1. The new element is the dependence of the performance index, terminal constraints, and dynamic equations on the time step Δ. Adjoin the constraints to the performance index with Lagrange multiplier vectors v and $\lambda(i)$ as follows:

$$\bar{J} = \phi + v^T \psi + \sum_{i=0}^{N-1} \lambda^T(i+1)\{f[x(i), u(i), i, \Delta] - x(i+1)\} + \lambda^T(0)[x_0 - x(0)] \,. \tag{4.5}$$

Define $H(i)$ and Φ as follows:

$$H(i) \stackrel{\Delta}{=} \lambda^T(i+1)f[x(i), u(i), i, \Delta] \,, \tag{4.6}$$

$$\Phi \stackrel{\Delta}{=} \phi + v^T \psi \,. \tag{4.7}$$

Using (4.6) and (4.7) in (4.5) and changing the index of summation on the last term in the summation yields

$$\bar{J} = \Phi - \lambda^T(N)x(N) + \sum_{i=0}^{N-1}[H(i) - \lambda^T(i)x(i)] + \lambda^T(0)x_0 \,. \tag{4.8}$$

Now consider differential changes in $\bar{J}$ due to differential changes in $u(i)$ and in Δ:

$$d\bar{J} = [\Phi_x - \lambda^T(N)]dx(N) + \Phi_\Delta d\Delta + \sum_{i=0}^{N-1}\{[H_x(i) - \lambda^T(i)]dx(i) + H_u(i)du(i) + H_\Delta(i)d\Delta\} \,. \tag{4.9}$$

We now choose $\lambda(i)$ so that the coefficients of $dx(i)$ in (4.9) vanish:

$$\lambda^T(i) = H_x(i) \equiv \lambda^T(i+1)f_x \,, \tag{4.10}$$

$$\lambda^T(N) = \Phi_x \equiv \phi_x + v^T \psi_x \,, \tag{4.11}$$

for $i = 0, \cdots, N - 1$. Equation (4.9) then becomes

$$d\bar{J} = \left[\Phi_\Delta + \sum_{i=0}^{N-1} H_\Delta(i) \right] d\Delta + \sum_{i=0}^{N-1} H_u(i) du(i) . \qquad (4.12)$$

For a stationary solution $d\bar{J} = 0$ for arbitrary $du(i)$ and arbitrary $d\Delta$; thus, necessary conditions for a stationary solution are

$$0 = H_u(i) , \quad i = 0, \cdots, N - 1 , \qquad (4.13)$$

$$0 = \Phi_\Delta + \sum_{i=0}^{N-1} H_\Delta(i) . \qquad (4.14)$$

ν must be found so that $\psi = 0$. Equation (4.13) is the same condition that occurs with fixed final time. The new necessary condition (4.14) is called a *transversality condition*. It is the additional condition that determines the optimal time step Δ.

Example 4.2.1—DTDP for Min Time to Specified (u_f, v_f, y_f)

This is the *dual problem* to Example 3.1.1 [max $u_f{}^1$ in time t_f to specified (v_f, y_f)]. If we measure (u, v) in units of $U \stackrel{\Delta}{=} u(t_f)$, the final time t_f in U/a and (x, y, h) in U^2/a, then the equations of motion are

$$u(i + 1) = u(i) + \Delta \cos \theta(i) , \quad v(i + 1) = v(i) + \Delta \sin \theta(i) ,$$

$$y(i + 1) = y(i) + \Delta v(i) + \tfrac{1}{2}\Delta^2 \sin \theta(i) ,$$

where all the initial states are zero, $u(N) = 1$, $v(N) = 0$, $y(N) = h$, and $\Delta = t_f/N$. For given N, we wish to find the $\theta(i)$ sequence that minimizes Δ so that

$$\phi = N\Delta , \quad \psi = [u(N) - 1 \ y(N) - h \ v(N)]^T .$$

The $H(i)$ sequence is then

$$H(i) = \lambda_u(i + 1)u(i + 1) + \lambda_v(i + 1)v(i + 1) + \lambda_y(i + 1)y(i + 1) ,$$

and the augmented performance index is

$$\Phi = N\Delta + \nu_u[u(N) - 1] + \nu_y[y(N) - h] + \nu_v v(N) .$$

The discrete EL equations are then

$$\lambda_u(i) = H_u(i) = \lambda_u(i + 1) ,$$
$$\lambda_v(i) = H_v(i) = \lambda_v(i + 1) + \Delta\lambda_y(i + 1) ,$$
$$\lambda_y(i) = H_y(i) = \lambda_y(i + 1) ,$$

---

[1]Note that u is a state variable here and θ is the control.

with terminal conditions

$$\lambda_u(N) = \Phi_u = \nu_u , \quad \lambda_v(N) = \Phi_v = \nu_v , \quad \lambda_y(N) = \Phi_y = \nu_y .$$

The optimality condition is

$$0 = \frac{H_\theta(i)}{\Delta} = -\lambda_u(i+1)\sin\theta(i) + \lambda_v(i+1)\cos\theta(i) + \frac{1}{2}\Delta\lambda_y(i+1)\cos\theta(i) .$$

The transversality condition is

$$0 = \Omega \overset{\Delta}{=} N + \sum_{i=0}^{N-1} H_\Delta(i) ,$$

where

$$H_\Delta = \lambda_u(i+1)\cos\theta(i) + \lambda_v(i+1)\sin\theta(i) + \lambda_y(i+1)[v(i) + \Delta\sin\theta(i)] .$$

The adjoint equations are easily solved in this case:

$$\lambda_u(i) = \nu_u , \quad \lambda_y(i) = \nu_y , \quad \lambda_v(i) = \nu_v + \nu_y(N-i)\Delta ,$$
$$\nu_u \tan\theta(i) = \nu_v + \nu_y(N-i-1/2)\Delta .$$

The four parameters ν_u, ν_v, ν_y, Δ must be determined to satisfy the four conditions $u(N) = 1$, $v(N) = 0$, $y(N) = h$, $\Omega = 0$.

If we assume symmetry of $v(i)$ about the midpoint [which yields $v(N) = 0$], this implies anti-symmetry of $\theta(i)$ about the midpoint $\Rightarrow \theta(N-1) = -\theta(0)$. This, in turn, implies

$$\nu_v = -\frac{N\Delta\nu_y}{2} ,$$

so that

$$\tan\theta(i) = \tan\theta(0)\left(1 - \frac{2i}{N-1}\right) ,$$

where $\tan\theta(0) \overset{\Delta}{=} (N-1)\Delta\nu_y/(2\nu_u)$. The two parameters $\theta(0)$ and Δ must be determined to satisfy the two conditions $u(N) = 1$, $y(N) = h$; then the condition $\Omega = 0$ and $\theta(0)$ determine ν_v, ν_y which, in turn, determines ν_v.

Problems

4.2.1 DVDP for Min Time to a Point with Gravity: This is the dual problem to Problem 3.1.1. Measure (x, y) in units of x_f, V in $\sqrt{gx_f}$, t in $\sqrt{x_f/g}$. We wish to find $\gamma(i)$ to minimize the time to go to normalized $y(N) = y_f$, $x(N) = 1$.

(a) Show that the necessary conditions for a stationary solution are the discrete equations of motion plus the following

$$\lambda_x(i) = v_x, \ \lambda_y(i) = v_y, \ \lambda_v(i) = \lambda_v(i+1) + v_x\Delta\cos\gamma(i) + v_y\Delta\sin\gamma(i), \ \lambda_v(N) = 0 \ ,$$
$$0 = \lambda_v(i+1)\cos\gamma(i) + v_x[-v(i)\sin\gamma(i) + (\Delta/2)\cos2\gamma(i)$$
$$+v_y[v(i)\cos\gamma(i) + \Delta\sin2\gamma(i)] \ ,$$
$$0 = 1 + \sum_{i=0}^{N-1}\{\lambda_v(i+1)\sin\gamma(i) + [v_x\cos\gamma(i) + v_y\sin\gamma(i)][v(i) + \Delta\sin\gamma(i)]\} \ .$$

(b) For normalized $y_f = 1/3$, use a code like FSOLVE in MATLAB to satisfy the necessary conditions in (a) and plot the optimal path. Use as parameters the $\gamma(i)$ sequence, v_x, v_y, and Δ.

4.2.2 DVDP for Min Time to a Point with u_c = Vy/h:

This is the dual problem to Problem 3.1.2. We wish to find $\theta(i)$ to minimize the time to go to $y(N) = y_f$, $x(N) = x_f$.

(a) Show that the optimal control sequence $\gamma(i)$ is a *linear tangent* law

$$\tan\theta(i) = \frac{v_y}{v_x} + \Delta(N - i + .5) \ ,$$

and the transversality condition is

$$0 = \Omega \triangleq N + \sum_{i=0}^{N-1}\{v_x[y(i) + \cos\theta(i)] + \sin\theta(i)[v_y + \Delta v_x(N - i)]\} \ .$$

(b) Use a code like FSOLVE in MATLAB to determine v_x, v_y, Δ in the optimal control law to satisfy the terminal conditions $y(N) = 0$, $x(N) = 12$, $\Omega = 0$ and plot the optimal path.

4.2.3 DVDP for Min Distance between Two Points on a Sphere:

This is the same as Problem 3.1.3 except that $s =$ distance along the path is used as the independent variable with latitude θ and longitude ϕ as state variables. The continuous EOM are

$$\frac{d\theta}{ds} = \sin\beta \ , \quad \frac{d\phi}{ds} = \cos\beta\sec\theta \ .$$

(a) With β held constant over the distance Δs, show that the discrete EOM are

$$\theta(i + 1) = \theta(i) + \sin\beta(i)\Delta s \ , \quad \phi(i + 1) = \phi(i) + \cot\beta(i)\{\Lambda[\theta(i + 1)] - \Lambda[\theta(i)]\} \ ,$$

where $\Lambda(x) \triangleq \log[\tan(x/2 + \pi/4)]$.

(b) Show that the discrete EL equations are

$$\lambda_\theta(i) = \lambda_\theta(i + 1) + v_\phi\frac{c_\beta}{s_\beta}\{\sec[\theta(i + 1)] - \sec[\theta(i)]\} \ , \quad \lambda_\theta(N) = v_\theta \ ,$$

$$0 = \lambda_\theta(i + 1)c_\beta\Delta s + v_\phi\left\{\frac{c_\beta}{s_\beta}\sec[\theta(i + 1)]c_\beta\Delta s - \frac{1}{s_\beta^2}\{\Lambda[\theta(i + 1)] - \Lambda[\theta(i)]\}\right\} \ ,$$

where $s_\beta \triangleq \sin\beta(i)$, $c_\beta \triangleq \cos\beta(i)$.

(c) Use a code like FSOLVE in MATLAB to find $\beta(i)$, $i = 0, \cdots, N-1$, and v_θ, v_ϕ for the min distance path from Tokyo to New York (see data in Problem 3.1.3).

4.2.4 DVDP for Min Time to a Point with $V = V_0(1 + y/h)$:   Use the MATLAB code CONSTR to find $\theta(i)$ to minimize the time to go to from the origin to the point $x_f = 2.3472$, $y_f = 0$, which is the dual to Problem 3.1.4.

4.2.5 DTDP for Min Time to a point with Specified (u_0, v_0):   Dual problem to Problem 3.1.5. Use $u_0 = v_0 = .5/\sqrt{2}$, $x_f = 1.4142$.

4.2.6 DTDP for Min Time to Specified (u_f, v_f, y_f):   Dual problem to 3.1.6. Using Example 4.2.1, calculate the min time path to $u_f = 1$, $v_f = 0$, $y_f = .2/(.6716)^2$.

4.2.7 DTDP for Min Time to Specified (u_f, v_f, y_f, x_f):   Dual problem to Problem 3.1.7. Calculate the min time path to $u_f = 1$, $v_f = 0$, $y_f = .2(.4874)^2$, $x_f = .15/(.4874)^2$.

4.2.8 DTDP for Min Time with Gravity with Specified $(x_f, y_f, V_0, \gamma_0)$:   Dual problem to Problem 3.1.8. Use $V_0 = .5$, $\gamma_0 = \pi/4$, $x_f = 1.6346$ to make it the exact dual of Problem 3.1.8.

4.2.9 DTDP for Min Time to Specified (u_f, v_f, y_f) with Gravity: Dual problem to Problem 3.1.9. Calculate the min time path to $u_f = 1$, $v_f = 0$, $y_f = .2/(.4930)^2$.

4.2.10 DTDP for Min Time to Specified (u_f, v_f, y_f, x_f) with Gravity: Dual problem to Problem 3.1.10. Calculate the min time path to $u_f = 1$, $v_f = 0$, $y_f = .2/(.4575)^2$, $x_f = .15/(.4575)^2$.

4.3 Numerical Solution with Gradient Methods

We can again use the parameter optimization algorithm POP of Section 1.3 by putting $y = [u(0) \ u(1) \ \cdots \ u(N-1) \ \Delta]^T$, and replacing (L, f) by (ϕ, ψ). The time and the controls should be scaled so that one unit of each is of comparable significance. Starting with a guess for $u(i)$ and Δ, we may calculate (ϕ, ψ) and their gradients with respect to y. Making these substitutions, the algorithm becomes

- Choose step-size parameters k and η.
- Guess t_f and $u(i)$ for $i = 0, \cdots, N-1$.
- (∗) Forward sequencing; $\Delta = t_f/N$; compute and store $x(i)$ for $i = 1, \cdots, N$ from (4.3).
- Evaluate ϕ, ψ, determine ϕ_Δ, ψ_Δ, ϕ_x, ψ_x, and put

$$\lambda^\phi(N) = \phi_x^T , \quad \lambda^\psi(N) = \psi_x^T .$$

- Backward sequencing. Compute and store $(1 + q)$ pulse response sequences $H_u^\phi(i)$ and $H_u^\psi(i)$ for $i = 0, \cdots, N - 1$ as in DOPC (Section 3.2), and the gradients $\bar{\phi}_\Delta$, $\bar{\psi}_\Delta$, where

$$\bar{\phi}_\Delta \stackrel{\Delta}{=} \phi_\Delta + \sum_{i=0}^{N-1} H_\Delta^\phi(i) , \quad \bar{\psi}_\Delta \stackrel{\Delta}{=} \psi_\Delta + \sum_{i=0}^{N-1} H_\Delta^\psi(i) ,$$

and

$$\left[H_u^\phi(i) \right]^T = f_u^T(i)\lambda^\phi(i + 1) , \quad \left[H_u^\psi(i) \right]^T = f_u^T(i)\lambda^\psi(i + 1) .$$

- Determine $\nu = -Q^{-1}g$, where

$$g = \bar{\psi}_\Delta \bar{\phi}_\Delta + H_u^\psi[H_u^\phi]^T , \quad Q = \bar{\psi}_\Delta \bar{\psi}_\Delta^T + H_u^\psi[H_u^\psi]^T .$$

- Compute $\delta u(i)$ and dt_f:

$$\delta u(i) = -k \left[H_u^\phi(i) + \nu^T H_u^\psi(i) \right]^T - \eta[H_u^\psi(i)]^T Q^{-1}\psi ,$$
$$\frac{dt_f}{N} = -k(\bar{\phi}_\Delta + \nu^T \bar{\psi}_\Delta) - \eta\bar{\psi}_\Delta^T Q^{-1}\psi .$$

- Compute δu_{avg}:

$$\delta u_{avg} = \sqrt{\frac{\delta u^T \delta u}{N}} .$$

- If $\max[\delta u_{avg} , |dt_f|] < tol$ then end.
- Compute new $u(i)$ and t_f:

$$u(i) \leftarrow u(i) + \delta u(i) , \quad t_f \leftarrow t_f + dt_f .$$

- Go to $(*)$.

For *min time*, $\phi = N\Delta$, so that $\phi_x = 0$ which implies that $H_u^\phi(i) = 0$.

A MATLAB Code for DOPT

A MATLAB code for implementing DOPT is listed in Table 4.1. It requires another file of subroutines for (1) the equations of motion, (2) the performance index and terminal constraints and their derivatives, and (3) the derivatives of the equations of motion with respect to the states, the controls, and the time step Δ.

Example 4.3.1—DTDP for Min Time to $v_f = 0$ with Specified u_f, y_f

This is a numerical solution of the problem treated analytically in Example 4.2.1. The required subroutine file is listed below. The value of y_f produces the exact dual of Example 3.2.1 since the max normalized $u_f = .6716$ in that example.

TABLE 4.1 MATLAB Code DOPT

```
function [u,s,tf,nu,la0]=dopt(name,u,s0,tf,k,tol,mxit,eta)
% Disc. OPTim. w. term. constr., Tf open, multiple controls;
    name must be in
% single quotes; function file 'name' should be in the Matlab
    path & compute
% the s(i) sequence for flg=1, perf. index and term. constr. & gradients
% (Phi,Phis,Phid) for flg=2, (fs,fu,fd) for flg=3; inputs: u(nc,N)=estimate
% of optimal control sequence; s0(ns,1)=initial state; tf=estimate of final
% time; k=step-size parameter; t and elements of u should be scaled so that
% one unit of each is of comparable significance; stopping criterion is
% max[norm(dua) pmag abs(dtf)]<tol or the number of iterations >= mxit;
% 0<eta<= 1 where d(psi)=-eta*psi is desired change in term. conditions;
% outputs (u,s,tf)=improved estimates; BASIC version '84;
    Matlab version '91;
% vector control version 9/96, rev. 1/8/98.
%
if nargin<8, eta=1; end; if nargin<7, mxit=10; end;
ns=length(s0); [nc,N]=size(u); N1=N+1; dum=zeros(nc,1);
Phi=feval(name,dum,s0,1,0,2); nt1=length(Phi); nt=nt1-1;
s=zeros(ns,N1); la=zeros(ns,nt1); Hu=zeros(nt1,nc,N);
nu=zeros(nt,1); dua=1; pmag=1; dtf=1; it=0; s(:,1)=s0;
disp('     Iter.     tf     phi     norm(psi)     dua     dtf')
while max([norm(dua) pmag abs(dtf)]) > tol,
 % Forward sequencing and store s(:,i):
   dt=tf/N; for i=1:N,
     s(:,i+1)=feval(name,u(:,i),s(:,i),dt,(i-1)*dt,1); end
 % Perf. index, term. constraints, & B.C.s for bkwd sequences:
   [Phi,Phis,Phid]=feval(name,dum,s(:,N1),dt,N*dt,2);
   phi=Phi(1); psi=Phi([2:nt1]); pmag=norm(psi)/sqrt(nt);
   if mxit==0, disp([it phi pmag]); break; end; la=Phis';
 % Backward sequencing and store Hu(:,:,i):
   for i=N:-1:1
     [fs,fu,fd]=feval(name,u(:,i),s(:,i),dt,(i-1)*dt,3);
     Hu(:,:,i)=la'*fu; Phid=Phid+la'*fd; la=fs'*la;
   end; phid=Phid(1); n2=[2:nt1]; psid=Phid(n2);
 % nu and la0:
   ga=Phid*phid; Qa=Phid*Phid'; n1=[1:nt1]; dua=zeros(nc,1);
   for i=1:N
     ga=ga+Hu(n1,:,i)*Hu(1,:,i); Qa=Qa+Hu(n1,:,i)*Hu(n1,:,i)';
   end; g=ga(n2); Q=Qa(n2,n2); nu=-Q\g; la0=la*[1;nu];
 % New u(i) and tf:
   for i=1:N, Huphi=Hu(1,:,i); Hupsi=Hu(n2,:,i);
     du(:,i)=-k*(Huphi'+Hupsi'*nu)-eta*Hupsi'*(Q\psi);
     dua=dua+norm(du(:,i))/sqrt(N);
   end
   u=u+du; dtf=-N*k*(phid+nu'*psid)-N*eta*psid'*(Q\psi);
   disp([it tf phi norm(psi) dua dtf])
   tf=tf+dtf; it=it+1; if it>=mxit, break, end
end
```

```
function [f1,f2,f3]=dtdpt(th,s,dt,t,flg)
% Subroutine for Example 4.3.1; DTDP for min tf to vf=0 and specified
% (uf,yf); t in units of uf/a; (u,v) in uf; (x,y) in uf^2/a;
% s=[u v y x]'=state vector; th=theta= control.
%
uf=1; vf=0; yf=.2/.6716^2; u=s(1); v=s(2); y=s(3); x=s(4); co=cos(th);
si=sin(th);
if flg==1,                                          % f1 = f
 f1=s+dt*[co; si; v+dt*si/2; u+dt*co/2];
elseif flg==2,                          % f1=Phi=[phi;psi]; f2=Phix
 f1=[t; u-uf; v-vf; y-yf];
 f2=[0 0 0 0; eye(3) zeros(3,1)];
 f3=[t/dt 0 0 0]';
elseif flg==3,                          % f1=fx; f2=fu; f3=fdt
 f1=[1 0 0 0; 0 1 0 0; 0 dt 1 0; dt 0 0 1];
 f2=dt*[-si; co; dt*co/2; -dt*si/2];
 f3=[co; si; v+dt*si; u+dt*co];
end
```

A code that uses the subroutine is listed below.

```
% Script e04_2_1.m; DTDP for min tf to vf=0 and specified (uf,yf);
% (u,v) in uf, (x,y) in uf^2/a, t in uf/a;
%
th=[1:-2/9:-1]; x0=[0 0 0 0]'; tf=1/.6716; k=50; tol=5e-5;
mxit=60; N=length(th);
[th,s,tf]=dopt('dtdpt',th,x0,tf,k,tol,mxit);
t=tf*[0:.1:1]; u=s(1,:); v=s(2,:); y=s(3,:); x=s(4,:);
```

The converged min time path is identical to the one calculated in Example 4.2.1 and is the exact dual of Example 3.2.1. The paths are identical to those in Figs. 3.2 and 3.3 except for the normalization of the variables.

Using DOPT requires choosing the step size parameter k. Guess some value of k and put 'mxit' equal to 2 or 3. Revise k based on 'duavg' and 'dtf'. With a few tries, values that start convergence can be obtained. To obtain accurate results with low values of 'tol', further refinement of k may be required. If convergence is slow, increase k by a factor of 3 or 10; if convergence occurs for a while and then divergence starts, this indicates the values are too large, so decrease k by a factor of 3 or 10.

Use of the MATLAB Code CONSTR

The CONSTR command in the MATLAB Optimization Toolbox (see Section 3.2) can be used as an alternate to DOPT. The final time t_f is simply added as another parameter to be determined.

Example 4.3.2—DTDP Using the MATLAB Code CONSTR

Here we solve the same problem solved in Example 4.3.1, using CONSTR instead of DOPT. The function file to calculate the performance index 'f' and the constraint vector 'g' is listed below.

```
function [f,g,u,v,x,y]=dtdpt_fg(p,yf)
% Subroutine for Example 4.3.2; DTDP for min tf to vf=0 and
% specified (uf,yf); t in units of uf/a; (u,v) in uf;
% (x,y) in uf^2/a; s = [u,v,x,y]' = state vector; th =
% theta = control;
%
N=length(p)-1; th=p([1:N]); tf=p(N+1);
u(1)=0; v(1)=0; x(1)=0; y(1)=0; dt=tf/N;
for i=1:N, co=cos(th(i)); si=sin(th(i));
 u(i+1)=u(i)+dt*co; v(i+1)=v(i)+dt*si;
 x(i+1)=x(i)+dt*u(i)+dt^2*co/2;
 y(i+1)=y(i)+dt*v(i)+dt^2*si/2;
end
f=tf;
g=[u(N+1)-1 v(N+1) y(N+1)-yf]';
```

The code that uses this subroutine is listed below.

```
% Script e04_2_2.m; DTDP for min tf to vf=0 and specified (uf,yf);
% (u,v) in uf, (x,y) in uf^2/a, t in uf/a, using CONSTR;
%
yf=.2/.6716^2; th=[1:-2/9:-1]; tf=1/.6716; p=[th tf];
optn(1)=1; optn(13)=3; optn(14)=100;
p=constr('dtdpt_fg',p,optn,[],[],[],yf);
[f,g,u,v,x,y]=dtdpt_fg(p,yf);
N=10; th=p([1:N]); tf=p(N+1); t=tf*[0:1/N:1];
```

Problems

4.3.1 to 4.3.10:   Do Problems 4.2.1 to 4.2.10 using DOPT, CONSTR, or FSOLVE.

4.3.11 DVDP for Min Time to a Point with Gravity and Thrust: Do the dual problem to Problem 3.2.11 using DOPT or CONSTR or FSOLVE.

4.3.12 DVDP for Min Time to a Point with Gravity, Thrust, and Drag: Do the dual problem to Problem 3.2.12 using DOPT or CONSTR or FSOLVE.

4.4 Continuous Dynamic Systems

Continuous problems with open final time are very similar to discrete step problems with open final time; the only difference is that the constraints are differential equations instead of difference equations.

Necessary Conditions for a Stationary Solution

We wish to find the control vector function $u(t)$ for $t_0 \leq t \leq t_f$ and the final time t_f to minimize

$$J = \phi[x(t_f), t_f] , \qquad (4.15)$$

subject to the constraints

$$\dot{x} = f(x, u, t) , \quad x(t_0) = x_0 , \qquad (4.16)$$

$$0 = \psi[x(t_f), t_f] . \qquad (4.17)$$

For *min time problems*, $\phi = t_f$.

Necessary conditions for a stationary solution can be obtained by an extension of Sections 2.3.1 and 3.3.1. The new element is the dependence of the performance index, the terminal constraints, and the dynamic equations on the final time t_f. Adjoin the constraints (4.16) and (4.17) to the performance index with Lagrange multiplier vectors v and $\lambda(t)$ as follows:

$$\bar{J} = \phi + v^T \psi + \int_{t_0}^{t_f} \lambda^T(t)\{f[x(t), u(t), t] - \dot{x}\}dt . \qquad (4.18)$$

Define $H(t)$ and Φ as follows:

$$H(t) \overset{\Delta}{=} \lambda^T(t)f[x(t), u(t), t] , \qquad (4.19)$$

$$\Phi \overset{\Delta}{=} \phi + v^T \psi . \qquad (4.20)$$

Using (4.19) and (4.20) in (4.18) and integrating the last term in the integral by parts yields

$$\bar{J} = \lambda^T(t_f)x(t_f) + \int_{t_0}^{t_f} [H(t) + \dot{\lambda}^T x(t)]dt + \lambda^T(t_0)x_0 . \qquad (4.21)$$

Now consider differential changes in $\bar{J}$ due to variations in $u(t)$ and a differential change in t_f:

$$d\bar{J} = [\Phi_x - \lambda^T(t_f)]dx(t_f) + \dot{\Phi}dt_f + \int_{t_0}^{t_f} \{[H_x(t) + \dot{\lambda}^T]\delta x + H_u\delta u\}dt , \qquad (4.22)$$

where

$$\dot{\Phi} \overset{\Delta}{=} \Phi_{t_f} + \Phi_x \dot{x}(t_f) . \qquad (4.23)$$

We now *choose* $\lambda(t)$ so that the coefficients of δx in (4.22) vanish:

$$\dot{\lambda}^T = -H_x \equiv -\lambda^T f_x , \qquad (4.24)$$

$$\lambda^T(t_f) = \Phi_x \equiv \phi_x + v^T \psi_x . \qquad (4.25)$$

Equation (4.22) then becomes

$$d\bar{J} = \dot{\Phi}dt_f + \int_{t_0}^{t_f} H_u \delta u dt . \qquad (4.26)$$

For a stationary solution $d\bar{J} = 0$ for arbitrary $\delta u(t)$ and arbitrary dt_f; thus necessary conditions for a stationary solution are

$$0 = H_u(t), \quad t_0 \le t \le t_f, \tag{4.27}$$

$$0 = \dot{\Phi}. \tag{4.28}$$

v must be found so that $\psi = 0$. Equation (4.27) is the same condition that occurs with fixed final time. The new necessary condition (4.28) is called a *transversality condition*. It is the additional condition that determines the optimal final time t_f.

$\dot{\Phi}$ is the total time derivative of Φ, i.e.,

$$\dot{\Phi} = \Phi_{t_f} + \Phi_x \dot{x}(t_f) = 0, \tag{4.29}$$

where $\Phi = \phi + v^T \psi$. The transversality condition (4.28) can thus be written in terms of the Hamiltonian $H = \lambda^T f$ at the final time, since $\lambda^T(t_f) = \Phi_x$, and $\dot{x} = f$:

$$\Rightarrow \dot{\Phi} = \Phi_{t_f} + H(t_f) = 0. \tag{4.30}$$

If neither ϕ nor ψ are explicit functions of t_f, then $\Phi_{t_f} = 0$, and the transversality condition becomes simply

$$H(t_f) = 0. \tag{4.31}$$

For *min time problems*, $\phi = t_f$. If $\psi_{t_f} = 0$, then the transversality condition becomes simply

$$1 + H(t_f) = 0. \tag{4.32}$$

Example 4.4.1—TDP for Min Time to Specified (u_f, v_f, y_f)

This is the dual problem to Example 3.3.1 [max u_f with specified (v_f, y_f, t_f)]. It is also the continuous version of the discrete numerical Example 4.2.1.

Let $U \overset{\Delta}{=} u(t_f)$ and measure t in U/a, (u, v) in U, and (x, y) in U^2/a. The problem is then to find $\theta(t)$ to minimize t_f subject to

$$\dot{u} = \cos\theta, \quad \dot{v} = \sin\theta, \quad \dot{y} = v,$$

where all the initial conditions are zero and $u(1) = 1$, $v(1) = 0$, $y(1) = y_f$. The Hamiltonian function $H(t)$ is then

$$H(t) = \lambda_u(t)\cos\theta(t) + \lambda_v(t)\sin\theta(t) + \lambda_y(t)v(t).$$

The augmented performance index is

$$\Phi = t_f + v_u[u(t_f) - 1] + v_y[y(t_f) - h] + v_v v(t_f).$$

The EL equations are then

$$\dot{\lambda}_u = -H_u = 0 , \quad \dot{\lambda}_v = -H_v = -\lambda_y , \quad \dot{\lambda}_y = -H_y = 0 ,$$

and the optimality condition is

$$0 = H_\theta = -\lambda_u \sin\theta + \lambda_v \cos\theta .$$

The terminal conditions are

$$\lambda_u(t_f) = \Phi_u = \nu_u , \quad \lambda_v(t_f) = \Phi_v = \nu_v , \quad \lambda_y(t_f) = \Phi_y = \nu_y .$$

The adjoint equations are easily integrated in this case:

$$\lambda_u(t) = \nu_u , \quad \lambda_y(t) = \nu_y , \quad \lambda_v(t) = \nu_v + \nu_y(t_f - t) ,$$

so that

$$\tan\theta(t) = \frac{\nu_v + \nu_y(t_f - t)}{\nu_u} .$$

The constants ν_u, ν_v, and ν_y must be determined to satisfy $u(t_f) = 1$, $y(t_f) = h$, and $v(t_f) = 0$. t_f must be determined to satisfy the transversality condition

$$0 = \Omega = 1 + H(t_f) ,$$

or

$$0 = 1 + \nu_u \cos\theta(t_f) + \nu_v \sin\theta(t_f) .$$

Since this is the dual problem of Example 3.3.1, it is no surprise that the solution is *exactly the same except for the normalization of the state variables*:

$$m = 2\sinh^{-1}\tan\theta(0) , \quad t = \frac{1}{m}[\tan\theta(0) - \tan\theta] ,$$

$$u = \frac{1}{m}[\sinh^{-1}\tan\theta(0) - \sinh^{-1}\tan\theta] , \quad v = \frac{1}{m}[\sec\theta(0) - \sec\theta] ,$$

$$y = \frac{1}{2m}[t\sec\theta(0) - v\tan\theta - u] , \quad x = \frac{1}{m}(v - u\tan\theta) .$$

Note that $v(t_f) = 0$ implies $\theta(t_f) = -\theta(0)$; $u(t_f) = 1$ then determines m and $\theta(0)$ is then determined implicitly by putting $y(t_f) = h$; finally $\theta(0)$ determines t_f (see Problem 4.4.6).

Problems

4.4.1 VDP for Min Time to a Point with Gravity: This is the dual problem to Problem 3.3.1. Since final time is being minimized, a different normalization is appropriate, so here we measure (x, y) in x_f, t in $\sqrt{x_f/g}$, V in $\sqrt{gx_f}$.

(a) Show that the solution may be written as

$$V = \frac{\sin(bt)}{b} , \quad x = \frac{1}{2b}\left[t - \frac{\sin(2bt)}{2b}\right], \quad y = \frac{1 - \cos(2bt)}{4b^2} ,$$

where (b, t_f) are determined by setting $x_f = 1$, $t = t_f$ in these relations.

(b) For $y_f = 1/3$, use a code like FSOLVE in MATLAB to find (b, t_f) and plot the optimal path $-y$ vs. x.

4.4.2 VDP for Min Time to a Point with $u_c = Vy/h$: This is the dual problem to Problem 3.3.2. The same normalization of variables can be used here so the solution follows directly from the dual. (θ_0, θ_f) are determined by solving the implicit equations for (x, y) with $x = x_f$, $y = y_f$. Using a code like FSOLVE in MATLAB determine (θ_0, θ_f) for $x_f = 12$, $y_f = 0$ and plot the optimal path.

4.4.3 VDP for Min Distance to a Point on a Sphere: This is the same as Problem 3.3.3 except for a different independent variable. Use distance along the path as the independent variable so that both θ and ϕ become state variables.

(a) Show that the optimal path is given explicitly as a function of distance along the path t as

$$\theta = \sin^{-1}[\sin\theta_m \sin(t + \alpha)] , \quad \cos\beta = \frac{\cos\theta_m}{\cos\theta} ,$$

$$\phi = \tan^{-1}[\cos\theta_m \tan(t + \alpha)] - \tan^{-1}[\cos\theta_m \tan\alpha] ,$$

where $\sin\alpha = \sin\theta_0/\sin\theta_m$. The parameters θ_m and t_f can be found using a code like FSOLVE in MATLAB putting $\theta = \theta_f$, $\phi = \phi_f$ in the relations above.

(b) Calculate and plot the min distance path from Tokyo to New York (see Problem 3.1.3 for longitude, latitude data). Watch out for quadrant problems for ϕ and β using MATLAB for plots.

4.4.4 VDP for Min Time to a Point with $V = V_0(1 + y/h)$: This is the dual problem to Problem 3.3.4. The same normalization of variables can be used here so the solution follows directly from the dual. θ_0, θ_f are determined by solving the implicit equations for (x, y) with $x = x_f$, $y = y_f$.

Using a code like FSOLVE in MATLAB determine (θ_0, θ_f) for $x_f = 2.5$, $y_f = 0$ and plot the optimal path.

4.4.5 TDP for Min Time to a Point: This is the dual of Problem 3.3.5. With constant thrust specific force a, we wish to find the thrust direction $\theta(t)$ (positive when below the x-axis) for min time to the point (x_f, y_f).

(a) Show that the equations of motion may be written as

$$\dot{u} = \cos\theta , \quad \dot{v} = \sin\theta , \quad \dot{x} = u , \quad \dot{y} = v ,$$

with $u(0) = \cos\gamma_0$, $v(0) = \sin\gamma_0$, $x(0) = y(0) = 0$, $x(t_f) = x_f$, $y(0) = 0$, where (u, v) are in units of V_0, (x, y) are in units of V_0^2/a, t is in units of V_0/a.

(b) Show that the min final time t_f is the smallest positive real root of the quartic equation

$$t_f^4 - 4t_f^2 + 8x_f \cos\gamma_0 t_f - 4(x_f^2 + y_f^2) = 0 ,$$

and the optimal θ is *constant* and given by

$$\theta = \tan^{-1}\frac{y_f - t_f \sin\gamma_0}{x_f - y_f - t_f \cos\gamma_0} .$$

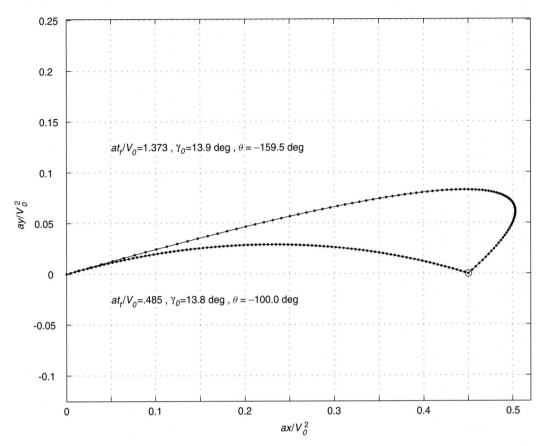

FIGURE 4.1 TDP for Min Time to a Point

(c) Calculate and plot the optimal paths for $x_f = .45$, $\gamma_0 = 13.8$ deg and $x_f = .45$, $\gamma_0 = 13.9$ deg. *Answer:* For the first path $\theta = -100.0$ deg and $t_f = .485$, while $\theta = -159.5$ deg and $t_f = 1.373$ for the second path. γ_0 is just enough larger for the second path that a direct trajectory is no longer possible and the rocket must go beyond the target and come back. Check your plot with Figure 4.1.

4.4.6 TDP for Min time to $v_f = 0$ and Specified (u_f, y_f): This is the dual problem to Problem 3.3.6, so the solution is exactly the same except for the normalization of variables (see Example 4.4.1). Make this change in normalization and use a code like FSOLVE in MATLAB to find θ_0 so that the solution is the exact dual of Problem 3.3.6. Note that max normalized $u_f = .6753$ in Problem 3.3.6.

4.4.7 TDP for Min Time to $v_f = 0$ and Specified (u_f, y_f, x_f): This is the dual problem to Problem 3.3.7, so the solution is the same except for the normalization of variables. Make this change in normalization and use a code like FSOLVE in MATLAB to find θ_0, θ_f so that the solution is the exact dual of Problem 3.3.7. Note that max normalized $u_f = .4876$ in Problem 3.3.7.

4.4.8 TDP for Min Time to a Point with Gravity: This is the dual problem to Problem 3.3.8, so the analytical solution is the same except for the normalization of variables. Make this change in normalization and use a code like FSOLVE in MATLAB for part (b).

(a) Show that the min final time t_f is the smallest positive real root of the quartic equation

$$(1 - g^2)t_f^4 + 4g \sin \gamma_0 t_f^3 - (4 + 4gy_f)t_f^2 + 8(x_f \cos \gamma_0 + y_f \sin \gamma_0)t_f - 4(x_f^2 + y_f^2) = 0 ,$$

and the optimal θ is *constant* and given by

$$\theta = \tan^{-1} \frac{-t_f \sin \gamma_0 + y_f + gt_f^2/2}{x_f - t_f \cos \gamma_0} .$$

(b) Calculate and plot the optimal path for $x_f = 5$, $y_f = .05$, $g = 1/3$, $\gamma_0 = 45$ deg. *Answer:* $\theta = -116.8$ deg and $t_f = 1.0782$.

4.4.9 TDP for Min time with Gravity to $v_f = 0$ and Specified (u_f, y_f):   This is the dual problem to Problem 3.3.9, so the solution is exactly the same except for the normalization of variables. Make this change in normalization and use a code like FSOLVE in MATLAB to find θ_0, θ_f. Note that the max normalized velocity in Problem 3.3.9 was .4935.

4.4.10 TDP for Min Time with Gravity to $v_f = 0$ and Specified (u_f, y_f, x_f):   This is the dual problem to Problem 3.3.10, so the solution is exactly the same except for the normalization of variables. Make this change in normalization and use a code like FSOLVE in MATLAB to find θ_0, θ_f. Note that the max normalized velocity in Problem 3.3.10 was .4577.

4.4.11 VDP for Min Time to a Point with Gravity and Thrust: Do the dual problem to Problem 3.3.11.

4.4.12 VDP for Min Time to a Point with Gravity, Thrust, and Drag: Do the dual problem to Problem 3.3.12.

4.4.13 VDP for Min Time to Enclose a Given Area in a Constant Wind: Do the dual problem of Problem 3.3.13 using distance along the path s as the independent variable so that both x and y become state variables.

4.4.14 VDP for Min Time to Cross a River with a Parabolic Current: Find the heading angle $\theta(t)$ to minimize the time to cross a river from $x = 0$, $y = -h$ to $x = 0$, $y = +h$ where the current is in the x direction with velocity

$$u = u_c \left(1 - \frac{y^2}{h^2}\right) .$$

(a) Show that min time paths are generated by the nonlinear feedback

$$\sec \theta = \sec \theta_0 + \frac{u_c}{V} \left(1 - \frac{y^2}{h^2}\right) .$$

(b) Find the min time path for the case $u_c/V = 1$, using code like FSOLVE in MATLAB to determine θ_0; use a code like ODE23 in MATLAB to integrate the EOM.

4.4.15   TDP for Min Time Orbit Transfer and Rendezvous; Small Δr:

(a) Do the min time orbit transfer problem for small Δr, which is the dual of Problem 3.3.17.

(b) Do the min time orbital rendezvous problem for small Δr, which is (a) with the addition of a terminal constraint on θ = the central angle around the attracting center. For $|r(t_f) - r(0)| \ll r(0)$ the additional EOM is

$$\dot{\theta} \approx u - r ,$$

and the final conditions (for the case where the target S/C is in a circular orbit at $r = r_f$) are

$$r(t_f) = r_f , \quad \theta(t_f) = \theta_f , \quad u(t_f) = -r_f/2 , \quad v(t_f) = 0 ,$$

where the variables are normalized as in Problem 3.3.17. Show that the optimal thrust direction program is given by

$$\tan \beta = \frac{\sin(t - \alpha) - 2b}{2\cos(t - \alpha) + 3bt + c} ,$$

where the parameters α, b, c, t_f are to be determined to satisfy the four final conditions.

(c) Use a code like FSOLVE in MATLAB to determine the parameters in (b) for the case $\Delta\theta_f = \pi$, $\Delta r_f = .2r_0$, $v_f = 0$ and the correct u_f for the new circular orbit. Plot the optimal path.

4.4.16   Max Time Glide to a Point in a Horizontal Plane (Ref. VYC):

A glider varies its angle-of-attack $\alpha(t)$ so as to stay at constant altitude and gradually slows down due to drag until it stalls ($\alpha = \alpha_{max}$). We wish to find the bank angle history $\sigma(t)$ to *maximize the time to glide to stall at a specified point*. Using a parabolic lift-drag polar, the normalized equations of motion are (Ref. VYC)

$$\dot{x} = V \cos \psi , \quad \dot{y} = V \sin \psi , \quad \dot{V} = -\frac{1}{2E\omega}\left[V^2 + \frac{\omega^2}{(V \cos \sigma)^2}\right], \quad \dot{\psi} = \frac{\tan \sigma}{V} ,$$

where V is in units of $V_0 \stackrel{\Delta}{=} V(0)$, (x, y) are in units of V_0^2/g, and t is in units of V_0/g, $E \stackrel{\Delta}{=}$ the max lift/drag (L/D) ratio of the glider, $\omega \stackrel{\Delta}{=} 2W/\rho S V_0^2 C_L^*$, W = weight of the glider, ρ = air density, S = reference area for C_L, and $C_L^* = C_L$ for max L/D. The lift coefficient (hence angle-of-attack) required to maintain constant altitude is

$$C_L = \frac{C_L^* \omega}{V^2 \cos \sigma} ,$$

so we require that C_L be less than $(C_L)_{max}$. Also the load factor $n \stackrel{\Delta}{=} L/W = \sec \sigma$ must be less than n_{max}.

(a) Show that there are four integrals of the max time problem:

$$\lambda_x(t) = c_1 \ , \quad \lambda_y(t) = c_2 \ , \quad \lambda_\psi(t) = c_1 y(t) - c_2 x(t) + c_3 \ , \quad H(t) = c_4 \ ,$$

where $c_1, \cdots c_4$ are constants.

(b) Show that the optimal bank angle is given by $\tan \sigma = EV\lambda_\psi/(\omega\lambda_V)$.

(c) Show that transversality requires $c_4 = 1$. Use this with (a) and (b) to show that optimal $\tan \sigma$ is the solution of the following quadratic equation:

$$\tan^2 \sigma + 2B \tan \sigma - C = 0 \ ,$$

where

$$B \triangleq \frac{V^2(c_1 \cos \psi + c_2 \sin \psi) - V}{c_1 y - c_2 x + c_3} \ , \quad C \triangleq \frac{V^4 + \omega^2}{\omega^2} \ .$$

Note that $\tan \sigma = \text{sign}(B)\sqrt{B^2 + C} - B$ so that $\sigma \to 0$ as $|B| \to \infty$.

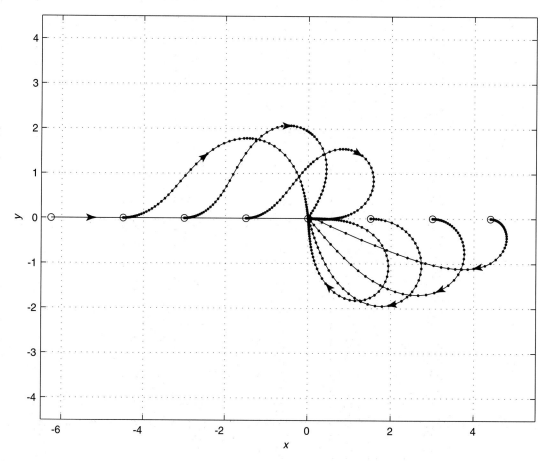

FIGURE 4.2　Max Time Glide Paths to $x = y = 0$ (Ref. VYC)

(d) With $E = 20$, $\omega = .23$, $y(0) = 0$, $V(0) = 1$, $\psi(0) = 0$ and $x(t_f) = y(t_f) = 0$, $V(t_f) = V_f$, determine the four parameters (c_1, c_2, c_3, t_f) to meet the three terminal conditions plus $\lambda_\psi(t_f) = 0$ with

$$x(0) = -6.26, \ -4.5, \ -3.0, \ -1.5, \ 0, \ 1.5, \ 3.0, \ 4.5 \ .$$

Take $(C_L)_{max} = 1.8C_L^*$, which yields $V_f = .3575$ since $\psi(t_f)$ open $\Rightarrow \lambda_\psi(t_f) = 0 \Rightarrow \sigma(t_f) = 0$. Show that the first value of $x(0)$ above is its min possible value and corresponds to a straight path with $\sigma(t) = 0$. The limit on positive $x(0)$ is load factor reaching n_{max} as the glider turns sharply to reverse its direction back toward $x = y = 0$. *Hint:* $\lambda_\psi(t_f) = 0 \Rightarrow c_3 = 0$, which reduces this to a three-parameter problem. Plot the max t_f paths in the (x, y) plane and the corresponding histories of σ vs. time. Compare your results with those in Figs. 4.2 and 4.3.

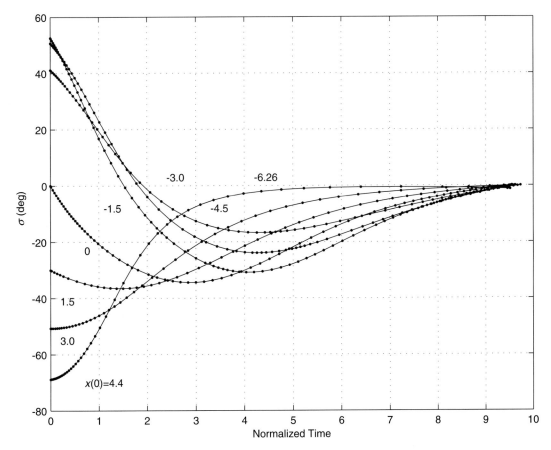

FIGURE 4.3   Bank Angle Histories for Max Time Glide Paths to $x = y = 0$ (Ref. VYC)

4.5 Numerical Solution with Gradient Methods

A gradient algorithm FOPT (for Function OPtimization with open final Time) is derived below using a function space extension of the parameter optimization algorithm POP derived in Section 1.3.

This algorithm may be interpreted as finding a small variation in the control functions $\Delta u(t)$, and a small change in the final time Δt_f to minimize a linear approximation of ΔJ subject to a quadratic penalty on $\Delta u(t)$ and on Δt_f, and to a linear approximation to coming closer to the terminal constraints $\psi = 0$, namely:

$$\min_{\Delta u(t),\ \Delta t_f} \Delta J \approx \dot\phi \Delta t_f + \int_0^{t_f} H_u^\phi \Delta u \, dt + \frac{1}{2k}\left[\int_0^{t_f} (\Delta u)^T \Delta u \, dt + (\Delta t_f)^2 \right] \quad (4.33)$$

subject to

$$\Delta \psi \approx \dot\psi \Delta t_f + \int_0^{t_f} H_u^\psi \Delta u \, dt = -\eta \psi \ , \quad (4.34)$$

where all the coefficients are evaluated at the current estimate of u and t_f, and $0 < \eta \le 1$. $\eta = 1$ asks for the constraints to be satisfied in one step, so $\eta < 1$ asks only to move part way toward satisfying the constraints. k is a scalar step-size parameter (see Section 2.2 for how to choose k). t and u should be scaled so that one unit of each is of comparable significance.

To solve this LQ problem, we adjoin the constraints (4.34) to the performance index (4.33) with a Lagrange multiplier vector $\bar v$:

$$\Delta \bar J \triangleq \Delta J + \bar v^T \left(\eta \psi + \dot\psi \Delta t_f + \int_0^{t_f} H_u^\psi \Delta u \, dt \right) . \quad (4.35)$$

Now take a differential of ΔJ:

$$d(\Delta \bar J) = \left(\dot\phi + \bar v^T \dot\psi + \frac{\Delta t_f}{k} \right) d(\Delta t_f) + \int_0^{t_f} \left(H_u^\phi + \bar v^T H_u^\psi + \frac{\Delta u^T}{k} \right) \delta(\Delta u) dt \ . \quad (4.36)$$

For a stationary solution, the coefficients of $d(\Delta t_f)$ and $\delta(\Delta u)$ must vanish, yielding

$$\Delta u = -k(H_u^\phi + \bar v^T H_u^\psi)^T \ , \quad \Delta t_f = -k(\dot\phi + \bar v^T \dot\psi) \ . \quad (4.37)$$

Substituting (4.37) into (4.34), we may solve for $\bar v$:

$$\bar v = v + \frac{\eta Q^{-1} \psi}{k} \ , \quad (4.38)$$

where $v = -Q^{-1} g$ and

$$Q = \dot\psi \dot\psi^T + \int_0^{t_f} H_u^\psi [H_u^\psi]^T \, dt \ , \quad g = \dot\psi \dot\phi + \int_0^{t_f} H_u^\psi [H_u^\phi]^T \, dt \ , \quad (4.39)$$

Substituting v into (4.37) gives the desired changes:

$$\Delta u = -k[H_u^\phi + v^T H_u^\psi]^T - \eta [H_u^\psi]^T Q^{-1} \psi \ , \quad (4.40)$$

$$\Delta t_f = -k(\dot\phi + v^T \dot\psi) - \eta \dot\psi^T Q^{-1} \psi \ . \quad (4.41)$$

Summary of FOPT Algorithm

- Choose k and η.
- Guess t_f and $u(t)$ for $0 \le t \le t_f$.

(∗) Forward integration. Compute and store $x(t)$ for $0 \le t \le t_f$.

- Evaluate ϕ, ψ and determine $\dot{\phi}$, $\dot{\psi}$ and

$$\lambda^\phi(t_f) = \phi_x^T , \quad \lambda^\psi(t_f) = \psi_x^T .$$

- Backward integration. Compute and store $(1+q)$ impulse response sequences $H_u^\phi(t)$ and $H_u^\psi(t)$ for $0 \le t \le t_f$:

$$\dot{\lambda}^\phi = -f_x^T \lambda^\phi , \quad (H_u^\phi)^T = f_u^T \lambda^\phi ,$$

and

$$\dot{\lambda}^\psi = -f_x^T \lambda^\psi , \quad (H_u^\psi)^T = f_u^T \lambda^\psi ,$$

- Determine $\nu = -Q^{-1}g$, where

$$g = \dot{\psi}\dot{\phi} + \int_0^{t_f} H_u^\psi(t)[H_u^\phi(t)]^T dt , \quad Q = \dot{\psi}\dot{\psi}^T + \int_0^{t_f} H_u^\psi(t)[H_u^\psi(t)]^T dt ,$$

- Compute $\delta u(t)$ for $0 \le t \le t_f$:

$$\delta u(t) = -k \left[H_u^\phi(t) + \nu^T H_u^\psi(t) \right]^T - \eta [H_u^\psi(t)]^T Q^{-1} \psi .$$

- Compute dt_f:

$$dt_f = -k(\dot{\phi} + \nu^T \dot{\psi}) - \eta \dot{\psi}^T Q^{-1} \psi .$$

- Compute δu_{avg}:

$$\Delta u_{avg} = \sqrt{\frac{1}{t_f} \int_0^{t_f} \delta u^T(t) \delta u(t) dt} .$$

- If $\delta u_{avg} < tol$ and $|dt_f| < tol$ then end.
- Compute new $u(t)$ and t_f:

$$u(t) \leftarrow u(t) + \delta u(t) , \quad t_f \leftarrow t_f + dt_f .$$

- Go to (∗) .

A MATLAB Code for FOPT

A MATLAB code that implements the FOPT algorithm for multiple controls is listed in Table 4.2. The remarks in Section 3.4 about FOPC also apply here.

TABLE 4.2   MATLAB Code FOPT

```
function [tu,ts,tf,nu,la0]=fopt(name,tu,tf,s0,k,told,tols,mxit,eta)
% Function OPtim. w. terminal constraints, Tf Open; name must be in single
% quotes; function file 'name' computes f(s,u)=sdot for flg=1, (Phi,Phis,
% Phit) for flg=2, [fs,fu] for flg=3; inputs: tu=[t(1:N1),1) u(1:N1,nc)]=
% init. estim. u(t); s0(ns,1)=init. state, tf=final time, k=step size
% param, told=reltol for ode23; stops when norm(dua)<tols or number of
% iterations > mxit; 0<eta<=1 where d(psi)=-eta*psi is desired change in
% term. constraints on next iteration; outputs: optimum tu and ts=[t(1:N,1)
% s(1:N,ns)]), la0=initial adjoint vector; BASIC version AEB '84; Matlab
% versions Sun H. Hur '90, Fred A. Wiesinger '91, AEB 11/94, 8/97, 2/2/98
%
if nargin<9, eta=1; end; if nargin<8, mxit=10; end;
[Nu,nc1]=size(tu); nc=nc1-1; it=0; dua=ones(1,nc);
disp('    Iter.      tf      phi     norm(psi)    dua     dtf');
options=odeset('reltol',told); pmag=1; dtf=1;
while max([norm(dua) pmag abs(dtf)])>tols,
  [t,s]=ode23('fopc_f',[0 tf],s0,options,tu,name); ts=[t s];
  Ns=length(t); ns=length(s0); u1=tu(Nu,[2:nc1]); s1=ts(Ns,[2:ns+1]);
  [Phi,Phis,Phit]=feval(name,u1',s1',tf,2); np=length(Phi);
  phi=Phi(1); psi=Phi([2:np]); pmag=norm(psi)/sqrt(np-1);
  Phid=Phit+Phis*feval(name,u1',s1',tf,1);
  phid=Phid(1); psid=Phid([2:np]);
  yf=[formm(Phis','c'); zeros(np*(np+1)/2,1)]; ny=length(yf);
  [tb,y]=ode23('fopc_b',[tf 0],yf,options,tu,ts,name,ns,np);
  Nb=length(tb); Qg=forms(y(Nb,[np*ns+1:ny])');
  Q=psid*psid'+Qg([2:np],[2:np]); g=psid*phid+Qg([2:np],1);
  nu=-Q\g; la0=formm(y(Nb,[1:np*ns])',ns)*[1; nu];
  du=zeros(Nb,nc); u=du;
  for i=1:Nb,
    u1=table1(tu,tb(i)); s1=table1(ts,tb(i));
    [fs,fu]=feval(name,u1',s1',tb(i),3);
    la=formm(y(i,[1:np*ns])',ns); HuPhi=la'*fu;
    Huphi=HuPhi(1,:); Hupsi=HuPhi([2:np],:);
    du(i,:)=-k*(Huphi+nu'*Hupsi)-eta*(psi'/Q)*Hupsi;
    u(Nb-i+1,:)=u1+du(i,:);
  end
  dtf=-k*(phid+nu'*psid)-eta*psid'*(Q\psi);
  for j=1:nc, dua(j)=norm(du(:,j))/sqrt(Nb); end;
  if mxit==0, disp([it tf phi pmag dua dtf]); break; end
  disp([it tf phi pmag dua dtf]);
  tf=tf+dtf; t1=zeros(Nb,1); for i=1:Nb, t1(i)=tb(Nb-i+1); end;
  tu=[t1*tf/t1(Nb) u]; Nu=Nb; tu(Nb,1)=tf*(1+1e-8);
  if it>=mxit, break, end; it=it+1;
end
```

Example 4.5.1—TDP for Min Time Transfer to Mars Orbit

This is the dual problem to the one treated in Example 3.4.1. The required subroutine for transfering from Earth orbit to the Mars orbit using directed low thrust is listed below.

```
function [f1,f2,f3]=mart(be,s,t,flg)
T=.1405; B=.07489; r=s(1); u=s(2); v=s(3); c=cos(be); si=sin(be);
a=T/(1-B*t); global rf;
if flg==1,
  f1=[u; v^2/r-1/r^2+a*si; -u*v/r+a*c; v/r];          % f1 = f
elseif flg==2,
  f1=[t; r-rf; u; v-1/sqrt(rf)];                      % f1 = Phi
  f2=[zeros(1,4); eye(3) zeros(3,1)];                 % f2 = Phis
  f3=[1 0 0 0]';                                      % f3 = Phit
elseif flg==3,
  f1=[0 1 0 0; -(v/r)^2+2/r^3   0 2*v/r 0; ...        % f1 = fs
       u*v/r^2  -v/r  -u/r  0;   0 0 0 0];
  f2=a*[0; c; -si; 0];                                % f2 = fu
end
```

A script that runs the example problem is listed below.

```
% Script e04_4_1.m; TDP for min time to Mars orbit;
%
name='mart'; be0=[.5:.25:5]; tf=3.33; N=length(be0)-1; t=tf*[0:1/N:1];
tu=[t' be0']; s0=[1 0 1 0]'; k=1.5; told=1e-4; tols=3e-3; mxit=20;
global rf; rf=1.5237;
[tu,ts,tf,nu,la0]=fopt(name,tu,tf,s0,k,told,tols,mxit); t1=ts(:,1);
r=ts(:,2); u=ts(:,3); v=ts(:,4); th=ts(:,5); N1=length(t1); t2=tu(:,1);
be=tu(:,2)*180/pi;
```

Figure 3.18 shows the converged min time path, along with the thrust direction at many of the computed points; the min time obtained here was used as the *specified time* in the max radius (dual) problem of Figure 3.18.

Use of the MATLAB Code CONSTR

This was discussed in Section 3.4; the only difference here is that the final time t_f is added to the parameter vector p. The advantage over FOPT is that the gradients f_s, f_u are not required; the disadvantage is that convergence is slow and the results are not as accurate.

Example 4.5.2—TDP for Min Time Transfer to Mars Orbit

This is the same problems treated in Example 4.5.1. A MATLAB script for solving the problem using CONSTR is listed below along with the required subroutine.

```
% Script e04_4_2.m; TDP for min time transfer to Mars orbit using CONSTR;
% s=[r,u,v,th]';
%
be0=[.4 .5 .6 .7 .8 .9 1.1 1.2 1.4 1.7 2.7 4.5 4.7 4.9 5.0 5.1 5.2 5.3...
    5.3 5.4 5.4]; tf=3.3155; p0=[be0 tf];    % Initial guess
N1=length(be0); N=N1-1;  un=ones(1,N1); vlb=.2*un; vub=6*un;
optn(1)=1; optn(13)=3; optn(14)=1000; name='mart_fg'; s0=[1 0 1 0]';
p=constr(name,p0,optn,vlb,vub,[],s0);
[f,g,s]=mart_fg(p,s0); be=p([1:N1]); tf=p(N1+1); t=tf*[0:1/N:1];
r=s(1,:); u=s(2,:); v=s(3,:); th=s(4,:);

function [f,g,s]=mart_fg(p,s0)
% Subroutine for Example 4.5.2; TDP for min time transfer to Mars orbit
% using CONSTR; th=s(4) is ignorable but included for plotting;
%
N=length(p)-1; be=p([1:N]); tf=p(N+1);
[t,s]=odeu('marc',be,s0,tf);
rf=s(1,N); uf=s(2,N); vf=s(3,N);
f=tf; g=[rf-1.5237 uf vf-1/sqrt(rf)];
```

Example 4.5.3—A/C Min Time to Climb

This example repeats a computation made in 1961 (Ref. DB) for the min time flight path for the F4 A/C to climb to 20 km, arriving with Mach 1 and level flight. The computation was tested in flight in January 1962 at the Patuxent River Naval Air Station. The copilot had a card with the optimal Mach number tabulated for every 1000 ft of altitude, which he read off to the pilot as they went through that altitude. The pilot then moved the stick forward or backward to get as close to this Mach number as he could. They got to the desired flight condition in 338 sec, where the predicted value was 332 sec.

The rigid-body attitude motions of an A/C take place in a few seconds. Hence, when analyzing performance problems lasting tens of seconds to minutes, the attitude motions can be approximated as quasi-instantaneous, i.e., the A/C can be treated as a point mass acted on by lift, drag, gravity, and thrust forces (see Figure 4.4). The EOM are

$$m\dot{V} = T\cos(\alpha + \epsilon) - D - mg\sin\gamma \, , \tag{4.42}$$

$$mV\dot{\gamma} = T\sin(\alpha + \epsilon) + L - mg\cos\gamma \, , \tag{4.43}$$

$$\dot{h} = V\sin\gamma \, , \tag{4.44}$$

$$\dot{m} = -f(V, h) \, , \tag{4.45}$$

$$\dot{x} = V\cos\gamma \, , \tag{4.46}$$

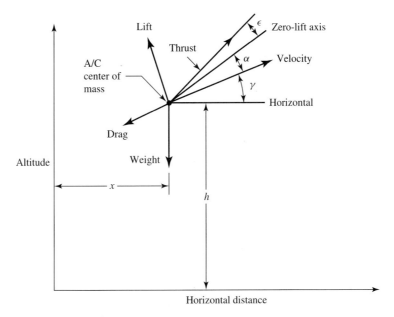

FIGURE 4.4   Nomenclature Used in the Point Mass Model of an Aircraft

where V = velocity, γ = flight-path-angle, h = altitude, m = mass, x = horizontal range, T = thrust, D = drag, L = lift, α = angle-of-attack, ϵ = thrust angle relative to the zero lift axis (a constant, usually a few degrees), and g = gravitational force per unit mass. The angle-of-attack $\alpha(t)$ was used as the control variable.

The manufacturer's data for lift, drag, and max thrust were used. Figure 4.5 shows the max thrust [divided by a reference A/C weight (34 klb)] vs. Mach number and altitude for the two J79 jet engines (courtesy of the General Electric Co.). Since partial derivatives of thrust with respect to V and h are required for the computation, an analytical fit was made to these data of the form

$$T_{\max} = [1 \; M \; M^2 \; M^3 \; M^4]Q[1 \; h \; h^2 \; h^3 \; h^4]^T , \qquad (4.47)$$

where $T_{\max}$ is in units of 1000 lb, M is the Mach number, and h is altitude in units of 10,000 ft. The resulting Q matrix was

$$Q = \begin{bmatrix} 30.21 & -.668 & -6.877 & 1.951 & -.1512 \\ -33.80 & 3.347 & 18.13 & -5.865 & .4757 \\ 100.80 & -77.56 & 5.441 & 2.864 & -.3355 \\ -78.99 & 101.40 & -30.28 & 3.236 & -.1089 \\ 18.74 & -31.60 & 12.04 & -1.785 & .09417 \end{bmatrix} . \qquad (4.48)$$

This analytical fit is also shown in Figure 4.5.

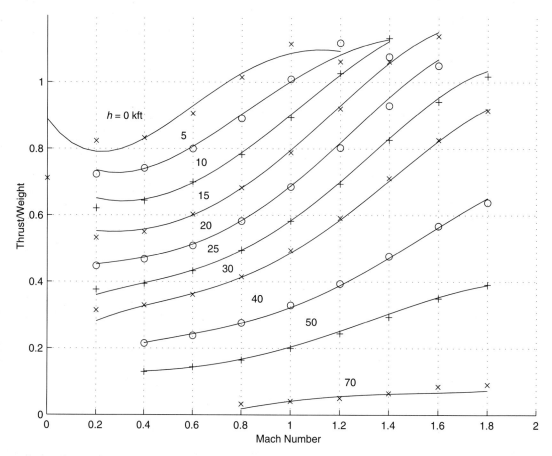

FIGURE 4.5   Thrust/Weight as a Function of Altitude h and Mach Number M for Two GE J79-8 Turbojet Engines; Reference Weight = 34.0 klb

Figure 4.6 shows the aerodynamic data of the F4 A/C (courtesy of McDonnell-Douglas) needed for the computation, namely, C_{L_α}, C_{D_o}, and κ as functions of Mach number, and the analytical fits to these data that were used in the calculations. The drag coefficient is approximated by

$$C_D = C_{D_o} + \kappa C_{L_\alpha}\alpha^2 , \tag{4.49}$$

and the lift coefficient by

$$C_L = C_{L_\alpha}\alpha . \tag{4.50}$$

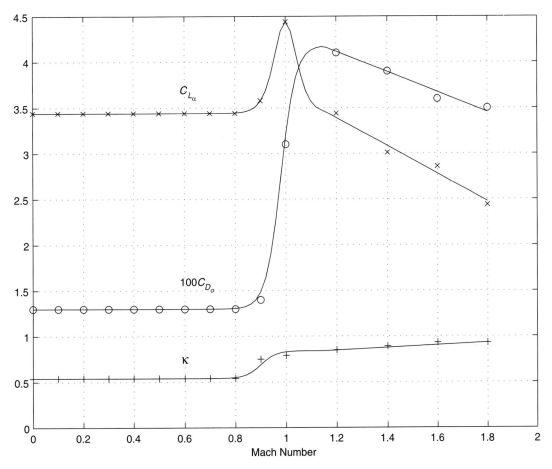

FIGURE 4.6 $dC_L/d\alpha$, C_{Do}, and κ as Functions of Mach number for the F4 Aircraft

The analytical fits for $M < 1.15$ are

$$C_{D_o} = .013 + .0144 \left[1 + \tanh \left(\frac{M - .98}{.06} \right) \right] , \tag{4.51}$$

$$C_{L_\alpha} = 3.44 + 1/\cosh^2 \left(\frac{M - 1}{.06} \right) , \tag{4.52}$$

$$\kappa = .54 + .15 \left[1 + \tanh \left(\frac{M - .9}{.06} \right) \right] . \tag{4.53}$$

For $M > 1.15$ the fits are

$$C_{D_o} = .013 + .0144 \left[1 + \tanh \left(\frac{.17}{.06} \right) \right] - .011(M - 1.15) , \tag{4.54}$$

$$C_{L_\alpha} = 3.44 + 1/\cosh^2 \left(\frac{.15}{.06} \right) - \frac{.96}{.63}(M - 1.15) , \tag{4.55}$$

$$\kappa = .54 + .15 \left[1 + \tanh \left(\frac{.25}{.06} \right) \right] + .14(M - 1.15) . \tag{4.56}$$

Fuel consumption [needed in Equation (4.45)] was modeled as

$$\dot{m} = -\frac{T}{cg} , \tag{4.57}$$

where T = thrust, g = force of gravity per unit mass, and $c = 1600$ sec.
    The subroutine required for use of FOPT is shown below.

```
function [f1,f2,f3]=f4_clmb(u,s,t,flg)
% F4H min time to climb to 20 km, M = 1, ga = 0, w. time-varying mass.
% s = [V ga h m x]'; u = alpha; lengths in lc, time in tc.
%
  Wo=.9888*41998; c=1600; S=530; g=32.2; rho=.002378;    % Parameters
  lc=2*Wo/(g*rho*S); Vc=sqrt(g*lc); tc=lc/Vc; c=c/tc; % Charac. 1,V,t
  Vf=968.1/Vc; hf=(20000/.3048)/lc;                   % Final b.c.'s
%
  V=s(1); ga=s(2); h=s(3); m=s(4);  x=s(5);  alp=u;
  cg=cos(ga);  sg=sin(ga);  ca=cos(alp);  sa=sin(alp);
%
% Atmospheric model - a(h), rhor(h); computation of Mach number:
  ao=1116/Vc;  hc=145820/lc;  lam=4.269; ha=20860/lc;  ht=36090/lc;
  if h<ht,
    a=ao*(1-h/hc)^.5;              rhor=(1-h/hc)^lam;
    dadh=-(.5*ao/(1-h/hc)^.5)/hc;  drhordh=-lam*(1-h/hc)^(lam-1)/hc;
  elseif h>=ht,
    a=968.1/Vc;                    rhor=((1-ht/hc)^lam)*exp(-(h-ht)/ha);
    dadh=0;                        drhordh=(-1/ha)*rhor;
  end
  M=V/a;
    dMdV=1/a;
    dMdh=-(V/a^2)*dadh;
%
% Aerodynamic model - cdo(M), ka(M), cla(M), & derivatives;
```

```
  if M<1.15, cdo=.013+.0144*(1+tanh((M-.98)/.06));
            dcdodM=(.0144/.06)/(cosh((M-.98)/.06))^2;
            cla=3.44+1.0/(cosh((M-1)/.06))^2;
    dcladM=-(2/.06)*sinh((M-1)/.06)/(cosh((M-1)/.06))^3;
            ka=.54+.15*(1+tanh((M-.9)/.06));
    dkadM=(.15/.06)/(cosh((M-.9)/.06))^2;
  else      cdo=.013+.0144*(1+tanh(.17/.06))-.011*(M-1.15);
       dcdodM=-.011;
            cla=3.44+1.0/(cosh(.15/.06))^2-(.96/.63)*(M-1.15);
    dcladM=-.96/.63;
            ka=.54+.15*(1+tanh((.25)/.06))+.14*(M-1.15);
    dkadM=.14;
  end;
    dcladV=dcladM*dMdV;
    dcladh=dcladM*dMdh;
  cd=cdo+ka*cla*alp^2;
    dcddM=dcdodM+(dkadM*cla+ka*dcladM)*alp^2;
    dcddV=dcddM*dMdV;
    dcddh=dcddM*dMdh;
    dcddalp=2*ka*cla*alp;
%
% Max thrust model - compute T(M,h), dT/dM, dT/dh:
  Q=[ 30.21      -.668  -6.877    1.951    -.1512; ...
      -33.80      3.347  18.13   -5.865     .4757; ...
      100.80    -77.56    5.441   2.864    -.3355; ...
      -78.99    101.40  -30.28    3.236    -.1089; ...
      18.74     -31.60   12.04   -1.785     .09417];
  h1=h*lc/1e4;    W1=Wo/1000;
  T=[1 M M^2 M^3 M^4]*Q*[1 h1 h1^2 h1^3 h1^4]'/W1;
    dTdM=[0 1 2*M 3*M^2 4*M^3]*Q*[1 h1 h1^2 h1^3 h1^4]'/W1;
    dTdh1=[1 M M^2 M^3 M^4]*Q*[0 1 2*h1 3*h1^2 4*h1^3]'/W1;
  dTdV=dTdM*dMdV;
  dTdh=dTdh1*lc/1e4+dTdM*dMdh;
%
% Outputs:
  if flg==1,                                     % f1 = f
    f1=[(T*ca-cd*rhor*V^2)/m-sg; ((T*sa+cla*rhor*V^2*alp)/m-cg)/V;...
        V*sg; -T/c; V*cg];
  elseif flg==2,
    f1=[t; V/Vf-1; 10*ga; h/hf-1];        % f1=Phi
    f2=[zeros(1,5); diag([1/Vf 10 1/hf]) zeros(3,2)];    % f2=Phis
    f3=[1 0 0 0]';      % f3=Phit
```

```
elseif flg==3,
  f1=[(dTdV*ca-rhor*(dcddV*V+2*cd)*V)/m    -cg ...
          (dTdh*ca-(drhordh*cd+rhor*dcddh)*V^2)/m ...
          -(T*ca-cd*rhor*V^2)/m^2    0; ...
      (dTdV*sa/V-T*sa/V^2+(cla+dcladV*V)*rhor*alp)/m+cg/V^2 sg/V...
          (dTdh*sa/V+(dcladh*rhor+cla*drhordh)*V*alp)/m ...
          -(T*sa+cla*rhor*V^2*alp)/m^2    0; ...
      sg    V*cg 0 0 0; ...
      -dTdV/c 0 -dTdh/c 0 0; ...
      cg    -V*sg  0 0 0]; % f1=fs, f2=fu
  f2=[(-T*sa-rhor*dcddalp*V^2)/m; (T*ca/V+cla*rhor*V)/m; 0; 0; 0];
end
```

A script that runs the problem, starting at the begin-climb point, is shown below.

```
% Script e04_4_3.m; min time-to-climb to 20 km, M=1, ga=0 for F4 A/C;
% starts from V=929 ft/sec and W=Wo*.9888 (where climb begins);
%
clear; name='f4_clmb';
u0=[.0632   .0507   .0324   .0201   .0148   .0144   .0166   .0195   .0224 ...
    .0243   .0247   .0224   .0173   .0105   .0057   .0081   .0164   .0252 ...
    .0314   .0343   .0341   .0324   .0298   .0272   .0249   .0231   .0219 ...
    .0212   .0207   .0205   .0206   .0207   .0209   .0210   .0212   .0211 ...
    .0211   .0209   .0205   .0201   .0197   .0193   .0193   .0196   .0207 ...
    .0228   .0260   .0307   .0366   .0437   .0510   .0579   .0634   .0655 ...
    .0634   .0569   .0467   .0341   .0195   .0006 -.0195  -.0390];
tf=37.2187;                                  % Converged u and tf
N=length(u0)-1; t=tf*[0:1/N:1]; tu=[t' u0']; W=.9888*41998; g=32.2;
rho=.002378; S=530; lc=2*W/(g*rho*S); Vc=sqrt(g*lc); Vo=929/Vc;
ho=50/lc; s0=[Vo 0 ho .9888 0]'; tc=lc/Vc; k=3e-5; told=1e-4;
tols=1e-3; mxit=10; c=180/pi;
[tu,ts,tf,nu,la0]=fopt(name,tu,tf,s0,k,told,tols,mxit);
t1=ts(:,1)/tf; V=ts(:,2)*Vc/1000; ga=c*ts(:,3); h=(lc/1000)*ts(:,4);
x=(lc/1000)*ts(:,6); t2=tu(:,1)/tf; alp=c*tu(:,2);
```

Figure 4.7 shows the calculated optimal path from the begin climb point. The sea-level acceleration from $M = .4$ to $M = .84$ takes 31 sec and the climb takes 296 sec for a total of 327 sec. This flight path was quite surprising in 1962, since it shows that the A/C should

- First accelerate just above ground level to $M = .84$, where the drag-rise begins (to show this we had to introduce an inequality constraint $h \geq 0$; see Problem 9.3.15).
- Then climb at nearly constant Mach number to about 30 kft.
- Then make a shallow dive to 24 kft followed by a slow climb back to 30 kft, increasing velocity until the energy approximately equals the desired final energy.
- Finally climb very rapidly to the desired final altitude of 65.7 kft (20 km).

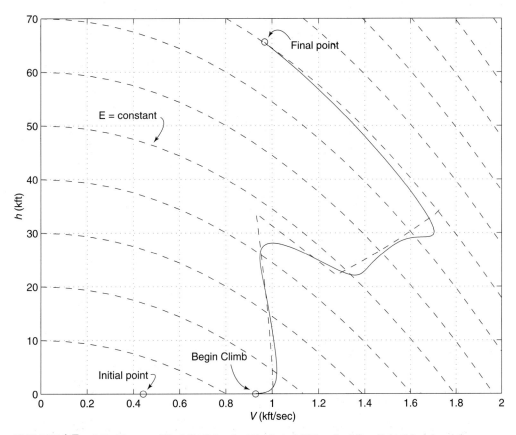

FIGURE 4.7   Min Time-to-Climb Path for the F4 Aircraft Using the Mass-Point Model and the Energy-State Model

Figure 4.8 shows angle-of-attack and flight-path-angle vs. time. There is a resonant pumping of α at the phugoid frequency, and a short segment of negative γ. The largest angles-of-attack occur at the sea-level begin-climb point and at the beginning of the final zoom climb. The largest load factor (specific lift force) also occurs at the beginning of the zoom climb but is surprisingly moderate (about 2 g's). The lowest angle-of-attack occurs right at the end, where a large negative angle-of-attack is used to level the A/C off to $\gamma(t_f) = 0$. Figure 4.9 shows altitude vs. range. The flight-path-angles are quite steep (between 30 and 40 deg) for an A/C of that period.

Example 4.5.4—A/C Min Time to Climb Using the Energy State Approximation

A few years after the test flight described above, a simpler A/C model using only two states, the energy per unit mass

$$E \overset{\Delta}{=} h + \frac{V^2}{2g} , \qquad (4.58)$$

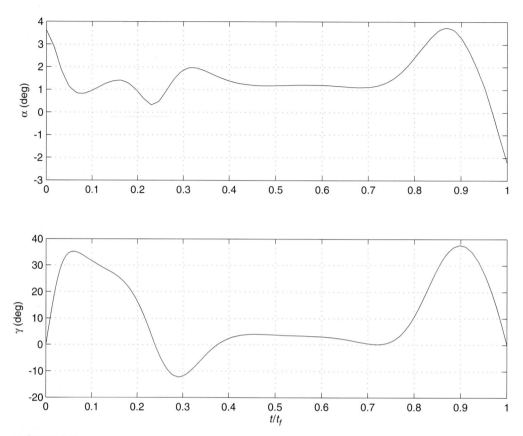

FIGURE 4.8   Angle-of-Attack and Flight-Path-Angle vs. Time for the F4 Min Time-to-Climb Flight Path

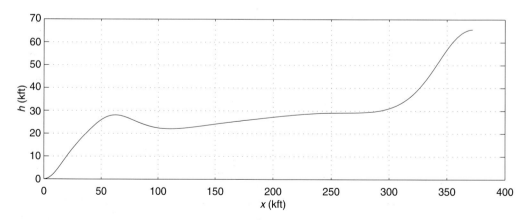

FIGURE 4.9   Altitude vs. Range for the F4 Min Time-to-Climb Flight Path

and A/C mass m, was used to calculate the optimal flight path as an initial value problem instead of a TPBVP (Ref. BDH). The results agreed quite well with the more accurate computation of Example 4.5.3.

The acceleration of the A/C perpendicular to the optimal flight path is almost always very small compared to 1 g. Hence it is a good approximation to ignore the D'Alembert force $mV\dot{\gamma}$ in (4.43). Furthermore, γ is almost always small compared to 1 radian, so that $\cos(\gamma) \cong 1$. (4.43) can then be used to find $\alpha(V, h, m)$:

$$T(V, h)\sin(\alpha + \epsilon) + L(\alpha, V, h) \cong mg \Rightarrow \alpha = \alpha(V, h, m) \ . \tag{4.59}$$

During World War II, Kaiser (Ref. Ks) suggested ways to take advantage of the then new jet engines for better climb performance. In the 1950s, Lush (Ref. Lu) and Rutkowski (Ref. Ru) introduced the concept of *energy climb,* noting that $\dot{E}$ does not depend explicitly on γ. Using (4.42) and (4.44) it follows that

$$\dot{E} = \frac{V(T - D)}{m} = \dot{E}(V, h, m, \alpha) \ . \tag{4.60}$$

α can be eliminated from (4.60) using (4.59), and we can express h in terms of (V, E) using (4.58). Thus (4.60) can be written as

$$\dot{E} = \dot{E}(E, m, V) \ . \tag{4.61}$$

This is a great simplification, since now there are only *two state variables, E and m, with V as the control variable.* h and α have become intermediate variables, which depend on E and V.

For the min time-to-climb problem, clearly we wish to increase E as rapidly as possible at all times, i.e., given E and m, find V that maximizes $\dot{E} = V(T - D)/m$; hence *this is no longer a TPBVP* but an initial value problem using the *nonlinear feedback* $V = V_{opt}(E, m)$.

Figure 4.10 shows $\dot{E} = V(T - D)$ as a function of altitude h and velocity V for the F4 A/C in level flight at a weight of 37.4 klb (the weight at the end of the climb maneuver in Figure 4.5). Also plotted are contours of constant E (parabolas). Thus an approximate solution to the min time-to-climb problem is to follow along the contours of constant E and locate the maxima of $V(T - D)$. A surprising feature of Figure 4.10 in 1962 was the appearance of a ridge of high $\dot{E}$ at supersonic velocities between 25 and 40 kft. This is a result of the large increase in thrust of the turbojet engines with velocity, which was shown in Figure 4.5. For $0 < E < 52$ kft, max $\dot{E}$ occurs near $M = .84$ with $0 < h < 30$ kft, which is the usual subsonic type of max climb rate behavior; the energy is added as *potential energy.* For $E > 52$ kft, max $\dot{E}$ occurs near 30 kft with $1 < M < 2$; the energy is added as *kinetic energy.* Kinetic and potential energy can be quickly exchanged in *zoom climbs* or *zoom dives.*

However, since there are significant variations in A/C mass during the flight, which result in changes of L and D, a more accurate solution can be obtained by integrating the nonlinear second-order system forward, finding the optimal V as a function of current E and m. A code for performing this *nonlinear feedback solution* is on the disc that accompanies this book and is not listed here. The result of this calculation is superimposed on Figure 4.5. Clearly this two-state (energy and mass) approximation is quite good.

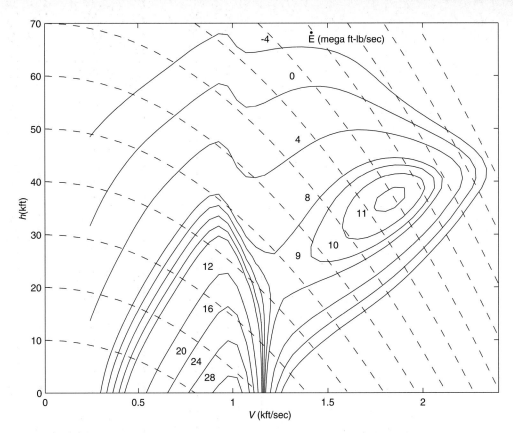

FIGURE 4.10 $\dot{E} = V(T - D)$ as a Function of Altitude h and Velocity V for the F4 Aircraft at $W = 37.4$ klb

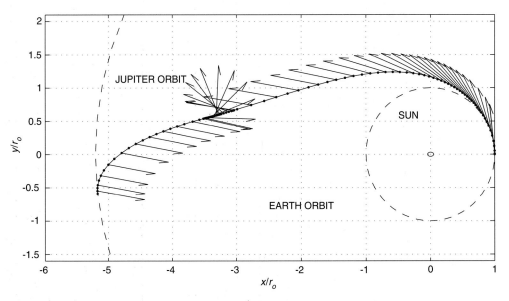

FIGURE 4.11 Min Time Path for Low-Thrust Earth-Jupiter Orbit Transfer

Problems

4.5.1 to 4.5.16:   Do Problems 4.4.1 to 4.4.16 numerically using FOPT or CONSTR.

4.5.17 TDP for Min Time Earth-Jupiter Orbit Transfer: Use FOPT with the S/C data from the example in this section to find the min time TDP to transfer from the Earth orbit to the Jupiter orbit ($r_{Jupiter}/r_{Earth} = 5.20$), and compare it with the solution shown in Figure 4.11 (from Ref. Wo).

4.5.18 TDP for Min Time Earth-Venus Orbit Transfer: Use FOPT with the S/C data from the example in this section to find the min time TDP to transfer from the Earth orbit to the Venus orbit ($r_{Venus}/r_{Earth} = .7233$), and compare it with the solution shown in Figure 4.12 (from Ref. Wo).

4.5.19 TDP for Min Time Orbit Injection from the Surface of the Moon: Use FOPT with constant specific thrust $a/g = 3$ to find the min time TDP for injection of a S/C into an orbit 100 nautical miles above the surface of the moon ($r_f/r_{Moon} = 1.1129$), starting from rest at the surface. Compare your solution with the solution shown in Figure 4.13.

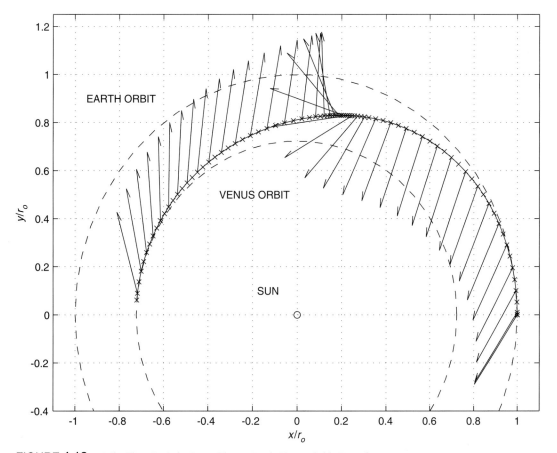

FIGURE 4.12   Min Time Path for Low-Thrust Earth-Venus Orbit Transfer

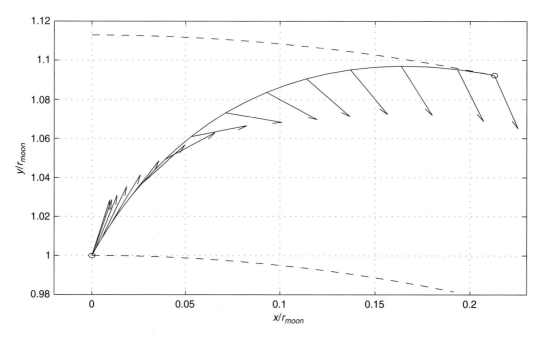

FIGURE 4.13   TDP for Min Time Orbit Injection from the Surface of the Moon

4.5.20 TDP for Min Time Mars Rendezvous: Use FOPT with the S/C data from the example in this section to find the min time TDP to rendezous with Mars, i.e., arrive at Mars with the correct orbital velocity. Note there is an optimum time to launch about every two years, which gives the min time (assumed in the example). Calculate the min times for launch 30 days before and 30 days after the optimum time. *Hint:* The final angular position around the Sun is now specified. Calculate backward and forward from the final angular position of the S/C around the Sun in the example, to find where Mars would be located 30 days before and after the optimum arrival time.

4.5.21 Solar Sail DP for Min Time to the Mars Orbit: In 1924 Tsander and Tsiolkovsky proposed the concept of propelling a vehicle in space using solar light pressure on a "sail" (Ref. Wr). While no fuel is required, a large sail is needed (on the order of 10,000 square meters). The sail force magnitude varies inversely with distance from the Sun and is proportional to $\cos^2 \theta$, where the sail angle θ is zero when the sail is normal to the Sun's rays. Since there is no "keel" it is impossible to obtain any component of force toward the Sun.

The normalized EOM are the same as for the low thrust example except that the thrust specific force a is replaced by the solar sail specific force a_{ss}, where

$$a_{ss} = \frac{\alpha \cos^2 \theta}{r^2},$$

and α is the ratio of max solar sail specific force to solar gravitational specific force. The components of this force are $a_{ss} |\cos \theta|$ in the radial direction and $a_{ss} \sin \theta$ in the tangential direction.

Using FOPT, find $\theta(t)$ to go from Earth orbit to the Mars orbit in min time. A reasonable value of α is on the order of .02 (Ref. Wr). However, to make the problem shorter (less

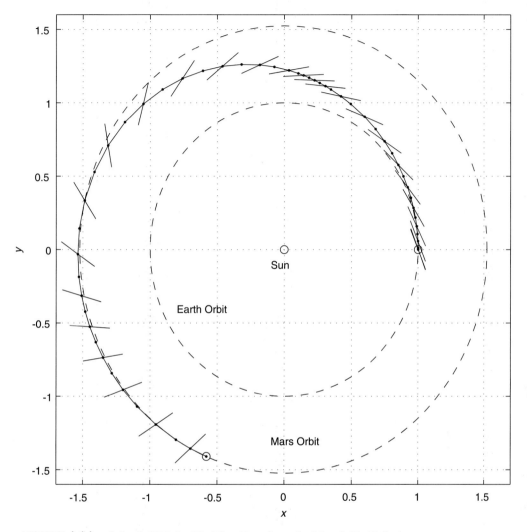

FIGURE 4.14   Solar Sail DP for Min Time Transfer to the Mars Orbit (Kelley)

integration time) use $\alpha = .17$ and compare your solution with Kelley's solution shown in Figure 4.14 (Ref. Le, p. 240).

4.5.22 Normal Force Programming (NFP) for Min Time Glide:

(a) Use FOPT to solve the dual problem of Problem 3.4.22, namely, find the min time glide path to lose 15 units of altitude with $V_f = 7$, starting from $V = 7$, $\gamma = 0$. Compare your solution with the energy-state solution of Problem 3.4.22(c).

(b) Add the terminal constraint $\gamma_f = 0$ to the problem in (a).

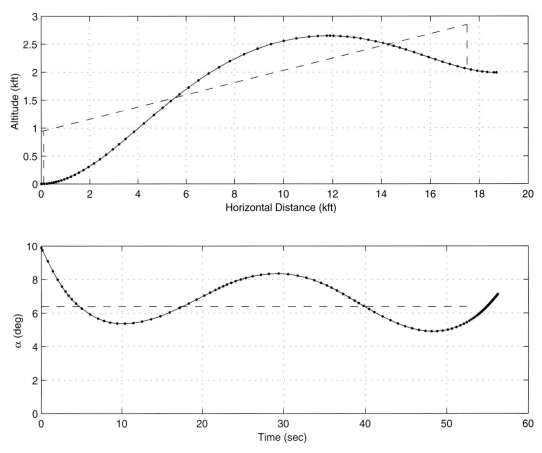

FIGURE 4.15   727 Min Time-to-Climb; Altitude vs. Horizontal Range and Angle-of-Attack vs. Time

4.5.23 NFP for Min Time Climb: For the generic A/C of Problem 3.4.23 with $\alpha_m = 1/12$, $\eta = 1/2$, $T = .2$, use FOPT to find

(a) The min time path from $V = 7$, $h = 0$, $\gamma = 0$ to $V = 7$, $h = 15$. Compare your solution with the energy-state solution of Problem 3.4.23(d).

(b) The min time path of (a) with the additional terminal constraint $\gamma_f = 0$.

(c) The max horizontal distance path from $V = 7$, $h = 0$, $\gamma = 0$ to $V = 7$, $h = 15$. Compare your solution with the energy-state solution of Problem 3.4.23(d).

(d) The max horizontal distance path of (c) with the additional terminal constraint $\gamma_f = 0$.

4.5.24 NFP for Min Time Climb of a 727 A/C: This is a dynamic version of Problem 1.3.21. For min time-to-climb, thrust is maximum at all times so the only control is angle-of-attack α. The equations of motion in normalized coordinates are

$$\dot{V} = T\cos(\alpha + \epsilon) - C_D V^2 - \sin\gamma \ , \quad V\dot{\gamma} = T\sin(\alpha + \epsilon) + C_L V^2 - \cos\gamma \ ,$$

$$\dot{h} = V\sin\gamma \ , \quad \dot{x} = V\cos\gamma \ .$$

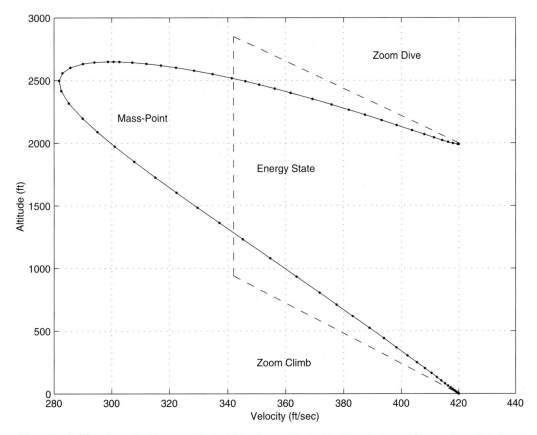

FIGURE 4.16   727 Min Time-to-Climb; Altitude vs. Velocity for Mass Point and Energy State Solutions

The expressions for $T(V)$, $C_D(\alpha)$, and $C_L(\alpha)$ are given in Problem 1.3.21. $h =$ altitude and $x =$ horizontal distance are in units of $\ell \stackrel{\Delta}{=} 2m/\rho S$, V is in units of $\sqrt{g\ell}$, and time is in units of $\sqrt{\ell/g}$.

(a) Use FOPT to find the min time path to climb 2000 ft (near sea-level where $\ell = 3243$ ft) starting and ending with $V = 420$ ft/sec and $\gamma = 0$. Such a flight path might be of interest for quick response to an air traffic control command. From Problem 1.3.21 the max steady climb rate is 37.6 ft/sec with $\alpha = 6.39$ deg and $V = 342.3$ ft/sec. Use this α as your initial guess for the optimal α history.

(b) Plot h vs. x in ft, α in deg vs. time in sec, and V in ft/sec vs. h in ft. On the last two plots, show the steady state optimal values as dashed curves. Check your solution against Figures 4.15 and 4.16.

4.5.25   NFP for Min Time Climbing Turn of a 727 A/C: Do the dual problem of Problem 3.4.25, i.e., find the two control histories $[\alpha(t), \sigma(t)]$ to minimize the time from normalized $V = 1.00$, $h = 0$, $\psi = 0$ to $V = .60$, $h = .5084$, $\psi = \pi/2$ and compare your results with Figures 3.20 and 3.21.

4.6 Direct Solution Methods for Discrete Systems

The NR and shooting algorithms developed here are almost identical to those in Chapter 3 except that the final time t_f is an additional parameter to be determined optimally.

A Discrete NR Algorithm

The NR code listed in Table 4.3 requires a good initial guess of the terminal Lagrange multipliers ν, and the final time t_f, and the control sequence $u(i)$ *at equal time steps,* such as one could obtain from a code like DOPT. Chapter 8 has another NR algorithm that requires analytical second derivatives and calculates the second variation using Riccati equations.

TABLE 4.3 A MATLAB Newton-Raphson Code DOPTN

```
function [f,s,la0]=doptn(p,name,s0)
% Disc. OPt. w. open Tf, Newton-raphson code; discrete Euler-Lagrange
% eqns. with plant eqns. forward, adjoint eqns. backward; use FSOLVE
% to iterate p=[u(i) nu' tf] to make f=[Hu(i) psi' Phid]=0; function
% file 'name' computes s(i+1)=f[s(i),u(i)] for flg=1, (Phi,Phis,Phid)
% for flg=2, and (fs,fu,fd) for flg=3 (same file used by DOPT);
% inputs: u = guess of control sequence (1 by N), nu = guess of
% terminal constraint Lagrange multipliers; tf = guess of final time;
% s0 = initial state (ns by 1); outputs: s = optimal state sequence,
% la0= initial lambda;
%
ns=length(s0); Phi=feval(name,0,s0,1,0,2); nt1=length(Phi);
N=length(p)-nt1; s=zeros(ns,N+1); la=s; Hu=zeros(1,N);
u=p([1:N]); nu=p([N+1:N+nt1-1]); tf=p(N+nt1); dt=tf/N;
%
% Forward sequencing and store x(:,i):
s(:,1)=s0; for i=1:N,
 s(:,i+1)=feval(name,u(:,i),s(:,i),dt,(i-1)*dt,1); end;
%
% Backward sequencing and store Hu(i) & Phid:
[Phi,Phis,Phid]=feval(name,0,s(:,N+1),dt,tf,2);
psi=Phi([2:nt1]); la=Phis'*[1;nu']; Phid=Phid'*[1;nu'];
for i=N:-1:1,
 [fs,fu,fd]=feval(name,u(:,i),s(:,i),dt,(i-1)*dt,3);
 Hu(i)=la'*fu; Phid=Phid+la'*fd; la=fs'*la;
end; la0=la;
f=[Hu psi' Phid];
```

Example 4.6.1—TDP for Min Time to $v_f = 0$ and Specified u_f, y_f

This is the same problem solved with DOPT in Example 4.3.1; the converged values of $\theta(i)$, ν, t_f from Example 4.3.1 were used as initial guesses here. The subroutine DTDPT is the same one used with FOPT in Example 4.3.1.

```
% Script e04_5_1.m; DTDP for min tf to vf=0 & spec. (uf,yf) using DOPTN;
% (u,v) in uf, (x,y) in uf^2/a, t in uf/a.
%
th=[1.1332 1.0293 .8709 .6191 .2332 -.2332 -.6191 -.8709 -1.0293 -1.1332];
 N=length(th); s0=[0 0 0 0]'; tf=1.4834; nu=[-.0389 .0924 -.1246]';
p0=[th nu' tf]; name='dtdpt'; optn(1)=1; optn(14)=100;
p=fsolve('doptn',p0,optn,[],name,s0); [f,s]=doptn(p,name,s0);
t=tf*[0:.1:1]; u=s(1,:); v=s(2,:); y=s(3,:); x=s(4,:);
```

The DOPT solution agrees with this more precise solution to three significant figures.

A Discrete Forward Shooting Algorithm

The code listed in Table 4.4 requires a good initial guess of the initial adjoint vector $\lambda(0)$, the terminal Lagrange multipliers ν, and the final time t_f, such as one would obtain from a code like DOPT. The subroutine DOPFU was listed in Table 2.8.

TABLE 4.4   A MATLAB Forward Shooting Code DOPTF

```
function [f,s,u,la]=doptf(p,name,u0,s0,N)
% Discrete OPtimization w. open Tf, Forward shooting; name must be in single
% quotes; function file 'name' must be in the Matlab path, giving s(i+1) for
% flg=1, (Phi,Phis,Phid) for flg=2, (fs,fu,fd) for flg=3; p=[la0,nu,tf]; u0=
% estim. init. control; s0=init. state; tf=estim. final time; N=no. steps.
%
ns=length(s0); s=zeros(ns,N+1); la=s; N1=length(p); la0=p([1:ns])';
nu=p([ns+1:N1-1])'; tf=p(N1); dt=tf/N; u=zeros(1,N); s(:,1)=s0;
la(:,1)=la0;
for i=1:N, t=(i-1)*dt;
  if i==1, u1=u0; else u1=u(i-1); end; l1=la(:,i); z0=[u1; l1];
  z1=fsolve('dopfu',z0,[],[],name,s(:,i),la(:,i),dt,t);
  u(i)=z1(1); la(:,i+1)=z1([2:ns+1]);
  s(:,i+1)=feval(name,u(i),s(:,i),dt,t,1);
  [fs,fu,fd]=feval(name,u1,s(:,i),dt,t,3); Hd(i)=la(:,i+1)'*fd;
end;
[Phi,Phis,Phid]=feval(name,u(N),s(:,N+1),dt,tf,2);
Phd=[1 nu']*Phid+sum(Hd); psi=Phi([2:N1-ns]);
f=[la(:,N+1)-Phis'*[1;nu]; psi; Phd];
```

Example 4.6.2—TDP for Min Time to $v_f = 0$ and Specified u_f, y_f Using DOPTF

This is the same problem solved above in Example 4.6.1 and uses the same subroutine DTDPT.

```
% Script e04_5_2.m; DTDP for min tf to vf=0 & spec. (uf,yf) using DOPTF;
(u,v) in uf, (x,y) in uf^2/a, t in uf/a;
%
N=10; la0=[-.3863 -.9257 -1.2434 0]'; th0=1.1366; s0=[0 0 0 0]';
tf=1.4890; nu=[-.3863 .9257 -1.2434]'; p0=[la0' nu' tf]; name='dtdpt';
optn(1)=1; optn(14)=100; p=fsolve('doptf',p0,optn,[],name,th0,s0,N);
[f,s,th,la]=doptf(p,name,th0,s0,N);
u=s(1,:); v=s(2,:); y=s(3,:); x=s(4,:);
```

As expected, the results agree almost exactly with those obtained using FOPTN.

A Discrete Backward Shooting Algorithm

The code listed in Table 4.5 requires a good initial guess of the final state vector $s(t_f)$, the terminal Lagrange multipliers v, and the final time t_f, such as one would obtain from a code like DOPT. The subroutine DOPBU was listed in Table 2.9.

TABLE 4.5   A MATLAB Backward Shooting Code DOPTB

```
function [f,s,u,la]=doptb(p,name,uf,s0,N)
% Discrete OPtimization w. term. constr., open Tf. Backward shooting;
% Name must be single quotes; function file 'name' must be in the
% Matlab path, giving s(i+1) for flg=1, (Phi,Phis,Phid) for flg=2,
% (fs,fu,fd) for flg=3; p=[sf,nu,tf]; uf=estim. final control;
% s0=initial state; tf=estim. final time; N=number steps;
%
ns=length(s0); s=zeros(ns,N+1); la=s; N1=length(p); sf=p([1:ns])';
nu=p([ns+1:N1-1])'; tf=p(N1); dt=tf/N; u=zeros(1,N); s(:,N+1)=sf;
[Phi,Phis,Phid]=feval(name,uf,sf,dt,tf,2);
psi=Phi([2:N1-ns]); la(:,N+1)=Phis'*[1;nu];
for i=N:-1:1, t=(i-1)*dt;
   if i==N, u1=uf; else u1=u(i+1); end; s1=s(:,i+1); z0=[u1; s1];
   z1=fsolve('dopbu',z0,[],[],name,s(:,i+1),la(:,i+1),dt,t);
   u(i)=z1(1); s(:,i)=z1([2:ns+1]);
   [fs,fu,fd]=feval(name,u(i),s(:,i),dt,t,3);
   la(:,i)=fs'*la(:,i+1); Hd(i)=la(:,i+1)'*fd;
end; Phd=[1 nu']*Phid+sum(Hd); f=[s(:,1)-s0; psi; Phd(1)];
```

Example 4.6.3—TDP for Min Time to $v_f = 0$ and Specified u_f, y_f Using DOPTB

This is the same problem solved above in Examples 4.6.1 and 4.6.2. The subroutine DTDPT is the same one used there. The results agree exactly with those in Examples 4.6.1 and 4.6.2.

```
% Script e04_5_3.m; DTDP for min tf to vf=0 & spec. (uf,yf) using DOPTB;
% (u,v) in uf, (x,y) in uf^2/a, t in uf/a.
%
N=10; sf=[1 0 .4390 .7417]'; thf=-1.1332; s0=[0 0 0 0]'; tf=1.4834;
nu=[-.3892 .9245 -1.2463]'; p0=[sf' nu' tf]; name='dtdpt'; optn(1)=1;
optn(14)=100; p=fsolve('doptb',p0,optn,[],name,thf,s0,N);
[f,s,th,la]=doptb(p,name,thf,s0,N); t=tf*[0:1/N:1]; u=s(1,:); v=s(2,:);
y=s(3,:); x=s(4,:);
```

Problems

4.6.1 to 4.6.12:   Do Problems 4.3.1 to 4.3.12 using DOPTN and either DOPTB or DOPTF. To do this you will first have to do them with DOPT to find good initial guesses for $\lambda(0)$ and ν. Note that Problem 4.6.12 is impossible with single shooting for $t_f >$ about 4; the optimal path is nearly steady state with transition segments of time duration $\Delta t \approx .25$ at both ends (see Problem 5.2.4 for an analytic example). FOPCN is for one control only so it cannot be used for in Table 4.3 4.6.25.

4.7 Direct Solution Methods for Continuous Systems

The NR and shooting algorithms developed here are almost identical to those in Chapter 3 except that t_f is an additional parameter to be determined optimally. The continuous codes FOPTN, FOPTB, and FOPTF are listed below, each with an example.

A Continuous NR Algorithm

The NR code listed in Table 4.6 requires a good initial guess of the control history $u(t)$, the terminal Lagrange multipliers ν, and the final time t_f, such as one would obtain from a code like FOPT. Chapter 8 has another NR algorithm that requires analytical second derivatives and calculates the second variation using Riccati equations. The subroutines ODEU and ODEHNU are fixed step RK codes included on the disc that accompanies this book.

Example 4.7.1—TDP for Min Time Transfer to Mars Orbit Using FOPTN

The Earth-Mars transfer orbit example is solved below, starting with $\lambda(0)$, ν, and t_f from the FOPT solution of Example 4.5.1.

TABLE 4.6   A MATLAB Newton-Raphson Code FOPTN

```
function [f,s,la0]=foptn(p,name,s0)
% Euler-Lagrange eqns: plant eqns fwd, adjoint eqns bkwd; a Newton-Raphson
% code since FSOLVE perturbs p=[u nu tf]; name must be in single quotes;
% function file 'name' computes sdot=f(s,u) for flg=1, perf. index phi,
% terminal constraints psi for flg=2, and fs,fu for flg=3; inputs: u =
% guess of control history (1 by N+1); nu = guess of terminal constraint
% Lagrange multipliers (1 by nt); tf = guess of final time; s0 = initial
% state (ns by 1); outputs: f=[Hu psi' Phid]; s = optimal state histories;
%
ns=length(s0); Phi=feval(name,0,s0,1,2); nt1=length(Phi); N1=length(p)-nt1;
s=zeros(ns,N1); Hu=zeros(1,N1); u=p([1:N1]); nu=p([N1+1:N1+nt1-1]);
 tf=p(N1+nt1); N=N1-1; [t,s]=odeu(name,u,s0,tf);
[Hu,phi,la0,psi]=odehnu(name,u,s,tf,nu);
[Phi,Phis,Phit]=feval(name,u(:,N1),s(:,N1),tf,2);
Phidot=Phit+Phis*feval(name,u(:,N1),s(:,N1),tf,1);
Phid=Phidot'*[1;nu']; f=[Hu psi' Phid];
```

```
% Script e04_6_1.m; TDP for min tf transfer to Mars orbit using FOPTN;
%
be0=[0.4332  0.5156  0.6132  0.7275  0.8594  1.0090  1.1766...
     1.3674  1.6209  2.6584  4.5315  4.7902  4.9363  5.0492...
     5.1447  5.2291  5.3057  5.3768  5.4442];
nu=[-3.7412  3.9796 -3.5747];  tf=3.3160; p0=[be0 nu tf];
name='mart'; s0=[1 0 1 0]'; optn(1)=1; optn(14)=200;
p=fsolve('foptn',p0,optn,[],name,s0); be=p([1:19]); tf=p(23);
[f,s]=foptn(p,name,s0); r=s(1,:); u=s(2,:); v=s(3,:);
th=s(4,:); t=tf*[0:1/18:1];
```

A diary of the execution of the solution is shown below:

```
% Diary e04_6_1.dia;
%
e04_6_1
f-COUNT        RESID      STEP-SIZE        GRAD/SD      LINE-SEARCH
   24 8.88211e-008              1     -1.78e-007
   50 1.59228e-008          0.576     -7.53e-008      incstep
   76 5.73634e-015              1      2.72e-016      int_step
Optimization Terminated Successfully
be=[0.4331  0.5154  0.6130  0.7273  0.8591  1.0086  1.1760
```

```
      1.3664  1.6200  2.6616  4.5322  4.7894  4.9359  5.0489
      5.1445  5.2289  5.3055  5.3767  5.4441];
nu=[-3.7419 3.9802 -3.5741]; tf=3.3160;
[f,s]=foptn(p,name,s0)
Hu=1e-7*[-.0681  -.0641  -.0697  -.0728  -.0778  -.0938  -.1216
         -.2025  -.2622   .1834  -.3114   .3463   .1936   .1495
          .1007   .1212   .0955   .0005  -.0702];
psi=1e-7*[.0791 .2072 .0880]; Phidot=1e-7*.0959;
```

Only three iterations were required (76 integrations) to bring $f = [H_u(t), \psi, \dot{\Phi}]$ to values less than $3 \cdot 10^{-8}$. The values in $p = [u(t), v, t_f]$ agree closely with those from FOPT but there are differences in the fourth and sometimes in the third significant figure.

A Forward Shooting Algorithm

The forward shooting code listed below requires a good initial guess of the initial adjoint vector $\lambda(t_0)$, the terminal Lagrange multiplier vector v, and the final time t_f, such as one would obtain from a code like FOPT.

FOPTF—A MATLAB Forward Shooting Code

Table 4.7 lists a MATLAB code that performs one step in the shooting algorithm described above. It uses the MATLAB command ODE23 which is a variable step-size RK code for integrating initial value problems for sets of first-order differential equations. It requires two subroutines, one for the EL equations, and the other to compute the terminal constraints.

TABLE 4.7   MATLAB Forward Shooting Code FOPTF

```
function [f,t,y]=foptf(p,name,s0,nc)
% Fcn. OPtim. w. open final Time using Fwd. shooting & FSOLVE;
% nc=no. controls
%
if nargin<4, nc=1; end; dum=zeros(nc,1); ns=length(s0); np=length(p);
n1=1:ns; la0=p(n1)'; nu=p([ns+1:np-1])'; tf=p(np); y0=[s0; la0];
optn=odeset('reltol',1e-4); [t,y]=ode23([name,'e'],[0 tf],y0,optn);
N=length(t); sf=y(N,n1); laf=y(N,[ns+1:2*ns])';
[Phi,Phis,Phit]=feval(name,dum,sf,tf,2); ydot=feval([name,'e'],tf,y(N,:));
Phidot=Phit+Phis*ydot(n1,1); ef=laf-Phis'*[1;nu]; psi=Phi([2:np-ns]);
Phd=[1 nu']*Phidot; f=[ef' psi' Phd];
```

Example 4.7.2—TDP for Min Time Transfer to Mars Orbit Using FOPTF

A script for solving the Earth-Mars transfer orbit problem using FOPTF is listed below, starting with $\lambda(0)$, ν, and t_f from the converged FOPT solution of Example 4.5.1. The first required subroutine is also listed; the second was listed in Example 4.5.1.

```
% Script e04_6_2.m; TDP for min time transfer to Mars orbit using FOPTF.
%
la0=[-5.23 -2.59 -5.62 0]; nu=[-3.75 3.99 -3.57]; tf=3.32; name='mart';
p=[la0 nu tf]; s0=[1 0 1 0]'; optn(1)=1; optn(14)=500; global rf;
rf=1.5237; p=fsolve('foptf',p,optn,[],name,s0);
[f,t,y]=foptf(p,name,s0);
be=(180/pi)*(atan2(y(:,5),y(:,6))+pi*ones(size(t)));
```

```
function yp=marte(t,y)
% Euler-Lagrange Eqns. for min tf, Earth to Mars; y=[r u v th lr lu lv lt]';
%
T=.1405; mdot=.07489; r=y(1); u=y(2); v=y(3); th=y(4); lr=y(5); lu=y(6);
lv=y(7); lt=y(8); a=T/(1-mdot*t); la=[lr lu lv lt]'; d=sqrt(lu^2+lv^2);
f=[u; v^2/r-1/r^2-a*lu/d; -u*v/r-a*lv/d; v/r];
fs=[0 1 0 0; -(v/r)^2+2/r^3 0 2*v/r 0; u*v/r^2 -v/r -u/r 0; 0 0 0 0];
yp=[f; -fs'*la];
```

An edited diary of the solution is shown below.

```
% Diary of forward shooting soln. min tf transfer to Mars orbit;
%
e04_6_2
f-COUNT        RESID    STEP-SIZE        GRAD/SD  LINE-SEARCH
    8    0.00308183            1        -0.00616
   19   0.000499837         1.44         0.00235  int_step
   29  4.28624e-008        0.993       -6.82e-008  incstep
   39  2.45056e-016            1       -5.36e-014  incstep
Optimization Terminated Successfully
f=1e-7*[-.0725 -.1158 -.0528 -.0116  .0143  .0116 -.0507]
```

It took only four iterations to satisfy the terminal constraints to an accuracy of 10^{-8}. The state and control histories are identical to those shown for the dual problem in Figures 3.21 and 3.23. The adjoint variable histories are shown in Figure 4.17; they are identical to those in Figure 3.22 *except for a negative scale factor*. Note that $\lambda_v > 0$ for a brief period near the middle of the path; this gives the rather unexpected result that the thrust vector has a *backward* component for a short time in the middle of this min time path; the thrust direction angle changes very rapidly during this time.

This solution is more accurate than the NR solution of Example 4.7.1 since (1) it used many more integration steps than the FOPTN code, and (2) the variable step-size feature

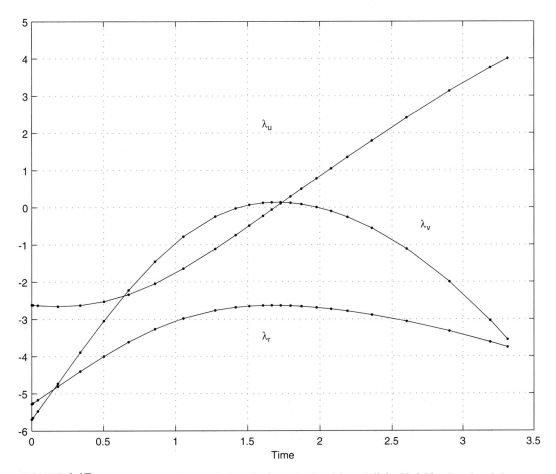

FIGURE 4.17   TDP for Min Time Orbit Transfer from Earth to Mars; Adjoint Variables (λ_r, λ_u, λ_v) vs. Time from Shooting Solution

of ODE23 causes it to use smaller steps where the control changes rapidly in the middle of the trajectory.

A Backward Shooting Algorithm

The backward shooting code listed below requires a good initial guess of the final state vector $s(t_f)$, the terminal Lagrange multiplier vector v, and the final time t_f, such as one would obtain from a code like FOPT.

FOPTB—A MATLAB Backward Shooting Code

Table 4.8 lists a MATLAB code that performs one step in a backward shooting algorithm. It uses the MATLAB command ODE23 and requires the same two subroutines required for FOPTF.

TABLE 4.8   MATLAB Backward Shooting Code FOPTB

```
function [f,t,y]=foptb(p,name,s0,nc)
% Fcn. OPtim. w. open Tf using Bkwd. shooting; p=[sf nu tf]; nc=no. controls;
%
if nargin<4, nc=1; end; dum=zeros(nc,1); ns=length(s0); sf=p([1:ns])';
n1=length(p); nu=p([ns+1:n1-1])'; global tf; tf=p(n1);
[Phi,Phis,Phit]=feval(name,dum,sf,tf,2);
yf=[sf; Phis'*[1;nu]]; psi=Phi([2:n1-ns]);
[t,y]=ode23([name,'e'],[tf 0],yf); N=length(t); e0=y(N,[1:ns])'-s0;
ydot=feval([name,'e'],0,yf); Phidot=[1 nu']*[Phit+Phis*ydot([1:ns])];
f=[e0; psi; Phidot];
```

Example 4.7.3—TDP for Min Time Transfer to Mars Orbit Using FOPTB

A script for solving the Earth-Mars transfer orbit problem using FOPTB is listed below, starting with s_f, v, and t_f from the converged FOPT solution of Example 4.5.1. The required subroutines were listed in Example 4.7.2.

```
% Script e04_6_3.m; TDP for min time to Mars orbit using FOPTB;
s=[r u v th]';
%
sf=[1.5237 0 .8101 0]; nu=[-2.8121 4.0011 -3.5518]; tf=3.3154; p=[sf nu tf];
global rf; rf=1.5237; name='mart'; s0=[1 0 1 0]'; optn(1)=1; optn(14)=500;
p=fsolve('foptb',p,optn,[],name,s0); [f,t,y]=foptb(p,name,s0);
be=(180/pi)*(atan2(y(:,6),y(:,7))+pi*ones(size(t))); N=length(t); tf=p(8);
```

Multishooting Algorithms

R. Bulirsch and his colleagues at the University of Munich have developed a robust multi-shooting code called BNDSCO (Refs. Bu and OG). This algorithm starts with guesses of s and λ at t_o and at several points between t_o and t_f; the EL equations are integrated over the intervals between these points and multiple interpolation is used iteratively to improve all the guesses until x and λ are continuous through the interior points. This avoids the overflow/underflow integration problems associated with single shooting and poor initial guesses of $\lambda(t_o)$ and v. This method works even for very long optimal paths where single shooting becomes impossible as in Problems 4.6.12 and 4.7.12.

Problems

4.7.1 to 4.7.25:   Do Problems 4.5.1 to 4.5.25 using FOPTN and either FOPTB or FOPTF. To do this you will first have to do them with FOPT to find good initial guesses for $\lambda(t_0)$ or $s(t_f)$ and v. Note that Problem 4.7.12 is impossible with single shooting for $t_f >$ about 4; the optimal path is nearly steady state with transition segments of time duration $\Delta t \approx .25$ at both ends (see Problem 5.2.4 for an analytic example). FOPCN is for one control only so it cannot be used for Problem 4.7.25.

4.8 Chapter Summary

Discrete System with Terminal Constraints and Final Time Open

In Mayer form, the problem is to choose the sequence $u(i)$, $i = 0, ..., N - 1$ and the time step Δ to minimize

$$J = \phi[x(N), t_f] , \quad t_f = N\Delta ,$$

subject to the difference equations

$$x(i + 1) = f[x(i), u(i), i\Delta] , \quad x(0) = x_o , \quad 0 = \psi[x(N), t_f] ,$$

where N is specified.

Necessary Conditions for a Stationary Path

Let

$$H(i) \stackrel{\Delta}{=} \lambda^T(i + 1)f[x(i), u(i), i, \Delta] , \quad \Phi \stackrel{\Delta}{=} \phi + \nu^T\psi .$$

Then necessary conditions for a stationary solution are

$$x(i + 1) = f[x(i), u(i), i, \Delta] , \quad x(0) = x_0 , \quad \lambda(i) = H_x^T(i) , \quad \lambda(N) = \Phi_x^T ,$$

where $u(i)$ is determined from

$$H_u(i) = 0 ,$$

and Δ is determined from the transversality condition

$$\Omega \stackrel{\Delta}{=} \Phi_\Delta + \sum_0^{N-1} H_\Delta(i) = 0 .$$

This forms a TPBVP for $x(i)$ and $\lambda(i)$, where we must also find ν and Δ to satisfy the terminal constraints $\psi = 0$ and $\Omega = 0$.

Continuous System with Terminal Constraints and Final Time Open

In Mayer form the problem is choose $u(t)$ and t_f to minimize

$$J = \phi[x(t_f)] ,$$

subject to

$$\dot{x} = f(x, u, t) , \quad x(t_o) = x_o , \quad 0 = \psi[x(t_f), t_f] .$$

Necessary Conditions for a Stationary Path

Let

$$H \overset{\Delta}{=} \lambda^T f , \quad \Phi \overset{\Delta}{=} \phi + v^T \psi .$$

Then necessary conditions for a minimum are:

$$\dot{x} = f(x, u, t) , \quad x(t_o) = x_o , \quad \dot{\lambda} = -H_x , \quad \lambda^T(t_f) = \Phi_x ,$$

where $u(t)$ is determined from

$$H_u(t) = 0 ,$$

and t_f is determined from the transversality condition

$$\Omega \overset{\Delta}{=} \Phi_t + H(t_f) = 0 .$$

This forms a TPBVP for $x(t)$ and $\lambda(t)$, where we must also find v and t_f to satisfy the terminal constraints $\psi = 0$ and $\Omega = 0$.

First Integral

If $H_t = 0$, an integral of the TPBVP is $H = $ constant.

Gradient Algorithms DOPT and FOPT

1. Guess $u(i)$ or $u(t)$ at N points and guess t_f.
2. Find $x(i)$ or $x(t)$ by forward sequencing or integration.
3. Evaluate ϕ, ψ, ϕ_x, ψ_x, $\dot{\phi}$, $\dot{\psi}$.
4. Find $H_u^\phi(i)$ and $H_u^\psi(i)$ or $H_u^\phi(t)$ and $H_u^\psi(t)$ by backward sequencing or integration.
5. Determine $v = -Q^{-1} g$ where Q, g are quadratic sums of $H_u^\phi(i)$, $H_u^\psi(i)$, $\bar{\phi}_\Delta$, $\bar{\psi}_\Delta$ or quadratic integrals of $H_u^\phi(t)$, $H_u^\psi(t)$, $\dot{\phi}$, $\dot{\psi}$.
6. $\delta u(i) = -k[H_u^\phi(i) + v^T H_u^\psi(i)]^T - \eta [H_u^\psi(i)]^T Q^{-1} \psi$, where $k > 0$, $0 < \eta \le 1$ or $t \leftarrow i$ for the continuous case.
7. $dt_f/N = -k(\bar{\phi}_\Delta + v^T \bar{\psi}_\Delta) - \eta \bar{\psi}_\Delta^T Q^{-1} \psi$ or $dt_f = -k(\dot{\phi} + v^T \dot{\psi}) - \eta \dot{\psi} Q^{-1} \psi$.
8. If $\max(\delta u_{avg}, |dt_f|) < \epsilon$ then stop.
9. $u(i) = u(i) + \delta u(i)$ or $u(t) = u(t) + \delta u(t)$ and $t_f = t_f + dt_f$.
10. Go to (2).

NR Algorithms DOPTN and FOPTN

1. Obtain a good approximation of $p = [u(t) \; v^T \; t_f]$ from DOPT or FOPT.
2. Calculate and store $s(t)$ by forward integration of the plant equations using $u(t)$.
3. At $t = t_f$ determine ψ and $\dot{\Phi} = \dot{\phi} + v^T \dot{\psi}$.
4. Calculate and store $H_u(t)$ by backward integration of the adjoint equations.
5. Evaluate the error vector $f = [H_u(t) \; \psi^T \; \dot{\Phi}]$.
6. Use FSOLVE to iterate p to make $f = 0$ to computer accuracy.

Forward Shooting Algorithms DOPTF and FOPTF

1. Obtain a good approximation of $p = [\lambda^T(t_0)\ v^T\ t_f]$ from DOPT or FOPT.
2. Find $s(t)$, $\lambda(t)$ by forward integration of the coupled EL equations, determining $u(t)$ by minimizing $H(s, \lambda, u, t)$ with respect to u.
3. Evaluate $f = [\lambda^T(t_f) - \phi_s - v^T\psi_s\ \psi^T\ \dot{\Phi}]$.
4. Use FSOLVE to iterate p to make $f = 0$ to computer accuracy.
5. This "single shooting" algorithm will FAIL for problems with energy dissipation and t_f large compared to characteristic times (see Problems 4.6.12, 4.7.12, and 5.2.4).

Backward Shooting Algorithms DOPTB and FOPTB

1. Obtain a good approximation of $p = [s^T(t_f)\ v^T\ t_f]$ from DOPT or FOPT.
2. Find $s(t)$, $\lambda(t)$ by backward integration of the coupled EL equations, determining $u(t)$ by minimizing $H(s, \lambda, u, t)$ with respect to u.
3. Evaluate $f = [s^T(t_0) - s_0^T\ \psi^T\ \dot{\Phi}]$.
4. Use FSOLVE to iterate p to make $f = 0$ to computer accuracy.
5. This "single shooting" algorithm will FAIL for problems with energy dissipation and t_f large compared to characteristic times (see Problems 4.6.12, 4.7.12, and 5.2.4).

CHAPTER 5

Linear-Quadratic
Terminal Controllers

5.1 Introduction

Many dynamic systems of interest are well approximated as *linear.* If *quadratic performance indices* and *linear constraints* are used to pose optimization problems for linear dynamic systems, the *optimal controls are linear feedbacks on the state variables.* Linear feedback is easy to implement and has been used for many years in classical control synthesis with frequency-domain or root-locus methods.

One of the main attractions of linear-quadratic (LQ) synthesis is that it produces graceful, coordinated feedback controls for multiple-input/multiple-output (MIMO) systems.

The use of state feedback requires that good estimates of the state variables be available. The states can be estimated from the available measurements by feeding back the difference between the measurements and the estimated measurements. It has been shown that feeding back this *estimated state* with optimal gains (the Kalman filter) constitutes the optimal LQ controller (see, e.g., Ref. FPE).

In this chapter we describe the synthesis of *terminal controllers,* which produces *time-varying feedback gains* even for time-invariant (TI) plants. Controllers with *soft terminal constraints* (quadratic penalty functions) are treated first, then controllers with *hard terminal constraints* (zero terminal error). In Chapter 6 we consider TI plants with $t_f \to \infty$; LQ synthesis then yields TI feedback gains.

5.2 Continuous Soft Terminal Controllers

A problem with many interesting applications is to find the control vector function $u(t)$ that minimizes

$$J = \frac{1}{2} e_f^T Q_f e_f + \frac{1}{2} \int_{t_o}^{t_f} \begin{bmatrix} x^T & u^T \end{bmatrix} \begin{bmatrix} Q & N \\ N^T & R \end{bmatrix} \begin{bmatrix} x \\ u \end{bmatrix} dt , \qquad (5.1)$$

subject to

$$\dot{x} = Ax + Bu , \quad x(t_o) = x_o , \quad e_f \overset{\Delta}{=} M_f x(t_f) - \psi , \qquad (5.2)$$

where M_f, Q_f are a given matrices and A, B, Q, N, R are given constant matrices or matrix time functions.

The control designer chooses Q_f, Q, N and R. One method of choosing these matrices is to choose $N = 0$ and

$$Q_f^{-1} = \text{diagonal matrix of maximum acceptable values of } [e_{fi}(t_f)]^2 , \qquad (5.3)$$

$$Q^{-1} = \text{diagonal matrix of maximum acceptable values of } T_{xi}[x_i(t)]^2 , \qquad (5.4)$$

$$R^{-1} = \text{diagonal matrix of maximum acceptable values of } T_{ui}[u_i(t)]^2 , \qquad (5.5)$$

where the T_i's are desired attenuation times. This makes the terms of the performance index dimensionless and of the same order of magnitude. The "costs" of terminal state deviations, in-flight output deviations, and in-flight controls are thus made roughly equal.

This is an LQ Bolza problem with Hamiltonian

$$H = \tfrac{1}{2}(x^T Q x + 2 x^T N u + u^T R u) + \lambda^T (Ax + Bu) . \qquad (5.6)$$

From Section 3.3 the EL equations are

$$\begin{bmatrix} \dot{x} \\ \dot{\lambda} \end{bmatrix} = \begin{bmatrix} A - B R^{-1} N^T & -B R^{-1} B^T \\ -Q + N R^{-1} N^T & -A^T + N R^{-1} B^T \end{bmatrix} \begin{bmatrix} x \\ \lambda \end{bmatrix} , \qquad (5.7)$$

$$u = -R^{-1}(N^T x + B^T \lambda) , \qquad (5.8)$$

with the two-point boundary conditions

$$x(t_0) = x_0 , \quad \lambda(t_f) = M_f^T Q_f [M_f x(t_f) - \psi] . \qquad (5.9)$$

x_0 and ψ are the only forcing functions in this TPBVP; the differential equations are *linear and homogeneous*.

Example 5.2.1—Lateral Intercept: Analytical Solution

Consider the double integrator system

$$\dot{y} = v , \quad \dot{v} = a , \tag{5.10}$$

with the performance index

$$J = \frac{1}{2}\left\{s_y[y(t_f)]^2 + s_v[v(t_f)]^2\right\} + \frac{1}{2}\int_0^{t_f} a^2 dt , \tag{5.11}$$

where $y(0) = y_o$, $v(0) = v_o$, and t_f is given.

Figure 5.1 shows an interpretation of this system as a lateral *intercept or rendezvous problem*. A pursuer has a closing velocity V relative to a nonmaneuvering target, parallel to an initial line-of-sight (ILOS) to the target, and a relative velocity $v(t)$ perpendicular to the ILOS. The relative acceleration and position perpendicular to the ILOS are $a(t)$ and $y(t)$. The current LOS makes an angle σ with the ILOS; hence

$$\sigma \approx \frac{y}{R} , \tag{5.12}$$

where R is the range-to-go and $T = -R/\dot{R}$ is the time-to-go.

It is straightforward to solve the linear EL equations (5.7) and (5.8) for this problem, yielding the "one-sample" solution

$$a = -v_y(t_f - t) - v_v , \tag{5.13}$$

$$v = \tfrac{1}{2}v_y[(t_f - t)^2 - t_f^2] - v_v t + v_o , \tag{5.14}$$

$$y = \frac{v_y}{6}[t_f^3 - (t_f - t)^3] - \frac{v_v}{2}t^2 + \left(v_o - \frac{v_y}{2}t_f^2\right)t + y_o , \tag{5.15}$$

where the constants v_y and v_v are given by

$$\begin{bmatrix} v_y \\ v_v \end{bmatrix} = \frac{1}{D}\begin{bmatrix} 1/s_v + t_f & -t_f^2/2 \\ -t_f^2/2 & 1/s_y + t_f^3/3 \end{bmatrix}\begin{bmatrix} v_o t_f + y_o \\ v_o \end{bmatrix} , \tag{5.16}$$

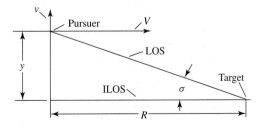

FIGURE 5.1   Nomenclature for Lateral Intercept/Rendezous Example

and

$$D \triangleq \frac{1}{s_v s_y} + \frac{t_f}{s_y} + \frac{t_f^3}{3s_v} + \frac{t_f^4}{12} \ . \tag{5.17}$$

The acceleration $a(t)$ is linear in time, giving velocity $v(t)$ that is quadratic in time, and position $y(t)$ that is cubic in time. At $t = 0$ the control a is a linear feedback on the initial states, and depends only on t_f:

$$a(t_f) = -k_v(t_f)v(0) - k_y(t_f)y(0) \ . \tag{5.18}$$

However, *any time t is a possible "initial" time on an optimal path* so we can replace t_f by $T = t_f - t$ time-to-go in (5.16) to (5.18), yielding a continuous *linear feedback*

$$a = -k_v(T)v - k_y(T)y \ , \tag{5.19}$$

with the *time-varying feedback gains*

$$k_v(T) = \frac{1}{D(T)} \left(\frac{1}{s_y} + \frac{T^2}{s_v} + \frac{T^3}{3} \right) \ , \tag{5.20}$$

$$k_y(T) = \frac{1}{D(T)} \left(\frac{T}{s_v} + \frac{T^2}{2} \right) \ , \tag{5.21}$$

where

$$D(T) \triangleq \frac{1}{s_v s_y} + \frac{T}{s_y} + \frac{T^3}{3s_v} + \frac{T^4}{12} \ . \tag{5.22}$$

A *perfect lateral intercept* $y(t_f) = 0$ is attained by letting $s_y \to \infty$, $s_v \to 0$. This gives

$$a = -\frac{3}{T^2} \begin{bmatrix} 1 & T \end{bmatrix} \begin{bmatrix} y \\ v \end{bmatrix} \equiv -3V\dot{\sigma} \ , \tag{5.23}$$

which is *proportional navigation*, used by many guided missiles. The acceleration normal to the ILOS is proportional to $\dot{\sigma}$, the time rate of change of the LOS angle with respect to inertial space. Figure 5.2 shows the state and control histories for this case. By definition, $y(0) = 0$ since the pursuer is on the ILOS at $t = 0$.

A *perfect lateral rendezvous* $y(t_f) = 0$ and $v(t_f) = 0$ is attained by letting $s_y \to \infty$, $s_v \to \infty$. This gives

$$a = -\frac{4}{T^2} \begin{bmatrix} 1.5 & T \end{bmatrix} \begin{bmatrix} y \\ v \end{bmatrix} \equiv -V \left(4\dot{\sigma} + \frac{2\sigma}{T} \right) \ . \tag{5.24}$$

Figure 5.3 shows the state and control histories for this case. The acceleration differs from the perfect lateral intercept case by having a change in the sign at $t = 2t_f/3$; this is needed to bring the velocity to zero at $t = t_f$.

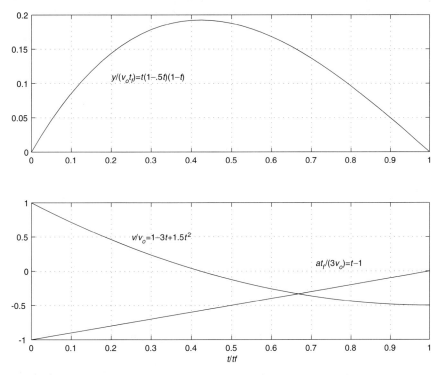

FIGURE 5.2 State and Control Histories for a Perfect Lateral Intercept

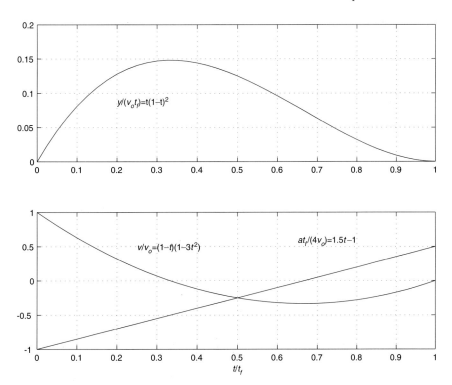

FIGURE 5.3 State and Control Histories for Perfect Lateral Rendezvous

Solution by Transition Matrix

The TPBVP can, in principle, be solved by finding a transition matrix for the EL equations, either forward or backward, then inverting partitions of this matrix to find the n missing boundary conditions; the coupled equations can then be integrated to get the solution; *iteration is not required.*

However, if $t_f - t_0 \gg$ characteristic times of the plant, the required integrations may overflow or underflow the computer so that the matrix to be inverted is ill-conditioned (nearly singular). The reason for this can be seen by considering the case where A, B, Q, N, R are constant matrices, since then the EL system is time-invariant; we show in Section 6.2 that eigenvalues of this system are symmetric across the $j\omega$-axis so that it is *unstable for integration in either direction.*

The forward transition matrix for the EL equations (5.7) can be partitioned as follows:

$$\begin{bmatrix} x(t) \\ \lambda(t) \end{bmatrix} = \begin{bmatrix} \Phi_1(t) & \Phi_2(t) \\ \Phi_3(t) & \Phi_4(t) \end{bmatrix} \begin{bmatrix} x(t_0) \\ \lambda(t_0) \end{bmatrix} , \tag{5.25}$$

Then, using the boundary conditions (5.9), the required value of $\lambda(t_0)$ can be determined:

$$\lambda(t_0) = [\Phi_4(t_f) - S_f \Phi_2(t_f)]^{-1} \left\{ [(S_f \Phi_1(t_f) - \Phi_3(t_f)] x(t_0) - a \right\} , \tag{5.26}$$

where

$$S_f \overset{\Delta}{=} M_f^T Q_f M_f , \quad a \overset{\Delta}{=} M_f^T Q_f \psi . \tag{5.27}$$

Having $\lambda(t_0)$, the EL equations (5.7) can be integrated forward to find $x(t)$, $\lambda(t)$; then $u(t)$ is obtained from (5.8). A MATLAB code TLQS for *time-invariant (TI) systems* is listed in Table 5.1.

Example 5.2.2—Lateral Intercept: Transition Matrix Solution

This problem was solved analytically in Example 5.2.1. Here it is solved numerically using TLQS.

```
% Script e05_2_2.m; trans. matrix soln. of intercept problem w. soft
% term. constr; t in tf, v in v0, y in v0*tf, a in v0/tf;
%
A=[0 1; 0 0]; B=[0 1]'; Q=zeros(2); N=[0 0]'; R=1; tf=1; x0=[0 1]';
Mf=[1 0]; Qf=3e4; psi=0; Ns=40;
[x,u,t]=tlqs(A,B,Q,N,R,tf,x0,Mf,Qf,psi,Ns); y=x(1,:); v=x(2,:); a=u;
```

The results agreed with the analytical solution to the expected accuracy.

Solution by Riccati Equation

A more reliable method of solution involves backward integration of a nonlinear matrix equation known as the *Riccati equation* plus an auxiliary vector equation. The boundary condition (5.9) at $t = t_f$ suggests a solution of the form

$$\lambda(t) = S(t)x(t) + g(t) , \tag{5.28}$$

TABLE 5.1   TLQS—Code for Soft LQ Terminal Controller

```
function [x,u,t]=tlqs(A,B,Q,N,R,tf,x0,Mf,Qf,psi,Ns)
% Time-varying LQ regulator w. Soft term. constraints; time-invariant
% plant xdot=A*x+B*u, 2J=ef'Qf*ef+int(0:tf)(x'*Q*x+2*x'*N*u+u'*R*u)dt ,
% ef=Mf*x(tf)-psi; Ns = number of steps in t; (x,u) are (state,control)
% histories with points at t.
%
% Find fwd TRANSITION MATRIX of EL eqns, t=tf to t=0:
Qb=Q-N*(R\N'); Ab=A-B*(R\N');
H=[Ab -B*(R\B'); -Qb -Ab'];                        % Hamiltonian matrix
Ph=expm(H*tf);                                     % Fwd transition matrix
%
% Partition transition matrix:
ns=length(x0); n1=[1:ns]; n2=[ns+1:2*ns];
F1=Ph(n1,n1); F2=Ph(n1,n2); F3=Ph(n2,n1); F4=Ph(n2,n2);
%
% Solve for lambda(0) in the TPBVP;
Sf=Mf'*Qf*Mf; a=Mf'*Qf*psi; la0=(F4-Sf*F2)\((Sf*F1-F3)*x0-a);
%
% Calculate optimal state and control histories:
y0=[x0; la0]; t=tf*[0:1/Ns:1];
for i=1:Ns+1, y(:,i)=expm(H*t(i))*y0;
 x(:,i)=y(n1,i); u(:,i)=-R\([N' B']*y(:,i));
end
```

where $S(t_f) = M_f^T Q_f M_f$ and $g(t_f) = -M_f^T Q_f \psi$. To obtain equations for $S(t)$ and $g(t)$ we differentiate (5.28) with respect to time

$$\dot{\lambda} = \dot{S}x + S\dot{x} + \dot{g} , \qquad (5.29)$$

and substitute for $\dot{x}$ and $\dot{\lambda}$ from (5.7), using (5.28) to eliminate λ:

$$-\bar{A}^T(Sx + g) - \bar{Q}x = \dot{S}x + S[\bar{A}x - \bar{B}(Sx + g)] + \dot{g} , \qquad (5.30)$$

where $\bar{A} \triangleq A - BR^{-1}N^T$, $\bar{B} \triangleq BR^{-1}B^T$, $\bar{Q} \triangleq Q - N^T R^{-1}N$. Since $x(t) \neq 0$, it follows that $S(t)$ and $g(t)$ are determined by the backward equations

$$\dot{S} = -S\bar{A} - \bar{A}^T S - \bar{Q} + S\bar{B}S , \quad S(t_f) = M_f^T Q_f M_f \triangleq S_f , \qquad (5.31)$$

$$\dot{g} = -[\bar{A} - \bar{B}S(t)]^T g , \quad g(t_f) = -M_f^T Q_f \psi . \qquad (5.32)$$

Since S_f and the Riccati equation (5.31) are symmetric, it follows that $S(t)$ is-symmetric.

This may be interpreted as 'sweeping' the coefficient matrix S_f of the terminal boundary condition in (5.9) backward, finding an equivalent coefficient matrix $S(t)$ relating $x(t)$ and $\lambda(t)$ at earlier times. If we sweep all the way back to $t = t_o$, we can find $\lambda(t_o)$ by using (5.28) and (5.9). We could then integrate the EL equations forward to find the solution. However, these equations are unstable, so numerical accuracy may be lost in such an integration. A stable way to find the solution is to regard the optimal control (5.8) as a feedforward/feedback law

$$u(t) = u_f(t) - K(t)x(t) \ , \quad u_f(t) \overset{\Delta}{=} -R^{-1}B^T g(t) \ , \quad K(t) \overset{\Delta}{=} R^{-1}[N^T + B^T S(t)] \ . \tag{5.33}$$

The feedforward control $u_f(t)$ and the feedback gain matrix $K(t)$ are calculated and stored on the backward integration of (5.31) and (5.32). The plant equations (5.2) are stable for forward integration using (5.33). Hence this is a possible real-time implementation of the optimal control. In practice this is often done with a discrete-step (digital) implementation, which is discussed in the next section.

Confirming the concept of the backward sweep of the terminal boundary conditions, it can be shown that (do Problem 5.2.2)

$$J_{\min} = \tfrac{1}{2}x_o^T S(t_o)x_o \ , \tag{5.34}$$

when $\psi = 0$. There are *three steps to the Riccati solution*:

1. The Riccati matrix history $S(t)$ and the vector $g(t)$ history are obtained using a differential equation solver (like ODE23 in MATLAB). If the solver handles vector only differential equations, the symmetric Riccati matrix $S(t)$ must be decomposed into a vector $y(t)$ with $n(n+1)/2$ components, where $n =$ the number of states. Inside the function file used by the solver for the backward integration of y, it is recomposed into S; then $\dot{S}$ is computed and decomposed into $\dot{y}$ (see Table 5.2 below for a MATLAB implementation).
2. The feedback gain history $K(t)$ is computed from $y(t)$ by recomposing it into $S(t)$ at each time step of the backward integration and using $K(t) = R^{-1}[N^T + B^T S(t)]$. Tables of $[t_k \ K]$ and $[t_k \ u_f]$ are formed where t_k is the time vector associated with the $S(t)$ integration. These tables must be passed on to step (3).
3. The closed-loop system $\dot{x} = [A - BK(t)]x + Bu_f(t)$ is integrated forward. Since this integration does not use the time vector t_k, $K(t)$ and $u_f(t)$ must be interpolated during the integration; this can be done in MATLAB using the TABLE1 command.

A MATLAB code TLQSR for *time-invariant systems* is listed in Table 5.2, along with four subroutines TLQS-RIC, TLQS-ST, FORMS, and FORMM (thanks to Sun Hur Diaz for the latter two codes and to Paul Y. Montgomery for the idea of using the TABLE1 command).

Example 5.2.3—Lateral Intercept Using a Riccati Solution

This problem was solved analytically in Example 5.2.1 and numerically using a transition matrix in Example 5.2.2. Here it is solved numerically using a Riccati solution that yields a *time-varying linear feedback solution*.

TABLE 5.2   TLQSR—Riccati Code for Controller w. STCs

```
function [x,u,t,tk,K,uf]=tlqsr(A,B,Q,N,R,tf,x0,Mf,Qf,psi,tol)
% Time-varying LQ ctrl w. Stc's - Riccati soln; time-invariant plant,
% xdot=A*x+B*u, x(0)=x0, 2J=ef'*Mf'*Qf*Mf*ef+int(0:tf)(x'*Q*x+2*x'*N*u
% +u'*R*u)dt, ef=Mf*x(tf)-psi; (x,u) are (state, control) histories
% at times t; K is fdbk gain at times tk.
%
Sf=Mf'*Qf*Mf; gf=-Mf'*Qf*psi; yf=[forms(Sf); gf];
optn=odeset('reltol',tol);
[tb,y]=ode23('tlqs_ric',[0 tf],yf,optn,A,B,Q,N,R);
N1=length(tb); tk=tf*ones(N1,1)-tb;
%
% Gain matrix K(t) and uf(t) from y(t):
[ns,nc]=size(B); K=zeros(N1,ns*nc); n1=ns*(ns+1)/2; ys=y(:,[1:n1]);
g=y(:,[1+n1:n1+ns])'; uf=zeros(N1,nc);
for i=1:N1, S=forms(ys(i,:)'); K(i,:)=formm(R\(N'+B'*S),'r');
 uf(i,:)=formm(-R\(B'*g(:,i)),'r');
end
K1=[tk K]; uf1=[tk uf]; disp('K(t) and uf(t) computed');
%
% Simulate closed-loop system:
[t,x]=ode23('tlqs_st',[0 tf],x0,options,A,B,K1,uf1);
N2=length(t); u=zeros(nc,N2);
for i=1:N2, Kt=formm(table1(K1,t(i)),ns);
 uft=table1(uf1,t(i)); u(:,i)=uft'-Kt*x(i,:)';
end;

function yp=tlqs_ric(t,y,flag,A,B,Q,N,R)
% Backward integ. S(t) and g(t); subroutine for TLQSR.
%
Qb=Q-N*(R\N'); Ab=A-B*(R\N'); [ns,ns]=size(A); n1=ns*(ns+1)/2;
ys=y([1:n1]); S=forms(ys); Sd=-S*Ab-Ab'*S-Qb+S*B*(R\(B'*S));
g=y([1+n1:n1+ns]); gd=-(A-B*(R\(N'+B'*S)))'*g; yp=-[forms(Sd); gd];

function yp=tlqs_st(t,y,flag,A,B,K1,uf1)
% Forward integ. of states; subroutine for TLQSR.
%
[ns,nc]=size(B); Kt=formm(table1(K1,t),ns); uft=table1(uf1,t);
yp=(A-B*Kt)*y+B*uft';

function a=forms(b)
% If b=column vector, a=symm. matrix; if b=symm. matrix, a=column vector;
%                                          Sun H. Hur 12/92, rev. AEB
```

(continued)

TABLE 5.2   (*continued*)

```
[nr,nc]=size(b);
if nc==1, n=sqrt(2*nr+1/4)-1/2; a=zeros(n);
  if n==1, a=b;
  else S1=zeros(n,n); a=S1;
  for i=1:n, j=1+(i-1)*(2*n+2-i)/2; k=i*(2*n+1-i)/2; S1(i,i:n)=b(j:k)';
  end; a=S1+S1'-diag(diag(S1));
end;
else n=nr; a=zeros(n*(n+1)/2,1);
  for i=1:n, j=(i-1)*(2*n+2-i)/2+1; k=i*(2*n+1-i)/2; a(j:k)=b(i,i:n)';
  end;
end;

function a=formm(b,d)
% If b is a matrix, a=(row,column) vector for d=('r','c'); if b is a
% (row,column) vector, a=matrix with d (columns,rows);
%                                          Sun H. Hur 11/91, rev. AEB
%
if nargin<2,error('Not enough input parameters!'),end
[m,n]=size(b);
if d=='c', a=zeros(m*n,1);
  for i=1:n, a(m*(i-1)+1:m*i)=b(:,i); end
elseif d=='r', a=zeros(1,m*n);
  for i=1:m, a(n*(i-1)+1:n*i)=b(i,:); end
elseif n==1, nc=length(b)/d; a=zeros(d,nc);
  for i=1:nc, a(:,i)=b(d*(i-1)+1:d*i); end
elseif m==1, nr=length(b)/d; a=zeros(nr,d);
  for i=1:nr, a(i,:)=b(d*(i-1)+1:d*i); end
end
```

```
% Script e05_2_3.m; lateral intercept using TLQSR;
%
A=[0 1; 0 0]; B=[0 1]'; Q=zeros(2); N=zeros(2,1); R=1;
tf=1; s0=[0 1]'; Mf=[1 0]; Qf=3e4; psi=0; tol=1e-4;
[x,u,t,tk,K]=tlqsr(A,B,Q,N,R,tf,s0,Mf,Qf,psi,tol);
```

The computed solution is very close to the analytic solution (5.13) to (5.17). Figure 5.4 shows the feedback gains $K(t)$ that are very close to the analytical solutions (5.20) and (5.21). The gain on velocity is smaller than the gain on position error and they both *peak shortly before the final time* and go to zero at the final time. This is typical of LQ problems using soft terminal constraints. With hard terminal constraints, the gains tend to infinity at the final time, which must be avoided when using noisy sensors (see Section 5.4).

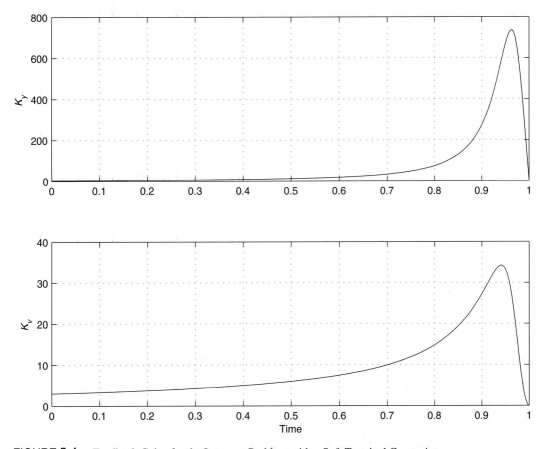

FIGURE 5.4 Feedback Gains for the Intercept Problem with a Soft Terminal Constraint

Minimum Integral-Square Control (MISC)

If there is no concern about in-flight outputs, the designer will put $Q = N = 0$ in the performance index (5.1), i.e., the control effort goes only into minimizing the terminal errors. The Riccati equation (5.31) then has no forcing function so that the differential equation for $P \stackrel{\Delta}{=} S^{-1}$ becomes *linear* (a Lyapunov equation):

$$\dot{P} = AP + PA^T - BR^{-1}B , \quad P(t_f) = S_f^{-1} , \tag{5.35}$$

which can be integrated readily if S_f is nonsingular.

If S_f is singular, we can still solve the problem with linear differential equations by letting

$$S_f = M_f^T Q_f M_f , \tag{5.36}$$

where the terminal output is $M_f x(t_f)$ and Q_f is a symmetric nonsingular weighting matrix. We then look for a solution of the form

$$\lambda(t) = M^T(t)Q(t)M(t)x(t) , \tag{5.37}$$

where $M(t_f) = M_f$ and $Q(t_f) = Q_f$ to satisfy (5.36).

Differentiating (5.37) with respect to time and substituting for $\dot{x}$ and $\dot{\lambda}$ from (5.7), it is straightforward to show that

$$\dot{M} = -MA , \quad M(t_f) = M_f , \tag{5.38}$$

$$\dot{Q} = QMBR^{-1}B^T M^T Q , \quad Q(t_f) = Q_f . \tag{5.39}$$

Moreover, (5.39) may be written as a *quadrature* for the inverse of Q:

$$\frac{d}{dt}(Q^{-1}) = -MBR^{-1}B^T M^T . \tag{5.40}$$

The *linear* matrix differential equations (5.38) and (5.40) can be integrated backward simultaneously to find $M(t)$ and $Q(t)$. The optimal *time-varying feedback gains* on the state $x(t)$ are then given by

$$K(t) = R^{-1}B^T M^T QM . \tag{5.41}$$

Example 5.2.4—Lateral Intercept: Analytical Solution Using MISC

This is the same problem considered in Examples 5.2.1 to 5.2.3. Using (5.38) we have

$$\dot{M} = -M \begin{bmatrix} 0 & 1 \\ 0 & 0 \end{bmatrix} , \quad M(t_f) = [1 \ 0] . \tag{5.42}$$

which is easily integrated, giving

$$M(T) = [1 \ T] , \quad T \stackrel{\Delta}{=} t_f - t . \tag{5.43}$$

The quadrature (5.40) with $Q_f = q_y$ yields

$$Q^{-1}(T) = \frac{1}{q_y} + \frac{T^3}{3} . \tag{5.44}$$

Using $B = [0 \ 1]^T$, (5.41) gives the feedback gains

$$K = \frac{[T \ T^2]}{1/q_y + T^3/3} , \tag{5.45}$$

which agrees with the limit of (5.20 to 5.22) as $s_v \to 0$. As $q_y \to \infty$

$$K \to 3\frac{[1 \ T]}{T^2} , \tag{5.46}$$

which is in agreement with (5.23) for a perfect lateral intercept.

MISC can also be used for numerical solutions and has the advantage over Riccati solutions that the differential equations are linear.

Following a Desired Output History

An interesting extension of the terminal controller algorithm is to consider *following a specified output history* as closely as possible using reasonable amounts of control. Consider the linear plant (5.2) with *a given desired output history* $y_d(t)$ where the actual output is $y = Cx + Du$. The performance index (5.1) is modified to

$$J = \frac{1}{2} e_f^T Q_f e_f + \frac{1}{2} \int_{t_o}^{t_f} [(y - y_d)^T Q_y (y - y_d) + u^T R_y u] dt , \tag{5.47}$$

It is straightforward to show (do Problem 5.2.1) that the EL equations are

$$\begin{bmatrix} \dot{x} \\ \dot{\lambda} \end{bmatrix} = \begin{bmatrix} A - BR^{-1}N^T & -BR^{-1}B^T \\ -Q + NR^{-1}N^T & -A^T + NR^{-1}B^T \end{bmatrix} \begin{bmatrix} x \\ \lambda \end{bmatrix} + \begin{bmatrix} 0 \\ C^T Q_y y_d \end{bmatrix} , \tag{5.48}$$

where

$$Q = C^T Q_y C , \quad N = C^T Q_y D , \quad R = R_y + D^T Q_y D . \tag{5.49}$$

The only difference of (5.48) from (5.7) is a forcing term proportional to $y_d(t)$. The boundary conditions are still (5.9).

It is straightforward to show that $S(t)$ in the sweep solution (5.28) still satisfies (5.31), while $g(t)$ is determined by

$$\dot{g} = -[A - BK(t)]^T g + C^T Q_y y_d , \quad g(t_f) = -M_f Q_f \psi , \tag{5.50}$$

where $K(t) = R^{-1} B^T S(t)$. The only difference from (5.32) is the forcing function $y_d(t)$. The form of the feedforward/feedback control law is identical to (5.33).

Rejecting a Known Disturbance History

A similar extension of the terminal controller algorithm is to consider *rejecting a known disturbance history* $w(t)$ as well as possible using reasonable amounts of control, where (5.2) is modified to

$$\dot{x} = Ax + Bu + \Gamma w . \tag{5.51}$$

This again involves use of control feedforward in addition to control feedback. It is straightforward to show that the only difference from (5.50) is in the forcing term

$$\dot{g} = -[A - BK(t)]^T g - S\Gamma w . \tag{5.52}$$

Problems

5.2.1 LQ Terminal Controller with $y = Cx + Du$: Another common form of the quadratic performance index (5.1) is

$$J = \frac{1}{2} e_f^T Q_f e_f + \frac{1}{2} \int_{t_o}^{t_f} (y^T Q_y y + u^T R_y u) dt ,$$

where $y = Cx + Du$. For this case, *show that*

$$Q = C^T Q_y C , \quad N = C^T Q_y D , \quad R = R_y + D^T Q_y D .$$

5.2.2 Derivation of $J_{\min} = \frac{1}{2}x^T(t_0)S(t_0)x(t_0)$ for $\psi = 0$: Derive (5.34) for the case $\psi = 0$ by showing that the integrand of the performance index in (5.1) is exactly equal to

$$-\frac{d}{dt}(x^T S x) ,$$

using the EL equations (5.7) and $\lambda = Sx$ from (5.8); note $g(t) \equiv 0$ if $\psi = 0$.

5.2.3 First-Order Plant Using MISC: Given the first-order system with quadratic criterion

$$\dot{x} = -x + u , \quad J = \frac{1}{2}s_f x_f^2 + \frac{1}{2}\int_0^{t_f} u^2 dt ,$$

where $x(0) = x_o$, $x_f \triangleq x(t_f)$ and all quantities are scalar.

(a) Show that the open-loop control history for min J is given by

$$u(t, t_f) = -\frac{\exp(-2t_f + t)}{1/s_f + (1/2)[1 - \exp(-2t_f)]}x_o .$$

(b) Show that the closed-loop control law for minimum J is given by

$$u(T) = -\frac{\exp(-2T)}{1/s_f + (1/2)[1 - \exp(-2T)]}x(t) ,$$

where $T = t_f - t = $ time-to-go.

(c) Find expressions for $x(t)$ for the open-loop case, and $M(T)$, $Q(T)$ for the closed-loop case.

(d) Show that $x(t_f) \to 0$ as $s_f \to \infty$.

5.2.4 First-Order Plant; Analytical and Numerical Solutions: For a first-order plant all the quantities are scalar. We let $B = R = M_f = 1$ with no loss in generality.

(a) Show that the optimal solution is

$$x = c_1 \sinh(at) + c_2 \sinh(aT) , \quad u = \dot{x} - Ax ,$$

where

$$c_2 \triangleq \frac{x_0}{\sinh(at_f)} , \quad c_1 \triangleq \frac{x_f + ac_2/Q_f}{\sinh(at_f) + [a\cosh(at_f) - \bar{A}\sinh(at_f)]/Q_f} .$$

and $a \triangleq \sqrt{\bar{A}^2 + \bar{Q}}$, $\bar{A} \triangleq A - N$, $\bar{Q} \triangleq Q - N^2$, $T \triangleq t_f - t$.

(b) Show that the transition matrix for the EL equations is

$$\begin{bmatrix} \Phi_1 & \Phi_2 \\ \Phi_3 & \Phi_4 \end{bmatrix} = \begin{bmatrix} \cosh(at) + \bar{A}\sinh(at)/a & -\sinh(at)/a \\ -\bar{Q}\sinh(at)/a & \cosh(at) - \bar{A}\sinh(at)/a \end{bmatrix} ,$$

(c) For $A = -1$, $Q = 4$, $N = 1$, $Q_f = 30$, $x_0 = 3$, $\psi \equiv x(t_f) = 1$, plot $x(t)$, $u(t)$ using the analytical solution from (a) for $t_f = 2$ and for $t_f = 20$. Do the same thing using the analytical transition matrix of (b), i.e., solve for $\lambda(0)$ from (5.26), then calculate and plot $x(t)$, $u(t)$. Check this solution using TLQS; note that both *fail to give the correct answer* for $t_f = 20$; explain why.

(d) Show that the Riccati matrix (a scalar here) and the feedforward control are given by

$$S(t) = \bar{A} + \frac{a}{\tanh(aT + \beta)} \;, \quad u_f(t) = \frac{Q_f x_f \sinh(aT + \beta)}{\sinh(\beta)} \;,$$

where $\beta \stackrel{\Delta}{=} \tanh^{-1}[a/(Q_f - \bar{A})]$, and that the feedback gain is $K(t) = N + S(t)$.

(e) Use TLQSR to check your solution for $K(t)$ and $u_f(t)$ in (d) using the data in (c). For $t_f = 20$ show that the numerical calculation still gives the correct answer; why doesn't it fail like the transition matrix solution? For $t_f = 20$ note that $x(t)$ and $u(t)$ are effectively zero over most of the path with 'boundary layers' near $t = 0$ and $t = t_f$.

5.2.5 Lateral Rendezvous: Derive the solution (5.13) to (5.22) of Example 5.2.1 in two ways:

(a) Using the transition matrix method on the EL equations (5.7).
(b) Using the linear matrix equation for $P = S^{-1}$ given in (5.35), i.e., using MISC.

5.2.6 Undamped Oscillator: Consider the second-order system and quadratic criterion

$$\dot{y} = v \;, \quad \dot{v} = -y + a \;, \quad J = \frac{1}{2}(s_y y_f^2 + s_v v_f^2) + \frac{1}{2}\int_0^{t_f} a^2 dt \;,$$

where $y(0) = y_0$, $v(0) = v_0$, $y_f \stackrel{\Delta}{=} y(t_f)$, $v_f \stackrel{\Delta}{=} v(t_f)$.

(a) Show that the optimal open-loop control history is to force the system at its resonant frequency as follows:

$$a(T, t_f) = -A_1(t_f)\sin T + A_2(t_f)\cos T \;, \quad T \stackrel{\Delta}{=} t_f - t \;,$$

where (A_1, A_2) are determined from the initial conditions by the linear equations

$$D\begin{bmatrix} A_1 \\ A_2 \end{bmatrix} = \begin{bmatrix} y(0) \\ v(0) \end{bmatrix} \;,$$

and

$$D = \begin{bmatrix} ct_f/s_y + t_f ct_f/2 - st_f/2 & st_f/s_v + t_f st_f/2 \\ st_f/s_y + t_f st_f/2 & -ct_f/s_y - t_f ct_f/2 - st_f/2 \end{bmatrix} \;.$$

where $ct_f \stackrel{\Delta}{=} \cos t_f$, $st_f \stackrel{\Delta}{=} \sin t_f$.

(b) Show that the optimal closed-loop control law is

$$a = -\begin{bmatrix} 0 & 1 \end{bmatrix} M^T(T) Q(T) M(T) \begin{bmatrix} y \\ v \end{bmatrix} ,$$

where

$$M(T) = \begin{bmatrix} \cos T & \sin T \\ -\sin T & \cos T \end{bmatrix} ,$$

$$[Q(T)]^{-1} = \begin{bmatrix} 1/s_y + T/2 - \sin 2T/4 & \sin^2 T/2 \\ \sin^2 T/2 & 1/s_v + T/2 + \sin 2T/4 \end{bmatrix} .$$

Verify that this is the same result as in (a).

(c) For $s_y \to \infty$, $s_v \to \infty$, $v(0) = 0$, find expressions for $y(t)$, $v(t)$, $a(t)$, and plot them for $t_f = 2\pi$ and $t_f = 10\pi$.

(d) Check your solutions in (c) using TLQS.

(e) Check your solutions in (c) using TLQSR.

5.2.7 Inverted Pendulum: Consider the unstable second-order system and quadratic criterion

$$\dot{y} = v , \quad \dot{v} = y + a , \quad J = \frac{1}{2}(s_y y_f^2 + s_v v_f^2) + \frac{1}{2} \int_0^{t_f} a^2 dt ,$$

where $y(0) = y_0$, $v(0) = v_0$, $y_f \triangleq y(t_f)$, $v_f \triangleq v(t_f)$.

(a) Show that the optimal open-loop control history is

$$a(T, t_f) = A_1(t_f) \sinh T + A_2(t_f) \cosh T , \quad T \triangleq t_f - t ,$$

where (A_1, A_2) are determined from the initial conditions by the linear equations:

$$D \begin{bmatrix} A_1 \\ A_2 \end{bmatrix} = \begin{bmatrix} y(0) \\ v(0) \end{bmatrix} ,$$

and

$$D \triangleq \begin{bmatrix} ct_f/s_y + t_f ct_f/2 - st_f/2 & st_f/s_v + t_f st_f/2 \\ -st_f/s_y - t_f st_f/2 & -ct_f/s_y - t_f ct_f/2 - st_f/2 \end{bmatrix} .$$

where $ct_f \triangleq \cosh t_f$, $st_f \triangleq \sinh t_f$.

(b) Show that the optimal closed-loop control law is

$$a = -\begin{bmatrix} 0 & 1 \end{bmatrix} M^T(T) Q(T) M(T) \begin{bmatrix} y \\ v \end{bmatrix} ,$$

where

$$M(T) = \begin{bmatrix} \cosh T & \sinh T \\ \sinh T & \cosh T \end{bmatrix},$$

$$[Q(T)]^{-1} = \begin{bmatrix} 1/s_y - T/2 + \sinh 2T/4 & \sinh^2 T/2 \\ \sinh^2 T/2 & 1/s_v + T/2 + \sinh 2T/4 \end{bmatrix}.$$

Verify that this is the same result as in (a).

(c) For $s_y \to \infty$, $s_v \to \infty$, $v(0) = 0$, find expressions for $y(t)$, $v(t)$, $a(t)$, and plot them for $t_f = 6$ and $t_f = 30$.

(d) As $T \to \infty$ show that $a \to -2y - v$, which stabilizes the system as a regulator.

(e) Check your solutions in (c) using TLQS.

(f) Check your solutions in (c) using TLQSR.

5.2.8 Triple Integrator Plant: Consider the third order plant and performance index

$$\dot{y} = v , \quad \dot{v} = a , \quad \dot{a} = u , \quad J = \frac{1}{2}(s_y y_f^2 + s_v v_f^2 + s_a a_f^2) + \frac{1}{2}\int_0^{t_f} u^2 dt ,$$

with $y(0) = y_o$, $v(0) = v_o$, $a(0) = a_o$, where $y_f \triangleq y(t_f)$, $v_f \triangleq v(t_f)$, $a_f \triangleq a(t_f)$.

(a) Show that the optimal control history is $u = A_1 + A_2 T + A_3 T^2$, where $T = t_f - t =$ time-to-go, and (A_1, A_2, A_3) are constants.

(b) Show that (A_1, A_2, A_3) can be calculated from

$$\begin{bmatrix} -T^3/6 - T^2/2s_a & -T^4/24 + T/s_v & -T^5/60 - 2/s_y \\ T^2/2 + T/s_a & T^3/6 - 1/s_v & T^4/12 \\ -T - 1/s_a & -T^2/2 & -T^3/3 \end{bmatrix} \begin{bmatrix} A_1 \\ A_2 \\ A_3 \end{bmatrix} = \begin{bmatrix} y(T) \\ v(T) \\ a(T) \end{bmatrix}.$$

(c) For s_y, s_v, s_a all tending to ∞, show that the min time-to-go T_o to bring the system from $y(T_o) = -1$, $v(T_o) = 0$, $a(T_o) = 0$ to $y_f = v_f = a_f = 0$ with $|u(t)| < 1$ is $T_o = 3.915$ and plot your solution.

(d) Check your solution in (c) using TLQS.

(e) Check your solution in (c) using TLQSR.

5.2.9 Cart with a Pendulum: Figure 5.5 shows a cart with a pendulum (or overhead crane), driven by an electric motor. We desire to move the cart from one point to another, starting and ending with the pendulum at rest while using as little electric power as possible (see Ref. Bw for a flexible S/C interpretation of the problem).

We designate the length of the pendulum as ℓ, the mass of the load (the pendulum bob) as m, the mass of the cart as M, the force on the cart as $u(t)$, the displacement of the cart as $y(t)$, and the deviation of the pendulum from the vertical as $\theta(t)$. The gravitational force per unit mass is g.

The problem is to choose $u(t)$ to minimize:

$$J = \frac{1}{2}(s_y y_f^2 + s_v v_f^2 + s_\theta \theta_f^2 + s_q q_f^2) + \frac{1}{2}\int_0^{t_f} u^2 dt ,$$

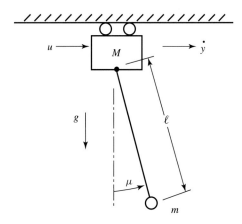

FIGURE 5.5   Nomenclature for a Cart with a
Pendulum

where $(\cdot)_f \overset{\Delta}{=} (\cdot)(t_f)$, subject to:

$$
\begin{bmatrix} \dot{y} \\ \dot{v} \\ \dot{\theta} \\ \dot{q} \end{bmatrix} = \begin{bmatrix} 0 & 1 & 0 & 0 \\ 0 & 0 & \epsilon & 0 \\ 0 & 0 & 0 & 1 \\ 0 & 0 & -1 & 0 \end{bmatrix} \begin{bmatrix} y \\ v \\ \theta \\ q \end{bmatrix} + \begin{bmatrix} 0 \\ 1 \\ 0 \\ -1 \end{bmatrix} u ,
$$

$$ y(0) = -1 , \quad v(0) = \theta(0) = q(0) = 0 , $$

where $\epsilon = m/(m + M)$, time is in units of $\sqrt{\ell M/g(m + M)}$, y in units of ℓ, and u in units of $(m + M)g$.

(a) Integrate the adjoint equations backward and show that the optimal control is a *linear function of time-to-go plus a sinusoidal function at the pendulum natural frequency* (normalized here to unity):

$$ u = A_1 \cos T + A_2 \sin T + A_3 T + A_4 , \quad T \overset{\Delta}{=} t_f - t , $$

where the $A_i s$ are constants to be determined by the boundary conditions.

(b) Substitute u from (a) into the equations of motion above and integrate to show that:

$$ \theta = -\tfrac{1}{2}A_1 T \sin T + \tfrac{1}{2}A_2 T \cos T - A_3 T - A_4 + A_5 \cos T + A_6 \sin T , $$

$$ y = A_1 \left[-\cos T + \epsilon \left(\cos T + \frac{1}{2}T \sin T \right) \right] + A_2 \left[-\sin T + \epsilon \left(\sin T - \frac{1}{2}T \cos T \right) \right] $$

$$ + \frac{A_3}{6}(1 - \epsilon)T^3 + \frac{A_4}{2}(1 - \epsilon)T^2 - A_5 \epsilon \cos T - A_6 \epsilon \sin T + A_7 T + A_8 , $$

where $A_5, \ldots, A_8$ are four more constants. Set up the linear relations for determining the eight $A_i s$ to meet the initial and final boundary conditions.

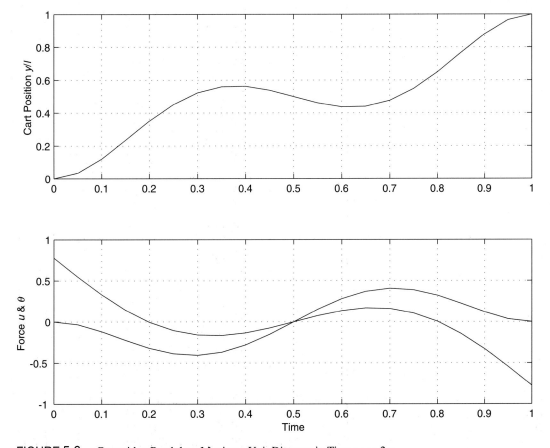

FIGURE 5.6   Cart with a Pendulum Moving a Unit Distance in Time $t_f = 2\pi$

(c) Calculate the A_is, then the optimal $y(t)$, $\theta(t)$, and $u(t)$ for the case $\epsilon = .5$, $y(0) = -1$, $v(0) = q(0) = \theta(0) = 0$, $S_f = 10^2 \cdot I$, $t_f = 2\pi$, and compare your results with Figures 5.6 and 5.7.

(d) Repeat (c) with $t_f = 10\pi$.

(e) Check your solutions in (c) and (d) using TLQS.

(f) Check your solutions in (c) and (d) using TLQSR.

In Figures 5.6 and 5.7 the force is initially positive, causing the pendulum bob to lag behind; then the force becomes negative so that at half time the pendulum is vertical and the force is zero; the second half of the path is a mirror reflection of the first half. The control is carefully phased so as to bring the pendulum to the vertical with zero angular velocity at the final time.

For the longer time $t_f = 10\pi$ (five periods), the specific force (the control) is a linear variation with time plus a sinusoidal variation at the resonant frequency; the phase of the sinusoidal part is chosen so as to produce decreasing oscillations of the pendulum during the

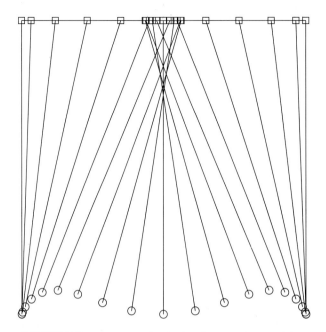

FIGURE 5.7   Stroboscopic Movie of Cart with a Pendulum
Moving a Unit Distance in Time $t_f = 2\pi$

first half of the time and increasing oscillations during the second half. The force magnitude is much smaller than in Figure 5.6 since the final time is longer.

5.2.10 Cart with an Inverted Pendulum: Figure 5.8 shows a cart with an inverted pendulum. The nomenclature and normalizations are the same as in Problem 5.2.9. For $|\theta| \ll 1$, the normalized equations of motion are

$$\begin{bmatrix} \dot{y} \\ \dot{v} \\ \dot{\theta} \\ \dot{q} \end{bmatrix} = \begin{bmatrix} 0 & 1 & 0 & 0 \\ 0 & 0 & -\epsilon & 0 \\ 0 & 0 & 0 & 1 \\ 0 & 0 & 1 & 0 \end{bmatrix} \begin{bmatrix} y \\ v \\ \theta \\ q \end{bmatrix} + \begin{bmatrix} 0 \\ 1 \\ 0 \\ -1 \end{bmatrix} u \ .$$

(a) Integrate the adjoint equations and *show that*

$$u = A_1 \exp(-t) + A_2 \exp[-(t_f - t)] + A_3 t + A_4 \ ,$$

where the A_is are constants to be determined by the boundary conditions.

(b) Substitute u from (a) into the equations of motion above and integrate to show that

$$\theta = \tfrac{1}{2} A_1 t \exp(-t) - \tfrac{1}{2} A_2 t \exp[-(t_f - t)] + A_3 t + A_4 + A_5 \exp(-t)$$
$$+ A_6 \exp[-(t_f - t)] \ ,$$

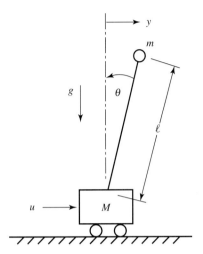

FIGURE 5.8 Nomenclature for a Cart with an Inverted Pendulum

$$y = (A_1 - \epsilon A_5)\exp(-t) + (A_2 - \epsilon A_6)\exp[-(t_f - t)] + \tfrac{1}{6}A_3(1 - \epsilon)t^3$$

$$+ \frac{1}{2}A_4(1 - \epsilon)t^2 - \frac{\epsilon}{2}A_1(2 + t)\exp(-t) - \frac{\epsilon}{2}A_2(2 - t)\exp[-(t_f - t)]$$

$$+ A_7 t + A_8 ,$$

where $A_5, \ldots, A_8$ are four more constants. Set up the relations for determining the eight A_is to meet the initial and final boundary conditions. Note the $A_i's$ are linearly proportional to the initial and final states.

(c) Calculate the optimal $y(t)$, $\theta(t)$, and $u(t)$ for the case $\epsilon = .5$, $x_0 = [-1\,0\,0\,0]^T$, $S_f = 10^4 \cdot I_4$, and $t_f = 6$, and compare your results with Figures 5.9 and 5.10. The cart is moved one pendulum length to the right, beginning and ending at rest with the pendulum vertical. The force is initially negative, backing the cart up to make the pendulum fall forward; then the force becomes positive to accelerate the system to the right so that at half time the pendulum is vertical and the force is zero; the second half of the path is a mirror reflection of the first half.

(d) Repeat (c) changing t_f to $t_f = 30$. The specific force (the control) is a linear variation with time plus exponential 'boundary layers' at both ends. The required force magnitude is much smaller than in (c) since the final time is longer.

(e) Solve (c) using TLQS and compare your results with Figure 5.10. Try (d) with TLQS and notice that it fails. Why?

(f) Solve (c) and (d) using TLQSR. Why does TLQSR work on (d) while TLQS does not?

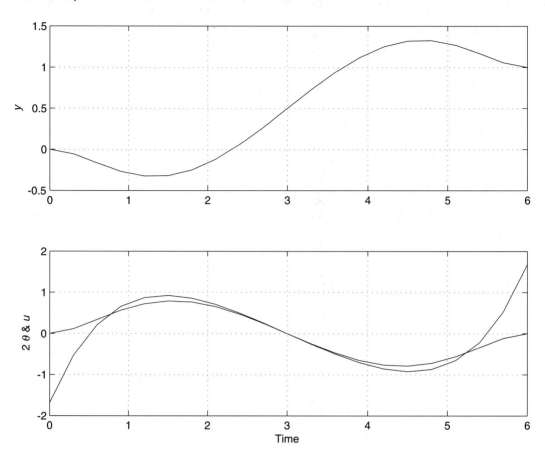

FIGURE 5.9   Cart with an Inverted Pendulum Moving a Unit Distance in Time $t_f = 6$

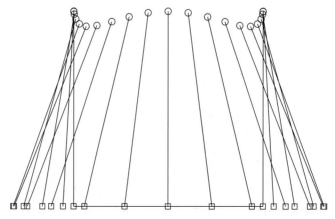

FIGURE 5.10   Stroboscopic Movie of Cart with an Inverted Pendulum Moving a Unit Distance in Time $t_f = 6$

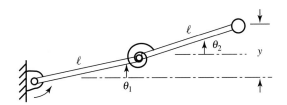

FIGURE 5.11   Nomenclature for a Simplified
Model of a Flexible Robot Arm

5.2.11 Flexible Robot Arm: Figure 5.11 shows a simplified model of a flexible robot arm (Ref. Sc). The first bending mode of the flexible arm is modeled by a pinned joint with a torsion spring with spring constant k. We wish to control the tip deflection y using a torque u at the shoulder. Taking the states as $s = [\theta_1 \; q_1 \; \theta_2 \; q_2]^T$, the EOM are (5.2) with

$$
A = \begin{bmatrix} 0 & 1 & 0 & 0 \\ -\epsilon & 0 & \epsilon & 0 \\ 0 & 0 & 0 & 1 \\ 1-\epsilon & 0 & -1+\epsilon & 0 \end{bmatrix}, \quad B = \begin{bmatrix} 0 \\ b \\ 0 \\ b-\epsilon \end{bmatrix},
$$

where $\epsilon = .4251$, $b = .2003$, $y = \theta_1 + \theta_2$ is in units of ℓ = half-length of the arm, u is in units of k, time is in units of $1/\omega$, and the angular velocities q_i are in units of ω, where $\omega = 2.782\sqrt{k/m\ell^2}$.

Use TLQS to determine the minimum integral square control to bring the tip from equilibrium at $y = 0$ to equilibrium at $y = .5$ in one period of the undamped vibration mode (normalized $t_f = 2\pi$). Use $M_f = I$, $\psi = [.25 \; 0 \; .25 \; 0]^T$ and compare your results with Figure 5.12.

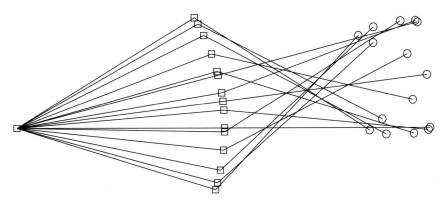

FIGURE 5.12   Stroboscopic Movie of Optimal Pick-and-Place Motion Using a Flexible
Robot Arm

5.2.12 Hovering Helicopter: A linear dynamic model (5.2) for the longitudinal motions of an OH-6A helicopter near hover is

$$A = \begin{bmatrix} -.0257 & .0130 & -.322 & 0 \\ 1.260 & -1.765 & 0 & 0 \\ 0 & 1 & 0 & 0 \\ 1 & 0 & 0 & 0 \end{bmatrix}, \quad B = \begin{bmatrix} .0860 \\ -7.41 \\ 0 \\ 0 \end{bmatrix},$$

where $x = \begin{bmatrix} u & q & \theta & y \end{bmatrix}^T$ and the control is δ_c. Here u = forward velocity, q = pitch rate, θ = pitch angle, y = longitudinal hover position, δ_c = longitudinal cyclic pitch. The units are ft, sec, centi-radians (crad), and deci-inches for δ_c.

With no in-flight penalties on output (Q=0), verify the optimal state and control histories shown in Figure 5.13 that move the helicopter from equilibrium at $y(0) = 0$ to equilibrium at $y(t_f) = 10$ ft with $t_f = 4$ sec.

5.2.13 Insertion of a S/C at the Earth-Moon L_1 Point: On the line between the Earth and the Moon there is a point called the L_1 point where a S/C would be in equilibrium,

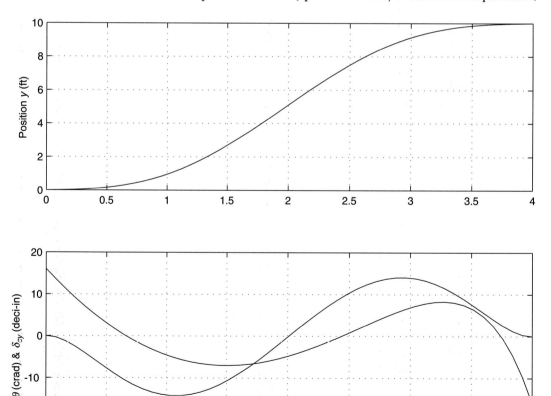

FIGURE 5.13   LQ Terminal Control of Position Change of a Hovering Helicopter

i.e., the gravitational pull of the Moon plus centrifugal force = gravitational pull of the Earth. However, it is an unstable equilbrium point, so that small thrusters operated by a feedback control system would be required to stabilize a S/C there.

If the thrusters are oriented along the Earth-Moon line, the linearized EOM in the orbit plane are

$$\ddot{x} = 2\dot{y} + (2\beta + 1)x + u \ , \quad \ddot{y} = -2\dot{x} - (\beta - 1)y \ ,$$

where $(x, \ y) = $ distance from L_1 (parallel, perpendicular) to the Earth-Moon line, $u = $ the thrust specific force, $\beta = 5.148$, time is in units of $1/n$, $n = $ orbital frequency of the Moon about the Earth, while distance is in units of the distance from Earth to the L_1 point.

Suppose the S/C is at $x = 0$, $y = -.01$ at $t = 0$. Find the MISC to bring it to $x = y = 0$ in one-eighth of a lunar orbit period ($t_f = 2\pi/8n$) using $S_f = 10^6 \cdot I$. Plot the optimal path and compare your solution with Figure 5.14.

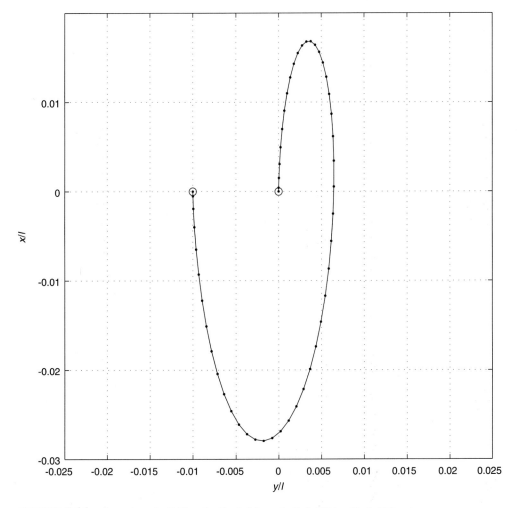

FIGURE 5.14   Insertion of a S/C at the Earth-Moon L_1 Point Using Radial Thrusters

5.2.14 Insertion of a Solar Sail S/C at the Sun-Earth L_2 Point: The Sun-Earth L_2 point is a point beyond the Earth on the Sun-Earth line where the gravitational forces of the Sun and the Earth are exactly balanced by centrifugal force. A S/C with a large solar sail is to be placed at an equilibrium point near L_2 where the centrifugal force is augmented by the sail force. This is an unstable equilibrium point, but the S/C position can be stabilized by small deviations of the sail angle θ that provides a small force perpendicular to the Sun-Earth line (see Figure 5.15). The EOM, linearized about this equilbrium point are

$$\ddot{x} = 2\dot{y} + b_1 x \ , \quad \ddot{y} = -2\dot{x} - b_2 y + c\theta \ ,$$

where (x, y) are in units of $1.51 \cdot 10^6$ km (the distance from Earth to the L_2 point), time is in units of $1/n$, $n =$ Earth angular velocity about the Sun, and b_1, b_2, c are positive constants. For one S/C considered in a design study (S. H. Hur and B. Pervan, 1991), these constants were found to be (12.762, 4.914, 1.948) with the equilibrium point 16% closer to Earth than the L_2 point.

   Suppose the S/C is at $x = 0$, $y = -.01$ at $t = 0$. Find the MISC to bring it to $x = y = 0$ in one-fourth of a year ($t_f = 2\pi/4n$) using $S_f = 10^6 \cdot I$. Plot the optimal path and compare it with Figure 5.16.

5.2.15 Dumping S/C Pitch Angular Momentum Using a Reaction Wheel: A S/C in circular orbit has a pitch reaction wheel which is actuated by an electric motor. The long axis of the S/C points downward, which is a stable attitude in pitch angle θ due to the gravity gradient torque (see Figure 5.17). The reaction wheel/motor combination has a time constant $1/\sigma$ that is short compared to the pitch libration period $2\pi/\omega_p$, where $\omega_p =$ the pitch libration frequency (approximately $\sqrt{3}$ times the orbit rate). The normalized EOM for small pitch angle θ are

$$
\begin{bmatrix} \dot{\theta} \\ \dot{\delta q} \\ \dot{\delta H} \\ \epsilon \dot{\theta}_w \end{bmatrix} =
\begin{bmatrix} 0 & 1 & 0 & 0 \\ -1 & -\sigma & \sigma/(1+\epsilon) & 0 \\ -1 & 0 & 0 & 0 \\ 0 & -1 & 1 & 0 \end{bmatrix}
\begin{bmatrix} \theta \\ \delta q \\ \delta H \\ \epsilon \theta_w \end{bmatrix} +
\begin{bmatrix} 0 \\ 1 \\ 0 \\ 0 \end{bmatrix} e \ ,
$$

where $\theta =$ deviation angle of the S/C from local vertical, $\delta q =$ deviation in S/C angular velocity, $\delta H =$ deviation in angular momentum of the S/C plus reaction wheel,

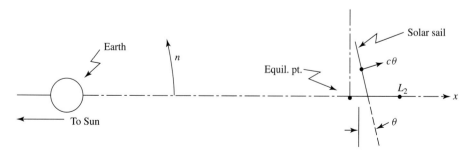

FIGURE 5.15   Nomenclature for Solar Sail S/C—Equilibrium Point Near the Sun-Earth L_2 Point

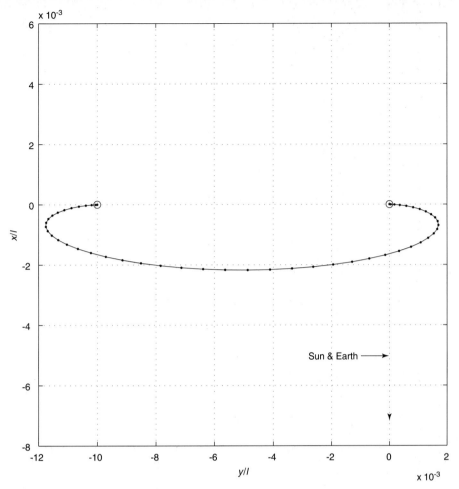

FIGURE 5.16 Insertion of a Solar Sail S/C at an Equilibrium Point Near the Sun-Earth L_2 Point

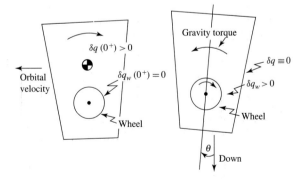

FIGURE 5.17 Pitch Control of a S/C Using a Reaction Wheel and Gravity Desaturation; On left—just after an impulsive disturbance torque; on right—using gravity torque to desaturate the reaction wheel

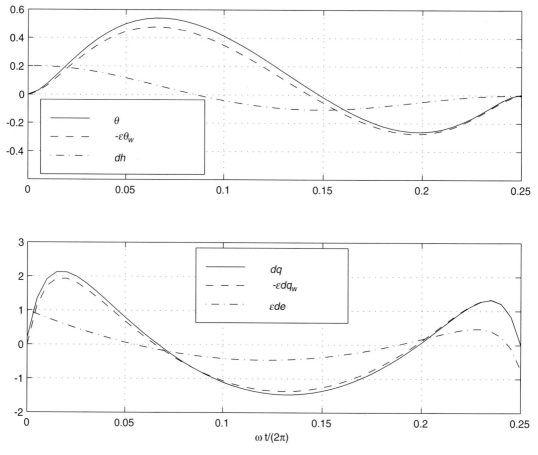

FIGURE 5.18   Dumping S/C Pitch Angular Momentum Using Pitch Gravity Torque and a Reaction Wheel

θ_w = deviation angle of a reference line on the reaction wheel from local vertical, e = armature voltage applied to the motor. $\epsilon = J/I_y$, where J = moment of inertia (MOI) of the reaction wheel and I_y = MOI of the S/C plus the reaction wheel. Time is in units of $1/\omega_p$, δH is in units of $I_y\omega_p$, while $(\delta q, \sigma)$ are in units of ω_p.

For $1/\sigma = .01(2\pi/\omega_p)$, $\epsilon = .02$ find the MISC $e(t)$ to bring the system from $\delta H = \delta q = .1$ to all states zero in time $t_f = 2\pi/(4\omega_p)$ = one-quarter of a libration period. This initial state has unwanted pitch angular momentum that must be "dumped" using the gravity-gradient torque. Compare your results with Figure 5.18. The problem solution on the disc displays a movie of the motion.

5.2.16 Dumping S/C Roll/Yaw Angular Momentum Using Reaction Wheels: A S/C in circular orbit has roll and yaw reaction wheels that are actuated by electric motors. The S/C is axially symmetric with its symmetry axis cross-track and is thin in the cross-track direction, so that it has two undamped libration modes, one primarily roll motion and the other primarily yaw motion due to the roll gravity gradient torque (see Figure 5.19).

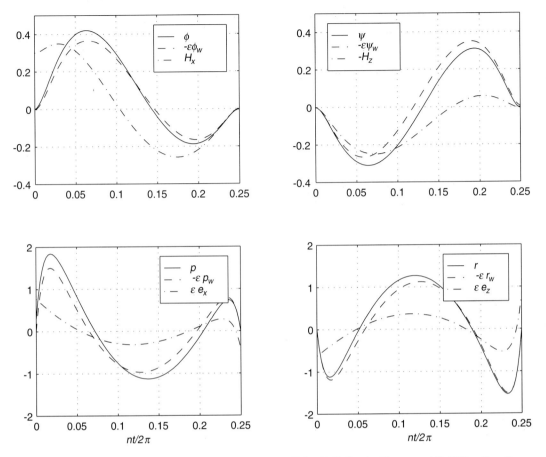

FIGURE 5.19   Dumping S/C Roll Angular Momentum Using Roll Gravity Torque and Roll/Yaw Reaction Wheels

The reaction wheel/motor combinations have the same time constant $1/\sigma$, which is short compared to the libration periods. The normalized EOM for small roll/yaw angles (ϕ, ψ) are

$$
\begin{bmatrix} \dot{\phi} \\ \dot{p} \\ \dot{H}_x \\ \dot{\psi} \\ \dot{r} \\ \dot{H}_z \\ \epsilon \dot{\phi}_w \\ \epsilon \dot{\psi}_w \end{bmatrix} = \begin{bmatrix} 0 & 1 & 0 & 1 & 0 & 0 & 0 & 0 \\ -3a & -\sigma & \sigma(1-\epsilon) & 0 & -a & 0 & 0 & 0 \\ -3a & 0 & 0 & 0 & -a & 0 & 0 & 0 \\ -1 & 0 & 0 & 0 & 1 & 0 & 0 & 0 \\ 0 & b & 0 & 0 & -\sigma & \sigma(1-\epsilon) & 0 & 0 \\ 0 & b & 0 & 0 & 0 & 0 & 0 & 0 \\ 0 & -1 & 1 & 0 & 0 & 0 & 0 & 0 \\ 0 & 0 & 0 & 0 & -1 & 1 & 0 & 0 \end{bmatrix} \begin{bmatrix} \phi \\ p \\ H_x \\ \psi \\ r \\ H_z \\ \epsilon \phi_w \\ \epsilon \psi_w \end{bmatrix} + \begin{bmatrix} 0 & 0 \\ 1 & 0 \\ 0 & 0 \\ 0 & 0 \\ 0 & 1 \\ 0 & 0 \\ 0 & 0 \\ 0 & 0 \end{bmatrix} \begin{bmatrix} e_x \\ e_z \end{bmatrix},
$$

where $(\phi, \psi) =$ the S/C (roll, yaw) angles, $(p, r) =$ (roll, yaw) angular velocities, $(H_x, H_z) =$ (roll, yaw) angular momenta of the S/C plus reaction wheels, $\phi_w, \psi_w) =$ reaction wheel deviation angles, $(e_x, e_z) =$ (roll, yaw) armature voltages applied to the motors. $\epsilon = J/I$, where $J =$ moment of inertia (MOI) of the reaction wheels and $I =$ MOI of the S/C plus the reaction wheel in roll and yaw. Time is in units of $1/n$, where $n =$ orbit rate; (H_x, H_z) are in units of In, while p, r, p_w, r_w are in units of n.

For $1/\sigma = .01(2\pi/n)$, $a = b = 1/2$, $\epsilon = .02$:

(a) Find the MISC $e_x(t)$, $e_z(t)$ to bring the system from $H_x = p = .3$, $\phi = \psi = H_z = r = 0$ to all states zero in time $t_f = 2\pi/(4n) =$ one-quarter of an orbit period. This initial state has unwanted roll angular momentum that must be "dumped" using the gravity-gradient torque. Compare your results with Figure 5.19. The problem solution on the disc displays a movie of the motion.

(b) Find the MISC $e_x(t)$, $e_z(t)$ to bring the system from $H_z = r = .2$, $\phi = \psi = H_x = p = 0$ to all states zero in time $t_f = 2\pi/(4n) =$ one-quarter of an orbit period. This initial state has unwanted yaw angular momentum that must be "dumped" using the gravity-gradient torque. Compare your results with Figure 5.20. The problem solution on the disc displays a movie of the motion.

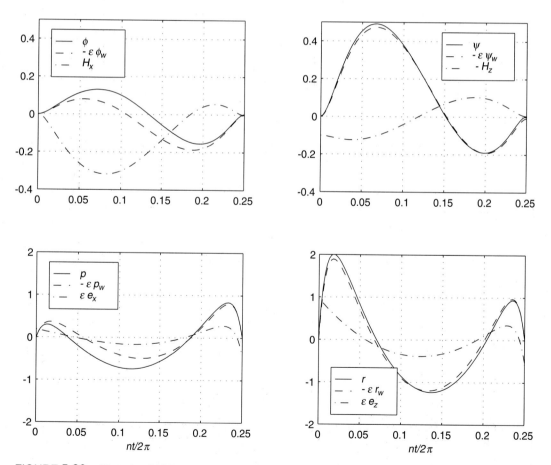

FIGURE 5.20   Dumping S/C Yaw Angular Momentum Using Roll Gravity Torque and Roll/Yaw Reaction Wheels

5.2.17 Altitude and Velocity Change for Navion A/C: A general aviation A/C (the Navion) is flying in cruise condition (176 ft/sec) near sea level. The perturbation longitudinal dynamics are given in Ref. Te and converted here to the form (5.2) with

$$A = \begin{bmatrix} -.045 & .036 & -.3220 & & \\ -.370 & -2.02 & 1.76 & 0 & 0 \\ .191 & -3.96 & -2.98 & 0 & 0 \\ 0 & 0 & 1 & 0 & 0 \\ 0 & -1 & 0 & 1.76 & 0 \end{bmatrix}, \quad B = \begin{bmatrix} 0 & 1 \\ -.282 & 0 \\ -11 & 0 \\ 0 & 0 \\ 0 & 0 \end{bmatrix},$$

where the state vector is $[u \ w \ q \ \theta \ h]^T$ and u = change in forward velocity, w = change in downward velocity, θ = pitch angle, $q = \dot{\theta}$, and h = change in altitude. The units are ft, sec, crad. The control vector is $[\delta e \ \delta T]^T$, where δe = change in elevator angle and δT = thrust specific force (proportional to throttle change).

(a) We wish to increase the altitude by 10 ft in 2 sec, with all the other final states zero using coordinated elevator and throttle histories. Compare your results with Figure 5.21.

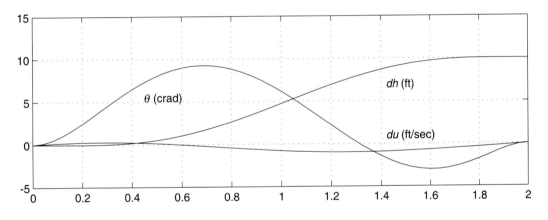

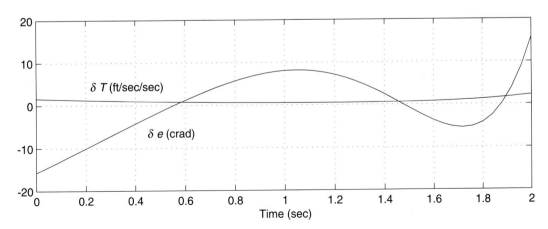

FIGURE 5.21 Navion Change in Altitude with Coordinated Elevator/Throttle

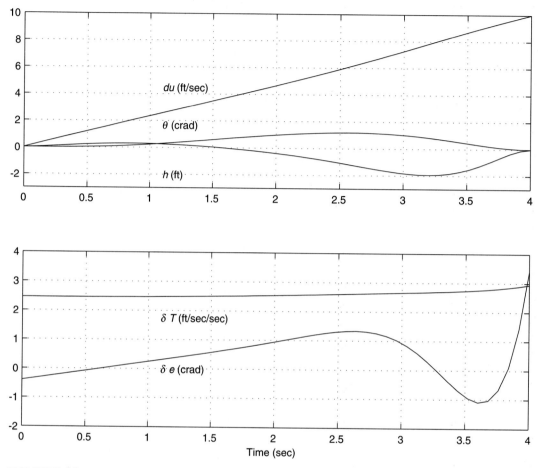

FIGURE 5.22   Navion Change in Velocity with Coordinated Elevator/Throttle

(b) We wish to increase the velocity by 10 ft/sec in 4 sec, with all other final states zero using coordinated elevator and throttle histories. Compare your results with Figure 5.22.

5.2.18 Coordinated Turn for Navion A/C: A general aviation A/C (the Navion) is flying in cruise condition (176 ft/sec) near sea level. The perturbation lateral dynamics are given in Ref. HJ and converted here to the form (5.2) with

$$
A = \begin{bmatrix}
-.254 & -1.76 & 0 & .322 & 0 \\
2.55 & -.76 & -.35 & 0 & 0 \\
-9.08 & 2.19 & -8.4 & 0 & 0 \\
0 & 0 & 1 & 0 & 0 \\
0 & 0 & 0 & 0 & 0]
\end{bmatrix}, \quad
B = \begin{bmatrix}
0 & .1246 \\
-.222 & -4.60 \\
29.0 & 2.55 \\
0 & 0 \\
0 & 0
\end{bmatrix},
$$

$$
C = \begin{bmatrix} -.254 & 0 & 0 & 0 & 0 \end{bmatrix}, \quad D = \begin{bmatrix} 0 & .1246 \end{bmatrix},
$$

where the state vector is $[v \ r \ p \ \phi \ \psi]^T$, the control vector is $[\delta_a \ \delta_r]^T$, and $v =$ sideslip velocity, $(r, p) =$ (yaw, roll) rate, $(\psi, \ \phi) =$ (yaw, roll) angle, $a_y =$ lateral specific force $= Cx + Du$, $\delta_a =$ aileron angle, $\delta_r =$ rudder angle. The units are ft, sec, crad.

Using the performance index

$$2J = e_f^T Q_f e_f + \int_0^{t_f} (Q_{ay} a_y^2 + \delta_a^2 + \delta_r^2) dt \ ,$$

with $Q_{ay} = 5^2$ and the initial and desired final states all zero except $\psi(t_f) = 10$ crad, find and plot the optimal state and controls histories for a coordinated turn (a_y small). Note that $D \neq 0$ so the performance index contains cross-product terms in the states and the controls.

5.2.19 Landing Flare of a 747 A/C: A 747 is approaching a runway along a glide slope of 2.5 deg in equilibrium flight at 221 ft/sec. At an altitude of 58 ft it begins its landing flare; the desired vertical velocity at impact is -2 ft/sec with a forward velocity of 206 ft/sec (i.e., losing 15 ft/sec during the flare) at time $t_f = 8$ sec from the beginning of the flare. The perturbation longitudinal dynamics are given in Ref. HJ and converted here to the form (5.2) with

$$A = \begin{bmatrix} -.021 & .122 & 0 & -.322 & 0 & 1 \\ -.209 & -.530 & 2.21 & 0 & 0 & -.044 \\ .017 & -.164 & -.412 & 0 & 0 & .544 \\ 0 & 0 & 1 & 0 & 0 & 0 \\ 0 & -1 & 0 & 2.21 & 0 & 0 \\ 0 & 0 & 0 & 0 & 0 & -.25 \end{bmatrix}, \quad B = \begin{bmatrix} .010 & 0 \\ -.064 & 0 \\ -.378 & 0 \\ 0 & 0 \\ 0 & 0 \\ 0 & 0 \\ 0 & .25 \end{bmatrix},$$

where the state vector is $[u \ w \ q \ \theta \ h \ \delta T]^T$, and $u =$ perturbation in velocity along the glide slope, $w =$ perturbation in velocity normal to the glide slope, $q =$ pitch rate, $\theta =$ perturbation in pitch angle, $h =$ altitude above the runway, $\delta T =$ perturbation in thrust specific force. The control vector is $[\delta e \ \delta T_c]^T$, where $\delta e =$ perturbation in elevator angle, $\delta T_c =$ perturbation in commanded thrust specific force (proportional to throttle change).

Using a MISC performance index with terminal penalties only on the deviations of forward velocity, vertical velocity, and altitude, find and plot the optimal state and control histories for the landing flare. Use $R = \text{diag}[1 \ 1/3]$ and note that vertical velocity $= -w + 2.21\theta$.

5.2.20 S-Turn of a 747 A/C (Ref. HB): A 747 on landing approach is a small distance off the centerline of the runway. An experienced pilot will make a coordinated S-turn (a coordinated turn is one with small lateral specific force a_y) to line up with the runway, and do it so gracefully that the passengers will hardly be aware of the maneuver. Well-designed LQR histories are close to the control histories used by experienced pilots, so here we synthesize time-varying control logic for this maneuver using two controls, aileron and rudder.

The lateral EOM for a 747 in landing configuration are (Ref. HJ) $\dot{x} = Ax + Bu$, and the controlled output vector is $z = Cx + Du$, where $x = [v \ r \ p \ \phi \ \psi \ y]^T$, $u = [\delta a \ \delta r]^T$,

v is the lateral velocity, r is the yaw rate, p is the roll rate, ϕ is the roll angle, ψ is the yaw angle, y is the lateral distance from the runway centerline, δa is the aileron angle, δr is the rudder angle, and $a_y = Cx + Du$ is the lateral specific force. In units of ft, sec, and crad, the system matrices are

$$A = \begin{bmatrix} -.089 & -2.19 & 0 & .319 & 0 & 0 \\ .076 & -.217 & -.166 & 0 & 0 & 0 \\ -.602 & .327 & -.975 & 0 & 0 & 0 \\ 0 & .15 & 1 & 0 & 0 & 0 \\ 0 & 1 & 0 & 0 & 0 & 0 \\ 1 & 0 & 0 & 0 & 2.19 & 0 \end{bmatrix}, \quad B = \begin{bmatrix} 0 & .0327 \\ .0264 & -.151 \\ .227 & .0636 \\ 0 & 0 \\ 0 & 0 \\ 0 & 0 \end{bmatrix},$$

$$C = \begin{bmatrix} -.089 & 0 & 0 & 0 & 0 & 0 \end{bmatrix}, \quad D = \begin{bmatrix} 0 & .0327 \end{bmatrix}$$

(a) Using the performance index

$$J = \int_0^{t_f} (Q a_{ay}^2 + \delta_a^2 + \delta_r^2) dt ,$$

with $Q_{ay} = 10$, $t_f = 10\,\text{sec}$, and the initial and final states all zero except $y(t_f) = 10\,\text{ft}$, verify Figures 5.23 and 5.24 for the S-turn maneuver. Note that the PI contains cross-product terms in the states and the controls since $D \neq 0$.

(b) Use the same performance index as in (a) but use rudder only with $t_f = 12\,\text{sec}$. Compare the resulting maneuver with the one in (a).

5.2.21 Helicopter Acceleration from Hover (Four Controls): Synthesize control logic to move a helicopter from hover to a slow forward velocity with no change in vertical velocity, lateral velocity, or yaw. Since the longitudinal and lateral motions are significantly coupled, this requires skillful coordination of all *four controls,* the collective stick δ_c, the (longitudinal, lateral) cyclic stick (δ_e, δ_a), and the tail rotor collective stick δ_r. We take the state vector to be $[u\ w\ q\ \theta\ v\ r\ p\ \phi\ \psi]^T$ and the control vector to be $[\delta_c\ \delta_e\ \delta_a\ \delta_r]^T$, where (u, v, w) is the velocity, (p, q, r) is the angular velocity, and (ϕ, θ, ψ) are the attitude angles. The EOM (Ref. He) are (5.2) with

$$A = \begin{bmatrix} -.0257 & .0113 & .013 & -.3216 & .0004 & -.0006 & -.0081 & 0 & 0 \\ -.0422 & -.3404 & .0001 & -.0093 & -.044 & .0147 & .0005 & .0171 & 0 \\ 1.26 & -.6 & -1.7645 & 0 & -.26 & .0719 & .3763 & 0 & 0 \\ 0 & 0 & .9986 & 0 & 0 & .0532 & 0 & 0 & 0 \\ .0158 & -.0194 & -.0084 & 0 & -.0435 & .0034 & -.0134 & .3216 & 0 \\ -2.62 & 3.1 & -.1724 & 0 & -.170 & -.8645 & -1.075 & 0 & 0 \\ .03 & -.19 & -1.136 & 0 & -4.620 & -.2873 & -4.92 & 0 & 0 \\ 0 & 0 & -.0015 & 0 & 0 & .0289 & 1 & 0 & 0 \\ 0 & 0 & 0 & 0 & 0 & 1 & 0 & 0 & 0 \end{bmatrix},$$

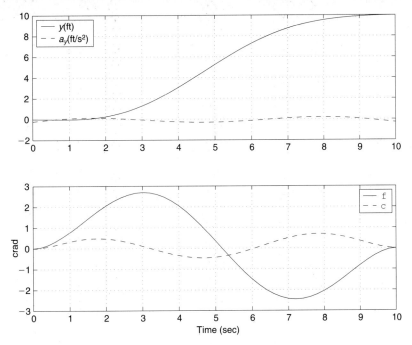

FIGURE 5.23   747 S-Turn; Lateral Distance/Specific Force and Roll/Yaw Angles

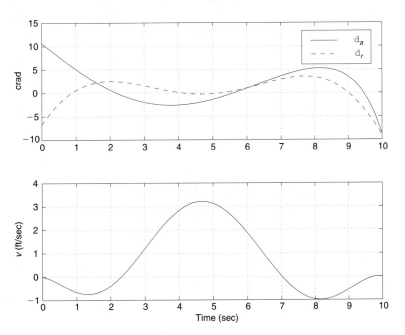

FIGURE 5.24   747 S-Turn; Aileron/Rudder Angles and Sideslip Velocity

$$B = \begin{bmatrix} .0216 & .086 & -.0028 & -.003 \\ -.7343 & -.0016 & .0011 & -.003 \\ -.785 & -7.408 & .35 & -.096 \\ 0 & 0 & 0 & 0 \\ .0057 & -.0038 & .0514 & .153 \\ 9.507 & .493 & 1.982 & -25.68 \\ 1.206 & 1.874 & 12.79 & -.781 \\ 0 & 0 & 0 & 0 \\ 0 & 0 & 0 & 0 \end{bmatrix}$$

where the units are ft, sec, crad for the states, and deci-inches for the controls.

Find the state and control histories to go from equilibrium at hover (all states zero) to equilibrium at $u = 10$ (all other states zero) in 3 sec ($t_f = 3$), putting an in-flight penalty on ψ so that it never exceeds 3 crad in magnitude. Compare your solution with Figures 5.25 and 5.26.

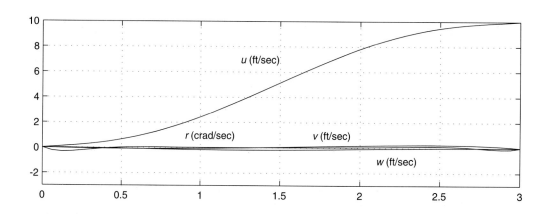

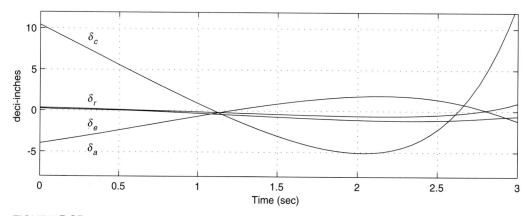

FIGURE 5.25   Helicopter Acceleration from Hover—Velocity, Yaw Rate, and Control Histories

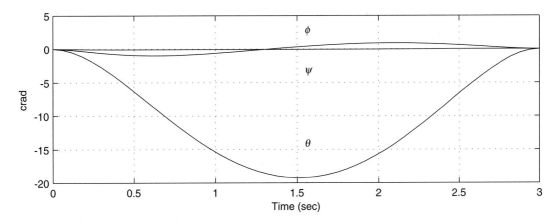

FIGURE 5.26 Helicopter Acceleration from Hover—Attitude Angles

5.2.22 Landing Flare of a STOL A/C (Two Controls): A STOL (Short Take-Off and Landing) A/C is approaching a runway along a glide slope of 7.0 deg (steep compared to conventional A/C) in equilibrium flight. At an altitude of 68 ft it begins its landing flare; the desired flight path angle at touchdown is -1 deg with no change in forward velocity during the flare, at time $t_f = 12$ sec from the beginning of the flare. The longitudinal dynamics are given in Ref. TB:

$$A = \begin{bmatrix} -.0397 & -.280 & -.282 & 0 & 0 \\ .135 & -.538 & .538 & .0434 & 0 \\ 0 & 0 & 0 & 1 & 0 \\ .0207 & .441 & -.441 & -1.41 & 0 \\ -.017 & 1.92 & 0 & 0 & 0 \end{bmatrix}, \quad B = \begin{bmatrix} -.0052 & -.102 \\ .031 & .037 \\ 0 & 0 \\ -1.46 & -.066 \\ 0 & 0 \end{bmatrix},$$

where the state vector is $[u \; \gamma \; \theta \; q \; h]^T$, and $u =$ perturbation in velocity along the glide slope, $\gamma =$ flight path angle (deg), $\theta =$ pitch angle (deg), $q =$ pitch rate (deg/sec), $h =$ altitude (ft) above the runway. The control vector is $[\delta e \; \delta n]^T$, where $\delta e =$ perturbation in elevator angle (deg), $\delta n =$ perturbation in nozzle angle (deg) of the deflected thrust. Initial $\theta = -4$ deg.

Using a MISC performance index with terminal penalties on the forward velocity, flight path angle, and altitude, find and plot the optimal state and control histories for the landing flare. Use $R = \text{diag}[1 \; 1/36]$.

5.2.23 Bicycle Turn - a Nonminimum Phase Problem: An approximate model of the lateral dynamics of a bicycle is

$$\dot{\phi} = p, \quad \dot{p} = a(\phi/c - \delta - eu), \quad \dot{\psi} = \delta, \quad \dot{\delta} = u,$$

where $\phi =$ roll angle, $p =$ roll rate, $\psi =$ direction of bike centerline with respect to an inertially fixed direction, $\delta =$ turn angle of the front wheel, and $u =$ turn rate of the front wheel. Time is measured in units of ℓ/V where ℓ = the wheelbase (distance between the ground contact points of the tires), V = velocity of the bike, p is in units of V/ℓ and the parameters are $a = m\ell h/I$, $e = b/\ell$, $c = V^2/g\ell$ where m = mass of the bike plus rider, g = force per unit mass due to gravity, h = height above the ground of the center of mass of the bike plus rider, I = roll moment-of-inertia of the bike plus rider about an axis through the ground contact points of the tires, b = horizontal distance from the center of mass to the ground contact point of the rear tire. For a turn of the bike centerline through an angle ψ_f in time t_f we take as the PI

$$J = \frac{1}{2} e_f^T Q_f e_f + \frac{1}{2} \int_0^{t_f} u^2 dt, \quad e_f = \psi(t_f) - \psi_f .$$

Calculate the optimal turn history for a bicycle plus rider where V = 12 mi/hr, ℓ = 40 in, h = 36 in, b = 20 in, $I = 4mh^2/3$, g = 32.2 ft/sec^2, ψ_f = 90 deg, and t_f = 25 sec. Plot ϕ, ψ, δ vs. time and note the nonminimum phase effect that $\psi(t)$ starts in the *opposite direction of the desired turn*. You can verify this by riding your bike on the edge of a sidewalk curb and trying to turn back onto the sidewalk; instead, you will go off the curb onto the street. The effect is also very noticeable on motorcycles at high speeds; you must first turn slightly left in order to turn right.

5.3 Discrete Soft Terminal Controllers

The corresponding problem for discrete-step systems is to find the control vector sequence $u(i)$ that minimizes

$$J = \frac{1}{2} e_f^T Q_f e_f + \frac{1}{2} \sum_{i=0}^{N-1} \begin{bmatrix} x^T(i) & u^T(i) \end{bmatrix} \begin{bmatrix} Q_d & N_d \\ N_d^T & R_d \end{bmatrix} \begin{bmatrix} x(i) \\ u(i) \end{bmatrix} \tag{5.53}$$

subject to

$$x(i+1) = \Phi x(i) + \Gamma u(i) , \quad x(0) = x_0 , \quad e_f = M_f x(N) - \psi , \tag{5.54}$$

for $i = 0, \ldots, N-1$, where M_f, Q_f are given and Φ, Γ, Q_d, N_d, R_d are given constant matrices or matrix functions of i.

   The control designer may choose Q_d, N_d, and R_d directly, or find them by converting a continuous quadratic performance index (QPI) using a code like CVRT (see Section 6.3.5). In general, $N_d \neq 0$ even if $N = 0$ in the corresponding continuous QPI. One method of choosing these matrices is as follows: take $N_d = 0$ and

$$Q_f^{-1} = \text{diagonal matrix of max acceptable values of } [e_{fi}(N)]^2 ,$$

$$Q_d^{-1} = \text{diagonal matrix of max acceptable values of } N_a[x_i(i)]^2 ,$$

$$R_d^{-1} = \text{diagonal matrix of max acceptable values of } N_a[u_i(i)]^2 ,$$

where N_a is the desired number of steps for attenuation. This makes the terms of the QPI dimensionless and of the same order of magnitude, i.e., the "costs" of terminal state deviations, in-flight output deviations, and in-flight controls are made roughly equal.

This is a discrete-step LQ Bolza problem with Hamiltonian sequence

$$H(i) = \tfrac{1}{2}[x^T(i)Q_d x(i) + 2x^T(i)N_d u(i) + u^T(i)R_d u(i)] + \lambda^T(i+1)[\Phi x(i) + \Gamma u(i)] . \quad (5.55)$$

Using the approach of Section 3.1, the discrete-step EL equations may be written as

$$\begin{bmatrix} \lambda(i) \\ 0 \\ x(i+1) \end{bmatrix} = \begin{bmatrix} Q_d & N_d & \Phi^T \\ N_d^T & R_d & \Gamma^T \\ \Phi & \Gamma & 0 \end{bmatrix} \begin{bmatrix} x(i) \\ u(i) \\ \lambda(i+1) \end{bmatrix} , \quad (5.56)$$

with the two-point boundary conditions

$$x(0) = x_0 , \quad \lambda(N) = M_f^T Q_f [M_f x(N) - \psi] . \quad (5.57)$$

x_0 and ψ are the only forcing functions; the differential equations are *linear and homogeneous*.

Solution by Transition Matrix

As in the continuous case, the TPBVP can, in principle, be solved by sequencing a transition matrix for the EL equations, either forward or backward, then inverting partitions of this matrix to find the n missing boundary conditions; the coupled equations can then be sequenced to get the solution; *iteration is not required.*

However, for many problems the required sequencing overflows or underflows the computer so that the matrix to be inverted is ill-conditioned (nearly singular). The reason for this can be seen by considering the case where $\Phi, \Gamma, Q_d, N_d, R_d$ are constant matrices, since then the EL system is time-invariant; we show in Section 6.3 that eigenvalues of this system are reciprocal in the unit circle so that it is unstable for sequencing in either direction.

To find a forward transition matrix for the EL equations (5.56), they must be put into the form

$$\begin{bmatrix} x(i+1) \\ \lambda(i+1) \end{bmatrix} = TM \begin{bmatrix} x(i) \\ \lambda(i) \end{bmatrix} . \quad (5.58)$$

After some algebraic manipulation we find

$$TM = \begin{bmatrix} \Phi - G_2 N_1 & -G_2 G_1 \\ -\Phi^{-T}(Q_d - N_2 N_1) & \Phi^{-T}(I + N_2 G_1) \end{bmatrix} . \quad (5.59)$$

where $G_1 \triangleq \Gamma_d^T \Phi^{-T}$, $R_1 \triangleq R_d - G_1 N_d$, $N_1 \triangleq N_d^T - G_1 Q_d$, $G_2 = \Gamma_d R_1^{-1}$, $N_2 = N_d R_1^{-1}$. It follows that

$$\begin{bmatrix} x(N) \\ \lambda(N) \end{bmatrix} = (TM)^N \begin{bmatrix} x(0) \\ \lambda(0) \end{bmatrix} . \quad (5.60)$$

We next partition $(TM)^N$ as in the continuous case (5.25) and use the boundary conditions (5.57) to obtain a relation similar to (5.26):

$$\lambda(0) = [\Phi_4(N) - S_f \Phi_2(N)]^{-1}\{[S_f \Phi_1(N) - \Phi_3(N)]x(0) - a\} , \quad (5.61)$$

TABLE 5.3   MATLAB Code TDLQS

```
function [x,u]=tdlqs(Ph,Ga,Qd,Nd,Rd,x0,Mf,Qf,psi,Ts,Ns)
% Time-varying Discrete LQ regulator w. Soft TC's for time-invariant
% plant x(k+1)=Ph*x(k)+Ga*u(k), ef=Mf*x(N)-psi, 2J=ef'*Qf*ef+sum(0:N)
% [x'(k)*Qd*x(k)+2x'(k)*Nd*u(k)+u'(k)*Rd*u(k)], Ns=number steps, Ts=
% sample time = tf/N; TRANSITION MATRIX solution.
%
% Finds fwd transition matrix Ph1 of disc. EL eqns:
[ns,nc]=size(Ga); G1=Ga'/Ph'; R1=Rd-G1*Nd;
N1=Nd'-G1*Qd; G2=Ga/R1; N2=Nd/R1;
TM=[Ph-G2*N1 -G2*G1; -Ph'\(Qd-N2*N1) Ph'\(eye(ns)+N2*G1)];
Ph1=TM^Ns;
%
% Partitions transition matrix:
n1=[1:ns]; n2=[ns+1:2*ns];
F1=Ph1(n1,n1); F2=Ph1(n1,n2); F3=Ph1(n2,n1); F4=Ph1(n2,n2);
%
% Solves for lambda(0) in the TPBVP;
Sf=(Mf)'*Qf*Mf; a=(Mf)'*Qf*psi; la0=(F4-Sf*F2)\((Sf*F1-F3)*x0-a);
%
% Calculates optimal state and control sequences
x=zeros(ns,Ns+1); la=x; u=zeros(nc,Ns);
yl=[x0; la0]; x(:,1)=yl(n1); la(:,1)=yl(n2);
for i=1:Ns,
 u(:,i)=-R1\(N1*yl(n1)+G1*yl(n2));
 yl=TM*yl; x(:,i+1)=yl(n1); la(:,i+1)=yl(n2);
end;
```

where

$$S_f \triangleq M_f^T Q_f M_f , \quad a \triangleq M_f^T Q_f \psi . \tag{5.62}$$

Having $\lambda(0)$, the EL equations (5.58) can be sequenced forward to find $x(i)$, $\lambda(i)$, and $u(i)$:

$$u(i) = -R_1^{-1}[N_1 x(i) + G_1 \lambda(i)] . \tag{5.63}$$

A MATLAB code TDLQS for *time-invariant systems* is listed in Table 5.3.

Example 5.3.1—Lateral Rendezvous: Transition Matrix Solution

This is a discrete version of Example 5.2.1. The discrete EOM are

$$\begin{bmatrix} y(i+1) \\ v(i+1) \end{bmatrix} = \begin{bmatrix} 1 & T \\ 0 & 1 \end{bmatrix} \begin{bmatrix} y(i) \\ v(i) \end{bmatrix} + \begin{bmatrix} T^2/2 \\ T \end{bmatrix} u(i) ,$$

where T is the sample time.

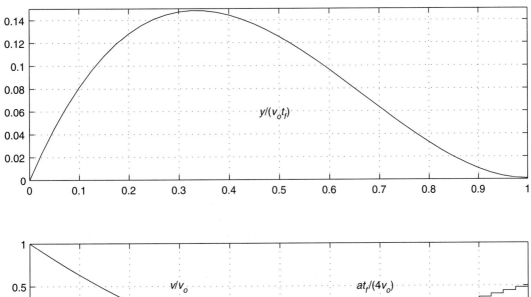

FIGURE 5.27 Discrete Lateral Rendezvous with $N = 40$, ; $s_y = s_v = 30,000$

A MATLAB script using the transition matrix code TDLQS is shown below.

```
% Script e05_3_1.m; lateral rendezvous using TDLQS; x =[y,v]';
% u = a; v in vo, t in tf, a in vo/tf, y in vo*tf;
%
tf=1; N=40; Ts=tf/N; Ph=[1 Ts; 0 1]; Ga=[Ts^2/2 Ts]'; psi=[0 0]';
Qd=zeros(2); Nd=zeros(2,1); Rd=1; x0=[0 1]'; Qf=3e4; Mf=eye(2);
[x,u]=tdlqs(Ph,Ga,Qd,Nd,Rd,x0,Mf,Qf,psi,Ts,N);
```

Figure 5.27 shows the results that are very close to the continuous solution shown in Figure 5.3.

Solution by Riccati Equation

The terminal boundary condition in (5.60) suggests a sweep solution for $\lambda(i)$ of the form

$$\lambda(i) = S(i)x(i) + g(i) , \qquad (5.64)$$

where $S(N) = M_f^T Q_f M_f$, $g(N) = -M_f^T Q_f \psi$ to satisfy (5.60).

To obtain an equation for $S(i)$ we write (5.64) as

$$\lambda(i+1) = S(i+1)x(i+1) + g(i+1) . \tag{5.65}$$

Substituting (5.65) into the first two equations of (5.56) and using the third equation of (5.56) to eliminate $x(i+1)$ gives

$$\begin{bmatrix} \lambda(i) \\ 0 \end{bmatrix} = \begin{bmatrix} Z_{xx}(i) & Z_{xu}(i) \\ Z_{xu}^T(i) & Z_{uu}(i) \end{bmatrix} \begin{bmatrix} x(i) \\ u(i) \end{bmatrix} + \begin{bmatrix} \Phi^T \\ \Gamma^T \end{bmatrix} g(i+1) , \tag{5.66}$$

where

$$\begin{bmatrix} Z_{xx}(i) & Z_{xu}(i) \\ Z_{xu}^T(i) & Z_{uu}(i) \end{bmatrix} \triangleq \begin{bmatrix} Q_d & N_d \\ N_d^T & R_d \end{bmatrix} + \begin{bmatrix} \Phi^T \\ \Gamma^T \end{bmatrix} S(i+1) \begin{bmatrix} \Phi & \Gamma \end{bmatrix} . \tag{5.67}$$

The second equation of (5.65) yields a feedforward plus a linear state feedback for $u(i)$:

$$u(i) = u_F(i) - K(i)x(i) , \quad u_F(i) \triangleq -Z_{uu}^{-1}(i)\Gamma^T g(i+1) , \quad K(i) \triangleq Z_{uu}^{-1}(i)Z_{xu}^T(i) . \tag{5.68}$$

Using (5.68) in the first equation of (5.56) gives:

$$\lambda(i) = [Z_{xx}(i) - Z_{xu}(i)Z_{uu}^{-1}(i)Z_{xu}^T(i)]x(i) + [\Phi - \Gamma K(i)]^T g(i+1) . \tag{5.69}$$

Finally, comparing (5.69) to (5.64) gives recursion relations for $S(i)$ and $g(i)$:

$$S(i) = Z_{xx}(i) - Z_{xu}(i)Z_{uu}^{-1}(i)Z_{xu}^T(i) , \quad g(i) = [\Phi - \Gamma K(i)]^T g(i+1) . \tag{5.70}$$

Since (5.70) and S_f are symmetric, it follows that $S(i)$ is symmetric.

As in the continuous case, this may be interpreted as sweeping the coefficient matrix of the terminal boundary condition of (5.57) backward, finding an equivalent matrix $S(i)$ relating $x(i)$ and $\lambda(i)$ at earlier steps. We also sweep $-M_f^T Q_f \psi$ backward, finding an equivalent vector $g(i)$ at earlier steps. If we sweep all the way back to $i = 0$, we can find $\lambda(0)$ by using (5.64) and the initial condition of (5.57). We could then sequence the EL equations forward to find the solution. However, if we store $K(i)$ and $u_F(i)$ from (5.68) during the backward sequencing, we have $u(i)$ as a *time-varying feedforward plus a linear state feedback* from (5.68). These results closely parallel the continuous results of Section 5.2.

As in the continuous case for $\psi = 0$, it can be shown that

$$J_{min} = \tfrac{1}{2} x_o^T S(0) x_o . \tag{5.71}$$

Table 5.4 is a MATLAB code TDLQSR for a TI plant with vector $u(i)$, that does the backward sequencing and then simulates the system (forward) for a specified initial condition.

TABLE 5.4   MATLAB Code TDLQSR

```
function [x,u,K,uf]=tdlqsr(Ad,Bd,Qd,Nd,Rd,x0,Mf,Qf,psi,Ts,Ns)
% Time-varying Discrete LQ controller with Soft TC's, Riccati solution;
% TI plant; x(k+1)=Ad*x(k)+Bd*u(k), x(0)=x0, ef=Qf*Mf*x(N)-psi;
% 2J=ef'*Qf*ef+sum(0:N)[x'(k)*Qd*x(k)+2x'(k)*Nd*u(k)+u'(k)*Rd*u(k)];
% Ns=number of steps; Ts=tf/Ns; computes bkwd Riccati matrix S(i) & g(i),
% stores fdbk gain vector K(i) & uf(i); simulates closed-loop system.
%
% Sequences S(i) & g(i) backward :
[ns,nc]=size(Bd); S=Mf'*Qf*Mf;
K=zeros(Ns,nc*ns); uf=zeros(nc,Ns); g=-Mf'*Qf*psi;
for i=Ns:-1:1,
 Zxx=Qd+Ad'*S*Ad; Zxu=Nd+Ad'*S*Bd; Zuu=Rd+Bd'*S*Bd;
 K1=Zuu\Zxu'; uf(:,i)=-Zuu\(Bd'*g);
 S=Zxx-Zxu*K1; g=(Ad'-K1'*Bd')*g;
 for j=1:nc, K(i,[1+(j-1)*ns:j*ns])=K1(j,:); end;
end;
%
% Sequences closed-loop system forward:
x(:,1)=x0; for i=1:Ns, for j=1:nc,
 u(j,i)=uf(j,i)-K(i,[1+(j-1)*ns:j*ns])*x(:,i); end;
 x(:,i+1)=Ad*x(:,i)+Bd*u(:,i);
end;
```

Example 5.3.2—Lateral Rendezvous; Riccati Solution

Example 5.3.1 was repeated with TDLQSR instead of TDLQS and, not surprisingly, the results agreed exactly. Figure 5.28 shows the feedback gains which are quite close to the continuous gains.

Problems

5.3.1 to 5.3.22:   Do discrete-step versions of the continuous soft terminal constraint problems 5.2.1 through 5.2.22. Use the code CVRT to convert the time-invariant continuous plant and QPI to a discrete-step time-invariant plant with an appropriate time step T_s.

5.4 Continuous Hard Terminal Controllers

In Section 5.2 the terminal constraints were "soft" in the sense that they were handled by quadratic penalty functions. While this leaves some small terminal errors, it also limits the control magnitudes near the terminal time; the feedback gains reach a peak just before the terminal time and decrease to zero at the terminal time. Therefore sensor noise is less likely to produce control saturation.

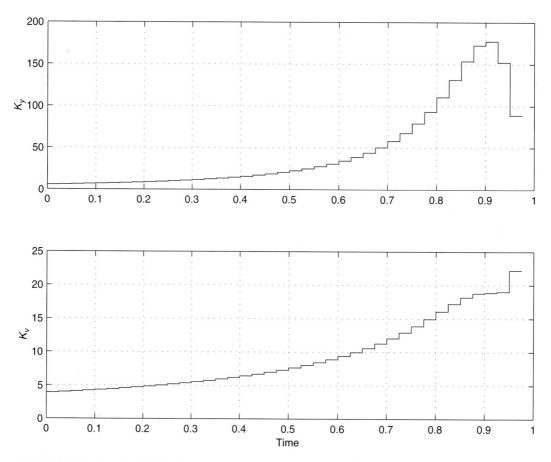

FIGURE 5.28   Feedback Gains for Discrete Lateral Rendezvous with Soft Terminal Constraints

In this section we consider the case of "hard" terminal constraints, i.e., *zero terminal errors*. Since the feedback gains tend to infinity at the terminal time we shall go open-loop for a short interval Δt_b just before the final time (called the "blind time" in missile engineering). Note that optimal control is used during the blind time based on the state at $t_f - \Delta t_b$, so one might call this "one-sample" feedback using a transition matrix solution.

We wish to design a terminal controller to bring certain linear combinations of $x(t_f)$ to specified numerical values ψ:

$$M_f x(t_f) = \psi \ , \tag{5.72}$$

This can be done by extending the problem of Section 5.2. We wish to find $u(t)$ to minimize the integral in (5.1) with the added constraints (5.72). This is a special case of the problems treated in Sections 3.3 and 3.4. The constraints (5.72) may be adjoined to the integral in

(5.1) by a Lagrange multiplier vector v:

$$\bar{J} = J + v^T[M_f x(t_f) - \psi] . \tag{5.73}$$

The EL equations are the same [see (5.7) to (5.8)] except for the terminal condition on $\lambda(t_f)$ in (5.9), which becomes

$$\lambda(t_f) = S_f x(t_f) + M_f^T v . \tag{5.74}$$

The problem is to find $u(t)$ and v to satisfy the EL equations and the constraints (5.72). Again this linear TPBVP problem can, in principle, be solved by the use of a transition matrix. As we have seen before, this is sometimes impossible because of overflow or underflow problems (see Problem 5.4.3).

Solution by Transition Matrix

Let

$$\begin{bmatrix} x(t) \\ \lambda(t) \end{bmatrix} = \begin{bmatrix} \Phi_1(t) & \Phi_2(t) \\ \Phi_3(t) & \Phi_4(t) \end{bmatrix} \begin{bmatrix} x(t_f) \\ \lambda(t_f) \end{bmatrix} . \tag{5.75}$$

The following matrix version of the EL equations is then integrated backward:

$$\begin{bmatrix} \dot{\Phi}_1 & \dot{\Phi}_2 \\ \dot{\Phi}_3 & \dot{\Phi}_4 \end{bmatrix} = \begin{bmatrix} A - BR^{-1}N^T & -BR^{-1}B^T \\ -Q + NR^{-1}N^T & -A^T + NR^{-1}B^T \end{bmatrix} \begin{bmatrix} \Phi_1 & \Phi_2 \\ \Phi_3 & \Phi_4 \end{bmatrix} , \tag{5.76}$$

with boundary conditions

$$\Phi_1(t_f) = \Phi_4(t_f) = I , \quad \Phi_2(t_f) = \Phi_3(t_f) = 0 . \tag{5.77}$$

At $t = t_o$ the following linear equations determine $x(t_f)$ and v:

$$x(t_o) = \Phi_1(t_o)x(t_f) + \Phi_2(t_o)[S_f x(t_f) + M_f^T v] , \tag{5.78}$$

$$\psi = M_f x(t_f) , \tag{5.79}$$

which, in turn, determine $\lambda(t_o)$

$$\lambda(t_o) = \Phi_3(t_o)x(t_f) + \Phi_4(t_o)[S_f x(t_f) + M_f^T v] . \tag{5.80}$$

Having $\lambda(t_o)$, the EL equations (5.7) can be integrated forward to find $x(t)$, $\lambda(t)$, $u(t)$. A MATLAB code TLQH for time-invariant systems is listed in Table 5.5.

Example 5.4.1—Lateral Intercept: Transition Matrix Solution

This is the same problem considered in Example 5.2.1 except that it has hard instead of soft terminal constraints. A MATLAB script is listed below that uses TLQH to solve the problem.

```
% Script e05_4_1.m; Lateral intercept and rendezvous with hard terminal
% constraints using TLQH; t in tf, v in vo, y in vo*tf, a in vo/tf.
%
A=[0 1; 0 0]; B=[0 1]'; Q=zeros(2); N=[0 0]'; R=1;
tf=1; x0=[0 1]'; Sf=zeros(2); Mf=[1 0]; psi=0; Ns=40;
[x,u,t]=tlqh(A,B,Q,N,R,tf,x0,Mf,psi,Ns);
```

TABLE 5.5   MATLAB Code TLQH

```
function [x,u,t]=tlqh(A,B,Q,N,R,tf,x0,Sf,Mf,psi,Ns)
% Time-varying LQ controller with Hard terminal constraints, TI plant,
% transition matrix method, xdot=Ax+Bu, x(0)=x0, psi=Mf*x(tf),
% 2J=x'(tf)*Sf*x(tf)+int(0:tf)(x'Q*x+2x'*N*u+u'R*u)dt, Ns = number of
% steps in t; (x,u) are (state,control) histories with points at t.
%
% Bkwd trans. matrix of EL eqns, t=tf to t=0:
Qb=Q-N*(R\N'); Ab=A-B*(R\N'); H=[Ab -B*(R\B'); -Qb -Ab'];
Phi=expm(-H*tf);                           % Bkwd trans. matrix
%
% Partition transition matrix:
n1=1:ns; n2=ns+1:2*ns; F1=Phi(n1,n1); F2=Phi(n1,n2);
F3=Phi(n2,n1); F4=Phi(n2,n2);
%
% Solve for xf, nu and la0 in the TPBVP;
[nt,ns]=size(Mf); Z=[F1+F2*Sf F2*Mf'; Mf zeros(nt)]; b=Z\[x0; psi];
xf=b(n1); nu=b([ns+1:ns+nt]); la0=F3*xf+F4*(Sf*xf+Mf'*nu);
%
% Calculate optimal state and control histories:
xa0=[x0; la0]; t=tf*[0:1/Ns:1];
for i=1:Ns+1, xa(:,i)=expm(H*t(i))*xa0;
 x(:,i)=xa(n1,i); u(:,i)=-R\(B'*xa(n2,i)+N'*x(:,i));
end
```

Solution by Riccati Equation

The zero terminal error problem may also be solved by using a backward Riccati equation (sweep method), which has significant advantages for numerical solution over the transition matrix method. This requires an extension of the sweep method, as presented in Section 5.2.

Since the problem is linear in $x(t_f)$ and v, we extend (5.28) assuming a sweep solution of the form

$$\begin{bmatrix} \lambda(t) \\ \psi \end{bmatrix} = \begin{bmatrix} S(t) & M^T(t) \\ M(t) & -Q(t) \end{bmatrix} \begin{bmatrix} x(t) \\ v \end{bmatrix} , \tag{5.81}$$

where, to satisfy (5.72) and (5.70),

$$S(t_f) = S_f , \quad M(t_f) = M_f , \quad Q(t_f) = 0 . \tag{5.82}$$

By differentiating (5.81) with respect to time (note ψ and v are constants) and substituting from the EL equations (5.7) for $\dot{x}$ and $\dot{\lambda}$, it is straightforward to show (as in Section 5.2) that S still satisfies the Riccati equation (5.31), and that

$$\dot{M} = -M[A - BR^{-1}(N^T + B^T S)] , \quad \dot{Q} = -MBR^{-1}B^T M^T . \tag{5.83}$$

Integrating (5.31) and (5.83) backward in time with the boundary conditions (5.82), we can then use (5.81) to determine ν and $\lambda(t_o)$ in terms of x_o and ψ:

$$\nu = \mathcal{Q}^{-1}[M(t_o)x_o - \psi] , \tag{5.84}$$

$$\lambda(t_o) = [S(t_o) + M^T(t_o)\mathcal{Q}^{-1}(t_o)M(t_o)]x_o - M^T(t_o)\mathcal{Q}^{-1}(t_o)\psi . \tag{5.85}$$

From (5.8) and (5.85) we have the initial control

$$u(t_o) = -R^{-1}\{N^T + B^T[S(t_o) + M^T(t_o)\mathcal{Q}^{-1}(t_o)M(t_o)]\}x_o + R^{-1}B^T M^T(t_o)\mathcal{Q}^{-1}(t_o)\psi . \tag{5.86}$$

However, *any time $t \le t_f$ is a possible initial time on an optimal path* (this is Bellman's Principle of Optimality, see Chapter 7), so we may replace t_o by t, giving us a *time-varying linear feedback/feedforward law*:

$$u(t) = -K(t)x(t) + K_\psi(t)\psi , \tag{5.87}$$

where

$$K(t) \stackrel{\Delta}{=} R^{-1}\{N^T + B^T[S(t) + M^T(t)\mathcal{Q}^{-1}(t)M(t)]\} , \tag{5.88}$$

$$K_\psi(t) \stackrel{\Delta}{=} R^{-1}B^T M^T(t)\mathcal{Q}^{-1}(t) . \tag{5.89}$$

For $\psi = 0$ it is straightforward to show that

$$J_{\min} = \frac{1}{2}x^T[S + M^T\mathcal{Q}^{-1}M]x\bigg|_{t=t_o} . \tag{5.90}$$

A MATLAB code TLQHR for time-independent systems is listed in Table 5.6, along with the subroutines TLQH-RIC and TLQH-ST.

TABLE 5.6   MATLAB Code TLQHR

```
function [x,u,t,t1k,K]=tlqhr(A,B,Q,N,R,tf,x0,Sf,Mf,psi,t1,tol)
% Time-varying LQ ctrl w. Htc's, Riccati soln, TI plant,
% xdot=Ax+Bu, x(0)=x0, Mf*x(tf)=psi, 2J=x'(tf)*Sf*x(tf)+int(0:tf)
% (x'Q*x+2x'*N*u+u'R*u)dt, Ns=number of steps in t; (x,u) are
% (state,control) histories with points at t; optimal open-loop
% control from t1 to tf.
%
[nt,ns]=size(Mf); Qcf=zeros(nt); yf=[forms(Sf); formm(Mf,'c');
forms(Qcf)]; options=odeset('reltol',tol);
[tb,y]=ode23('tlqh_ric',[0 tf],yf,options,A,B,Q,N,R,Mf);
Nk=length(tb); un=ones(Nk,1); tk=tf*un-tb;
[dum,N1]=min(abs(tk-t1*un)); t1=tk(N1); N1k=[N1:Nk];
t1k=tk(N1k);
%
% Gain matrix K(t) and uf(t) from y(t):
[ns,nc]=size(B); K=zeros(Nk-N1+1,ns*nc); uf=zeros(Nk-N1+1,nc);
n1=ns*(ns+1)/2; n2=nt*ns; n3=nt*(nt+1)/2; ys=y(N1k,[1:n1]);
```

(*continued*)

TABLE 5.6   (*continued*)

```
ym=y(N1k,[1+n1:n1+n2]); yq=y(N1k,[1+n1+n2:n1+n2+n3]);
for i=1:Nk-N1+1,
 S=forms(ys(i,:)'); M=formm(ym(i,:)',nt); Qc=forms(yq(i,:)');
 K(i,:)=formm(R\(N'+B'*(S+M'*(Qc\M))),'r');
 uf(i,:)=(R\(B'*M'*(Qc\psi)))';
end
K1=[t1k K]; uf1=[t1k uf]; disp('K(t) and uf(t) computed');
%
% Simulate closed-loop system:
[ta,xa]=ode23('tlqh_st',[0 t1],x0,options,A,B,K1,uf1);
N2=length(ta); ua=zeros(nc,N2);
for i=1:N2, Kt=formm(table1(K1,ta(i)),ns);
 uft=table1(uf1,ta(i)); ua(:,i)=uft'-Kt*xa(i,:)';
end;
Ns=round(200*(tf-t1)/tf);
[xc,uc,tc]=tlqh(A,B,Q,N,R,tf-t1,xa(N2,:)',Sf,Mf,psi,Ns);
N3=length(tc); un=ones(1,N3); tc=tc+t1*un;
x=[xa; xc']; u=[ua uc]; t=[ta; tc'];

function yp=tlqh_ric(t,y,flag,A,B,Q,N,R,Mf)
% Integrates S(t), M(t), & Qc(t) for TLQHR.
%
[nt,ns]=size(Mf); Ab=A-B*(R\N'); Bb=B*(R\B');
Qb=Q-N*(R\N'); n1=ns*(ns+1)/2; ys=y([1:n1]);
S=forms(ys); Sd=-S*Ab-Ab'*S-Qb+S*Bb*S;
n2=nt*ns; M=(formm(y([1+n1:n1+n2]),nt));
Md=-M*(Ab-Bb*S); Qcd=-M*Bb*M';
yp=-[forms(Sd); formm(Md,'c'); forms(Qcd)];

function yp=tlqh_st(t,y,flag,A,B,K1,uf1)
% Forward integration of states for TLQHR.
%
[ns,nc]=size(B); Kt=formm(table1(K1,t),ns);
uft=table1(uf1,t); yp=(A-B*Kt)*y+B*uft';
```

Example 5.4.2—Lateral Intercept: Riccati Solution

This is the same problem considered in Example 5.4.1 except that it is solved here with
TLQHR. A MATLAB script is listed below.

```
% Script e05_4_2.m; Lateral intercept using TLQHR;
% t in tf, v in vo, y in vo*tf, a in vo/tf.
%
A=[0 1; 0 0]; B=[0 1]'; Q=zeros(2); N=[0 0]'; R=1;
tf=1; x0=[0 1]'; Sf=zeros(2); Mf=[1 0]; psi=0;
```

```
t1=.85*tf; tol=1e-5;
[x,u,t,t1k,K]=tlqhr(A,B,Q,N,R,tf,x0,Sf,Mf,psi,t1,tol)
```

The solution is very close to the one obtained with TLQH.

MISC and Controllability

A special case of interest is when $S_f = \psi = Q = N = 0$. We are then simply minimizing the integral of a quadratic form in the control:

$$J = \frac{1}{2} \int_{t_o}^{t_f} u^T R u \, dt \; . \tag{5.91}$$

The Riccati equation (5.31) with $S_f = Q = N = 0$ has the simple solution

$$S(t) = 0 \; , \tag{5.92}$$

while the first equation of (5.83) simplifies to

$$\dot{M} = -MA \; . \tag{5.93}$$

The feedback gain (5.88) can then be written as

$$K(t) = R^{-1} B^T M^T(t) Q^{-1}(t) M(t) \; . \tag{5.94}$$

From (5.90) it follows that the min value of the performance index is

$$J_{\min} = \frac{1}{2} x^T M^T Q^{-1} M x \bigg|_{t=t_o} \; . \tag{5.95}$$

Clearly, if $Q(t_o)$ is singular, $J_{\min} \to \infty$ and one or more of the terminal constraints is *uncontrollable*. Thus the rank of $Q(t)$ is a measure of how many of the terminal constraints can be controlled at t_f starting at t.

Example 5.4.3—First-Order System: MISC Analytical Solution

Consider the unstable first-order system

$$\dot{x} = x + u \; , \quad x(0) = x_o \; , \quad x(t_f) = 0 \; ,$$

with the integral square performance index

$$J = \frac{1}{2} \int_0^{t_f} u^2 dt \; ,$$

where x_o and t_f are given. The solution to (5.83) with $A = B = R = M_f = 1$ and boundary conditions (5.82) is easily found

$$M = \exp(t_f - t) \; , \quad Q = \tfrac{1}{2} \{ \exp[2(t_f - t)] - 1 \} \; .$$

Finally, the feedback gain, from (5.88) is:

$$K = \frac{M^2}{Q} \equiv \frac{2}{1 - \exp[-2(t_t - t)]} \ .$$

Substituting the feedback law into the original system gives

$$\dot{x} = -\frac{x}{\tanh(t_f - t)} \ .$$

Thus, for $t_f - t \gg 1$, $\dot{x} = -x$ while the coefficient of $-x$ tends to infinity as $t \to t_f$.

Problems

5.4.1 First-Order System—MISC Analytical Solution: Consider the first-order system and quadratic performance index

$$\dot{x} = -ax + u \ , \quad J = \frac{1}{2} \int_0^{t_f} [u(t)]^2 dt \ ,$$

where all quantities are scalar and $x(0) = x_0$, $x(t_f) = x_f$.

(a) Show that the state and control histories that minimize J are

$$x(t, t_f) = \frac{\sinh(aT)x_0 + \sinh(at)x_f}{\sinh(at_f)} \ ,$$

$$u(t, t_f) = -a\frac{\exp(-aT)x_0 - \exp(at)x_f}{\sinh(at_f)} \ ,$$

where $T \overset{\Delta}{=} t_f - t$.

(b) Show that the optimal closed-loop control law is

$$u(t) = -a\frac{\exp(-aT)x(t) - x_f}{\sinh(aT)} \ .$$

(c) Show that these results can also be obtained as a special case of Problem 5.2.4.

5.4.2 A General Linear TPBVP: Consider the linear TPBVP

$$\dot{x} = F(t)x + w(t) \ , \quad Ax(t_0) = a \ , \quad Bx(t_f) = b \ .$$

where x is an n vector, a is an $(n - k)$ vector, b is a k vector, and $F(t)$, $w(t)$, A, B, a, and b are given.

(a) Show that a forward sweep solution may be obtained by letting

$$Ax(t) = S(t)Bx(t) + m(t) ,$$

where

$$\dot{S} = C_1 S - SC_4 - SC_3 S + C_2 , \quad S(t_0) = 0 ,$$
$$\dot{m} = (C_1 - SC_3)m + (A - SB)w , \quad m(t_0) = a ,$$

and

$$\begin{bmatrix} C_1 & C_2 \\ C_3 & C_4 \end{bmatrix} = \begin{bmatrix} A \\ B \end{bmatrix} F(t) \begin{bmatrix} A \\ B \end{bmatrix}^{-1} .$$

Note that S is an $(n - k) \times k$ matrix, while m is an $(n - k)$ vector. *Hint*: Change state vector to y and partition y into $[y_1; y_2]$ where $y_1 \triangleq Ax$, $y_2 \triangleq Bx$.

Integrating the differential equations for S and m *forward* generates a family of solutions that satisfy the initial conditions. At $t = t_f$, we have

$$Ax(t_f) = S(t_f)b + m(t_f) , \quad Bx(t_f) = b ,$$

which are n equations for $x(t_f)$. Having $x(t_f)$, the original equations may be integrated *backward* to obtain the desired solution. However, for a stable backward integration, integrate only $y_2 \equiv Bx$, eliminating $y_1 = Ax$ by the feedback $y_1 = Sy_2 + m$; this requires storing $S(t)$ and $m(t)$ during the forward integration.

(b) Show that a similar procedure could be used for a *backward sweep* solution by letting $Bx(t) = Q(t)Ax(t) + n(t)$.

Note: This procedure was suggested by G. B. Rybicki and P. D. Usher at the Smithsonian Astrophysical Lab, Harvard University, in the 1960s for solving a TPBVP to estimate the radial temperature distribution inside a star. Thanks to Prof. Max Krook, who made the author aware of their work. R. Bellman inspired this idea with his concept of "invariant embedding," which we call the sweep method.

5.4.3 to 5.4.22: Do problems 5.2.3 through 5.2.22 with hard instead of soft terminal constraints.

5.5 Discrete Hard Terminal Controllers

This case is the same as Section 5.3 except that we replace the soft terminal constraints by the hard terminal constraints

$$M_f x(N) = \psi , \tag{5.96}$$

This is a special case of Sections 3.1 and 3.2. These constraints may be adjoined to (5.53) by a Lagrange multiplier vector v:

$$\bar{J} = J + v^T [M_f x(N) - \psi] , \tag{5.97}$$

The EL equations are the same as for the soft terminal controller (5.58) with boundary conditions (5.96), $x(0) = x_0$, and

$$\lambda(N) = S_f x(N) + M_f^T v . \tag{5.98}$$

The resulting linear TPBVP can, in principle, be solved by the use of transition matrices. As before, for problems with large N this may fail because of numerical overflow or underflow.

Solution by Transition Matrix

The derivation is the same as the derivation in Section 5.3 except for the boundary conditions (5.96) and (5.98), which become, in terms of the symbols used in Section 5.3,

$$\begin{bmatrix} M_f \Phi_2 & 0 \\ \Phi_4 - S_f \Phi_2 & -M_f^T \end{bmatrix} \begin{bmatrix} \lambda(0) \\ v \end{bmatrix} = \begin{bmatrix} -M_f \Phi_1 & I \\ -\Phi_3 + S_f \Phi_1 & 0 \end{bmatrix} \begin{bmatrix} x_0 \\ \psi \end{bmatrix} . \tag{5.99}$$

This set of linear equations is solved for $\lambda(0)$ and v. Having $\lambda(0)$ the EL equations can be sequenced forward to find the optimal states and controls as in Section 5.3. Table 5.7 is a MATLAB code for this case.

Example 5.5.1—Cart with a Pendulum: Transition Matrix Solution

The continuous version of this problem is described in Problem 5.2.9. Here we solve the discrete version using TDLQH. A MATLAB script for solving it is listed below for the case $\epsilon = .5$, $t_f = 2\pi$, $N = 32$.

```
% Script e05_5_1.m; cart w. pendulum using TDLQH;
% M = cart mass, m = pendulum mass, l = pendulum length, u = force on
% cart, y=displacement of cart, th=angle pendulum; x=[y ydot th thdot]';
% t in sqrt[Ml/(M+m)g]/2*pi, u in (M+m)g, y in l, ep = m/(m+M); xdot=
% Ax+Bu; x(0)=xo; psi=Mf*x(N).
%
tf=2*pi; ep=.5; x0=[0 0 0 0]'; N=32; Ts=tf/N; psi=[1 0 0 0]';
A=[0 1 0 0; 0 0 ep 0; 0 0 0 1; 0 0 -1 0]; B=[0 1 0 -1]';
[Ph,Ga]=c2d(A,B,Ts);                    % Converts cont. to discrete
Qd=zeros(4); Nd=zeros(4,1); Rd=1; Sf=zeros(4); Mf=eye(4);
[x,u]=tdlqh(Ph,Ga,Qd,Nd,Rd,x0,Sf,Mf,psi,Ts,N);
```

The solution is very close to the continuous solution shown in Fig. 5.6 except for the 'staircase' control history due to the zero-order-hold.

TABLE 5.7   MATLAB Code TDLQH

```
function [x,u]=tdlqh(Ph,Ga,Qd,Nd,Rd,x0,Sf,Mf,psi,Ts,N)
% Discrete LQ controller w. HTC's - trans. matrix soln; TI plant,
    Mf*x(N)=psi,
% x(i+1)=Ph*x(i) +Ga*u(i); 2J=x'(N)*Sf*x(N)+sum(0:N)[x'(i)Qd*x(i)
% +2x'(i)*Nd*u(i)+u'(i)Rd*u(i)]; N=number of steps; Ts=sample time=tf/N;
% simulates w. initial state x0 & final psi; finds fwd transition
    matrix Ph1
% of EL eqns.
%
[nt,ns]=size(Mf); G1=Ga'/Ph'; R1=Rd-G1*Nd; N1=Nd'-G1*Qd; G2=Ga/R1;
    N2=Nd/R1;
TM=[Ph-G2*N1 -G2*G1; -Ph'\(Qd-N2*N1) Ph'\(eye(ns)+N2*G1)]; Ph1=TM^N;
%
% Partitions transition matrix:
n1=[1:ns]; n2=[ns+1:2*ns]; F1=Ph1(n1,n1); F2=Ph1(n1,n2); F3=Ph1(n2,n1);
F4=Ph1(n2,n2);
%
% Solves for lambda(0) and nu in the TPBVP;
M1=[Mf*F2 zeros(nt); F4-Sf*F2 -Mf']; M2=[-Mf*F1 eye(nt)
Sf*F1-F3 zeros(ns,nt)]; ln=M1\(M2*[x0;psi]); la0=ln(n1); nu=ln(ns+1:ns+nt);
%
% Calculate optimal state and control sequences
x=zeros(ns,N+1); la=x; [ns,nc]=size(Ga); u=zeros(nc,N); yl=[x0; la0];
x(:,1)=yl(n1); la(:,1)=yl(n2);
for i=1:N,
yl=TM*yl; x(:,i+1)=yl(n1); la(:,i+1)=yl(n2);
end;
for i=1:N,
u(:,i)=-Rd\(Nd'*x(:,i)+Ga'*la(:,i+1));
end;
```

Solution by Riccati Equation

As in the soft terminal constraint case (Section 5.3), the problem can be solved by the sweep method, which has significant advantages for numerical accuracy over the transition matrix method. This requires an extension of the sweep method, as presented in Section 5.3.

Since the problem is linear in $x(N)$ and v, we assume a sweep solution of the form

$$\begin{bmatrix} \lambda(i) \\ \psi \end{bmatrix} = \begin{bmatrix} S(i) & M^T(i) \\ M(i) & -Q(i) \end{bmatrix} \begin{bmatrix} x(i) \\ v \end{bmatrix} , \tag{5.100}$$

where, to satisfy (5.96) and (5.98),

$$S(N) = S_f , \quad M(N) = M_f , \quad Q(N) = 0 . \tag{5.101}$$

By considering (5.100) with i replaced by $i + 1$, and using the EL equations (5.58) to write $\lambda(i + 1)$ and $x(i + 1)$ in terms of $\lambda(i)$ and $x(i)$, we obtain

$$u(i) = -Z_{uu}^{-1}[Z_{xu}^T x(i) + \Gamma^T M^T(i + 1)v] , \tag{5.102}$$

and

$$\begin{bmatrix} \lambda(i) \\ \psi \end{bmatrix} = \begin{bmatrix} Z_{xx} - Z_{xu} Z_{uu}^{-1} Z_{xu}^T & (\Phi - \Gamma Z_{uu}^{-1} Z_{xu}^T)^T M^T(i + 1) \\ M(i + 1)(\Phi - \Gamma Z_{uu}^{-1} Z_{xu}^T) & -Q(i + 1) - M(i + 1)\Gamma Z_{uu}^{-1} \Gamma^T M^T(i + 1) \end{bmatrix} \begin{bmatrix} x(i) \\ v \end{bmatrix} , \tag{5.103}$$

where

$$\begin{bmatrix} Z_{xx} & Z_{xu} \\ Z_{xu}^T & Z_{uu} \end{bmatrix} \triangleq \begin{bmatrix} Q_d & N_d \\ N_d^T & R_d \end{bmatrix} + \begin{bmatrix} \Phi^T \\ \Gamma^T \end{bmatrix} S(i + 1) \begin{bmatrix} \Phi & \Gamma \end{bmatrix} . \tag{5.104}$$

Comparing the coefficients of (5.103) with those of (5.100) yields recursion relations for $S(i)$, $M(i)$, and $Q(i)$ in terms of $S(i + 1)$, $M(i + 1)$, and $Q(i + 1)$:

$$S(i) = Z_{xx} - Z_{xu} Z_{uu}^{-1} Z_{xu}^T , \tag{5.105}$$

$$M(i) = M(i + 1)(\Phi - \Gamma Z_{uu}^{-1} Z_{xu}^T) , \tag{5.106}$$

$$Q(i) = Q(i + 1) + M(i + 1)\Gamma Z_{uu}^{-1} \Gamma^T M^T(i + 1) . \tag{5.107}$$

Sequencing (5.105) to (5.107) backward with the boundary conditions (5.99), we can then use (5.100) to determine v and hence $\lambda(0)$ in terms of x_o:

$$v = Q^{-1}(0)[M(0)x_o - \psi] , \tag{5.108}$$

$$\lambda(0) = \left[S(0) + M^T(0)Q^{-1}(0)M(0)\right] x_o - M^T(0)Q^{-1}(0)\psi . \tag{5.109}$$

Hence, from (5.102) we have the initial control

$$u(0) = -Z_{uu}^{-1}[Z_{xu}^T + \Gamma^T M^T(1)Q^{-1}(0)M(0)]x_o + Z_{uu}^{-1}\Gamma^T M^T(1)Q^{-1}(0)\psi .$$

However, any step $i \leq N - n_f$ is a possible initial step on an optimal path, where $n_f \geq$ number of terminal constraints. Thus we may replace 0 by i, giving a *time-varying linear feedback/feedforward law*:

$$u(i) = -K(i)x(i) + K_\psi(i)\psi , \tag{5.110}$$

where

$$K(i) = Z_{uu}^{-1}[Z_{xu}^T + \Gamma^T M^T(i + 1)Q^{-1}(i)M(i)] , \tag{5.111}$$

$$K_\psi(i) = Z_{uu}^{-1}\Gamma^T M^T(i + 1)Q^{-1} . \tag{5.112}$$

When less than n_t steps remain, where $n_t = $ the number of terminal constraints, the system is uncontrollable with one control. Hence we choose some $n_f \leq n_t$ and use optimal open-loop control for the last n_f steps using the transition matrix method described above. If no disturbances occur during these last n_f steps, this will take the system to the desired final conditions while minimizing the performance index. This is done in the MATLAB code TDLQHR which is listed in Table 5.8. This code, for a time-invariant plant, does the backward sequencing and then simulates the system for specified x_0 and ψ. The subroutine FORMM was listed in Table 5.2.

TABLE 5.8   MATLAB Code TDLQHR

```
function [x,u,K,Kf]=tdlqhr(Ph,Ga,Qd,Nd,Rd,x0,Sf,Mf,psi,Ts,N,nf)
% Discrete LQ controller w. HTC's - Riccati soln; TI plant; x(i+1)=Ph*x(i)
% +Ga*u(i), 2J=x'(N)*Sf*x(N)+sum(0:N)[x'(i)Qd*x(i)+2x'(i)*Nd*u(i)
% +u'(i)*Rd*u(i)], Mf*x(N)=psi; N=number of steps; Ts=tf/N; computes bkwd
% Riccati matrix S(i) & aux. matrices M(i), Qc(i); stores fdbk gain matrix
% K(i), fdfwd gain matrix Kf(i); simulates closed-loop system with open-
% loop control for last nf steps.
%
% Bkwd sequencing of S(i), M(i), Qc(i) matrices:
[ns,nc]=size(Ga); [nt,ns]=size(Mf); S=Sf; Qc=zeros(nt);
K=zeros(N-nf,ns*nc); Kf=zeros(N-nf,nt*nc); M1=Mf;
for i=N:-1:1,
 Zxx=Qd+Ph'*S*Ph; Zxu=Nd+Ph'*S*Ga; Zuu=Rd+Ga'*S*Ga;
 Qc=Qc+M1*Ga*(Zuu\Ga')*M1'; M=M1*(Ph-Ga*(Zuu\Zxu'));
 if i<=N-nf,
  K(i,:)=formm(Zuu\(Zxu'+Ga'*M1'*(Qc\M)),'r');
  Kf(i,:)=formm(Zuu\Ga'*M1'/Qc,'r');
 end;
 S=Zxx-Zxu*(Zuu\Zxu'); M1=M;
end;
%
% Sequences closed-loop system, i=1 to N-nf:
x1=zeros(ns,N-nf+1); x1(:,1)=x0; u1=zeros(nc,N-nf);
for i=1:N-nf,
 K1=formm(K(i,:),ns); Kf1=formm(Kf(i,:),nt);
 u1(:,i)=-K1*x1(:,i)+Kf1*psi;
 x1(:,i+1)=Ph*x1(:,i)+Ga*u1(:,i);
end;
%
% Sequences open-loop system, i=N-nf+1 to N, based on x1(:,N-nf+1):
[x2,u2]=tdlqh(Ph,Ga,Qd,Nd,Rd,x1(:,N-nf+1),Sf,Mf,psi,Ts,nf);
x=[x1 x2(:,[2:nf+1])]; u=[u1 u2];
```

Example 5.5.2—Cart with a Pendulum: Riccati Solution

This the same as Example 5.5.1 except that TDLQHR is used to solve it instead of TDLQH. The solutions using these two very different algorithms agree to four significant figures.

MISC and Controllability

A special case of interest is when $S_f = Q_d = N_d = 0$ in the performance index, so that we are simply minimizing the sum of a quadratic form in the control:

$$J = \frac{1}{2} \sum_{i=0}^{N-1} u^T(i) R_d u(i) . \tag{5.113}$$

The Riccati equation then has the simple solution

$$S(i) = 0 \text{ for all } i , \tag{5.114}$$

while (5.112) simplifies to

$$M(i) = M(i+1)\Phi . \tag{5.115}$$

The feedback gain can then be written as

$$K(i) = R_d^{-1} \Gamma^T M^T(i+1) Q^{-1}(i) M(i) . \tag{5.116}$$

It is straightforward to show that the min value of the performance index is

$$J_{\min} = x_o^T M^T(0) Q^{-1}(0) M(0) x_o . \tag{5.117}$$

If $Q(0)$ is singular, $J_{\min} \to \infty$, which implies that one or more of the terminal constraints is *uncontrollable*. Thus the rank of $Q(i)$ is a direct measure of how many of the terminal constraints can be controlled at $i = N$ starting at i.

Problems

5.5.1 Discrete LQ Hard Terminal Constraints (HTC) as a Parameter Optimization Problem: Consider the min integral-square terminal control problem of this section with a scalar control sequence $u(i)$, namely:

$$\min_{u(i)} J = \frac{1}{2} \sum_{i=0}^{N-1} [u(i)]^2 ,$$

subject to

$$x(i+1) = \Phi x(i) + \Gamma u(i) , \quad x(0) = x_o , \quad M_f x(N) = 0 .$$

(a) Show that this may be written as the following parameter optimization problem:

$$\min_{U} J = \tfrac{1}{2} U^T U \; ,$$

subject to

$$A x_o + B U = 0 \; ,$$

where

$$U \triangleq \begin{bmatrix} u(0) & \cdots & u(N-1) \end{bmatrix}^T \; , \quad A \triangleq M_f \Phi^N \; , \quad B \triangleq M_f \begin{bmatrix} \Phi^{N-1} \Gamma & \cdots & \Gamma \end{bmatrix} \; .$$

(b) Show that the solution is

$$U = -K x_o \; ,$$

where

$$K = B^T (B B^T)^{-1} A \; .$$

Note that the existence of $(B B^T)^{-1}$ ensures *controllability* of $M_f x(N)$ with $u(i)$.

(c) Write a code to calculate K using software like MATLAB.

(d) Using your code from (d), find K for the discrete double-integrator plant

$$\begin{bmatrix} y(i+1) \\ v(i+1) \end{bmatrix} = \begin{bmatrix} 1 & T \\ 0 & 1 \end{bmatrix} \begin{bmatrix} y(i) \\ v(i) \end{bmatrix} + \begin{bmatrix} T^2/2 \\ T \end{bmatrix} u(i) \; ,$$

with $M_f = \begin{bmatrix} 1 & 0 \end{bmatrix}$, $T = .1$, $N = 20$.

5.5.2 Discrete Lateral Intercept/Rendezvous: Consider Example 5.2.2. Using TDLQH with $N = 20$ make plots for discrete-step control with hard terminal constraints that are comparable to the soft terminal constraint plots of Figs. 5.2 and 5.3.

5.5.3 to 5.5.22:   Using TDLQH or TDLQHR, do Problems 5.3.3 to 5.3.22 with hard terminal constraints.

5.6 Chapter Summary

LQ Soft Terminal Controllers—Continuous

Find $u(t)$ to minimize

$$J = \frac{1}{2} e_f^T(t_f) Q_f e_f + \frac{1}{2} \int_{t_o}^{t_f} \begin{bmatrix} x^T & u^T \end{bmatrix} \begin{bmatrix} Q & N \\ N^T & R \end{bmatrix} \begin{bmatrix} x \\ u \end{bmatrix} dt \; ,$$

subject to

$$\dot{x} = Ax + Bu \; , \quad x(t_o) = x_o \; , \quad e_f = M_f x(t_f) - \psi \; ,$$

where (t_o, t_f) are specified and $R > 0$.

Necessary Conditions for a Stationary Path

Stationary paths must satisfy the EL equations

$$
\begin{bmatrix} \dot{x} \\ \dot{\lambda} \end{bmatrix} = \begin{bmatrix} \bar{A} & -\bar{B} \\ -\bar{Q} & -\bar{A}^T \end{bmatrix} \begin{bmatrix} x \\ \lambda \end{bmatrix} ,
$$

where $\bar{A} \triangleq A - BR^{-1}N^T$, $\bar{B} \triangleq BR^{-1}B^T$, $\bar{Q} \triangleq Q - NR^{-1}N^T$, with the two-point boundary conditions

$$
x(t_o) = x_o , \quad \lambda(t_f) = M_f^T Q_f [M_f x(t_f) - \psi] .
$$

The optimal control is

$$
u = -R^{-1}(B^T \lambda + N^T x) .
$$

Numerical Solution of the EL Equations

If $t_f - t_0 \gg$ characteristic times of the plant, the solution of the EL equations by transition matrices may overflow or underflow the computer since the EL equations are unstable for both forward and backward integration. A stable solution is obtained by sweeping the coefficient matrix S_f backward to the initial time using a nonlinear matrix differential equation called the *Riccati equation*:

$$
\dot{S} = -S\bar{A} - \bar{A}^T S - \bar{Q} + S\bar{B}S , \quad S(t_f) = M_f^T Q_f M_f \triangleq S_f ,
$$

along with a linear vector equation for $g(t)$ if $\psi \neq 0$:

$$
\dot{g} = -[\bar{A} - \bar{B}S(t)]^T g , \quad g(t_f) = -M_f^T Q_f \psi .
$$

The optimal control involves a feedforward control proportional to $g(t)$ and a linear feedback on the state $x(t)$:

$$
u(t) = u_f(t) - K(t)x(t) , \quad u_f(t) \triangleq -R^{-1}B^T g(t) , \quad K(t) \triangleq R^{-1}[N^T + B^T S(t)] .
$$

The state equations can be integrated forward stably using this control law. The Riccati matrix $S(t)$ has the property (for $\psi = 0$) that

$$
J_{\min} = \tfrac{1}{2} x_o^T S(t_o) x_o .
$$

LQ Soft Terminal Controllers—Discrete

Choose $u(i)$ to minimize

$$
J = \frac{1}{2} e_f^T Q_f e_f + \frac{1}{2} \sum_{i=0}^{N-1} \begin{bmatrix} x^T(i) & u^T(i) \end{bmatrix} \begin{bmatrix} Q_d & N_d \\ N_d^T & R_d \end{bmatrix} \begin{bmatrix} x(i) \\ u(i) \end{bmatrix} ,
$$

subject to

$$
x(i+1) = \Phi x(i) + \Gamma u(i) , \quad x(0) = x_o , \quad e_f = M_f x(N) - \psi ,
$$

for $i = 0, \ldots, N - 1$, where $R_d > 0$.

Necessary Conditions for a Stationary Path

Stationary paths must satisfy the following linear difference equations (the discrete EL equations):

$$
\begin{bmatrix} \lambda(i) \\ 0 \\ x(i+1) \end{bmatrix} = \begin{bmatrix} Q_d & N_d & \Phi^T \\ N_d^T & R_d & \Gamma^T \\ \Phi & \Gamma & 0 \end{bmatrix} \begin{bmatrix} x(i) \\ u(i) \\ \lambda(i+1) \end{bmatrix} ,
$$

with the two-point boundary conditions

$$
x(0) = x_o , \quad \lambda(N) = M_f^T Q_f [M_f x(N) - \psi] .
$$

Numerical Solution of the EL Equations

If $N \gg$ characteristic step numbers of the plant, the solution of the EL equations by transition matrices may overflow or underflow the computer since the EL equations are unstable for both forward and backward sequencing. A stable solution is obtained by sweeping the coefficient matrix S_f backward to the initial time using a nonlinear matrix difference equation called the *discrete Riccati equation:*

$$
S(i) = Z_{xx}(i) - Z_{xu}(i) Z_{uu}(i)^{-1} Z_{ux}(i) , \quad i = N-1 , \dots, 0 ,
$$

where $S(N) = M_f^T Q_f M_f$ and

$$
\begin{bmatrix} Z_{xx} & Z_{xu} \\ Z_{ux} & Z_{uu} \end{bmatrix} \triangleq \begin{bmatrix} Q_d & N_d \\ N_d^T & R_d \end{bmatrix} + \begin{bmatrix} \Phi^T \\ \Gamma^T \end{bmatrix} S(i+1) \begin{bmatrix} \Phi & \Gamma \end{bmatrix} .
$$

If $\psi \neq 0$ a linear vector difference equation for $g(i)$ must also be sequenced backward along with the Riccati equation

$$
g(i) = [\Phi - \Gamma K(i)]^T g(i+1) , \quad g(N) = -M_f^T Q_f \psi .
$$

where $K(i) \triangleq Z_{uu}^{-1}(i) Z_{xu}^T(i)$. The optimal control is a feedforward proportional to $g(i)$ and a linear feedback on the state $x(i)$:

$$
u(i) = u_f(i) - K(i) x(i) , \quad u_f(i) \triangleq -Z_{uu}^{-1}(i) \Gamma^T g(i+1) .
$$

The state equations can be sequenced forward stably using this control law. The Riccati matrix $S(i)$ has the property (for $\psi = 0$) that

$$
J_{\min} = \tfrac{1}{2} x_o^T S(0) x_o .
$$

LQ Hard Terminal Controllers—Continuous

Same as the soft terminal controller except the terminal cost is $\frac{1}{2}x^T(t_f)S_f x(t_f)$ and there are hard terminal constraints $M_f x(t_f) = \psi$. The sweep solution has the same Riccati equation as for soft terminal constraints with $S(t_f) = S_f$, and the following additional linear matrix differential equations:

$$\dot{M} = -M[A - BR^{-1}(N^T + B^T S)] , \quad \dot{Q} = -MBR^{-1}B^T M^T ,$$

with boundary conditions $M(t_f) = M_f$, $Q(t_f) = 0$. The optimal control is

$$u(t) = -K(t)x(t) + K_\psi \psi ,$$

where

$$K(t) = R^{-1}\{N^T + B^T[S(t) + M^T(t)Q^{-1}(t)M(t)]\} ,$$
$$K_\psi(t) = R^{-1}B^T M^T(t)Q^{-1}(t) ,$$

and, for $\psi = 0$

$$J_{\min} = \frac{1}{2}x_o^T[S(t_o) + M^T(t_o)Q^{-1}(t_o)M(t_o)]x_o .$$

The gains tend $\to \infty$ as $t \to t_f$ which is not practical for implementation, so we use optimal open-loop control for a short time before $t = t_f$.

LQ Hard Terminal Controllers—Discrete

Same as the soft terminal controller except the terminal cost is $\frac{1}{2}x^T(N)S_f x(N)$ and there are hard terminal constraints $M_f x(N) = \psi$. The sweep solution has the same Riccati equation as for soft terminal constraints with $S(N) = S_f$, and the following additional linear matrix difference equations:

$$M(i) = M(i+1)(\Phi - \Gamma Z_{uu}^{-1}Z_{ux}) , \quad Q(i) = Q(i+1) + M(i+1)\Gamma Z_{uu}^{-1}\Gamma^T M^T(i+1) ,$$

with boundary conditions $M(N) = M_f^T$, $Q(N) = 0$. The optimal control is

$$u(i) = -K(i)x(i) + K_\psi \psi ,$$

where

$$K(i) = Z_{uu}^{-1}[Z_{ux} + \Gamma^T M^T(i+1)Q^{-1}(i)M(i)] ,$$
$$K_\psi(i) = Z_{uu}^{-1}\Gamma^T M^T(i+1)Q^{-1}(i) ,$$

and, for $\psi = 0$,

$$J_{\min} = \frac{1}{2}x_o^T[S(0) + M^T(0)Q^{-1}(0)M(0)]x_o .$$

The gains tend $\to \infty$ as $i \to N$, which is not practical for implementation, so we use optimal open-loop control for a few steps before $i = N$.

CHAPTER 6

Linear-Quadratic Regulators

6.1 Introduction

The feedback gains of a terminal controller for a TI plant with in-flight penalties on the output, i.e., $Q = $ constant $\neq 0$, tend to constant values as the time-to-go becomes large. In other words the controller effectively becomes a *regulator;* a regulator is a TI feedback controller that keeps a plant within an acceptable deviation from a reference condition using acceptable amounts of control for the anticipated level of disturbances.

The only disturbances we consider are nonzero initial conditions. We also assume perfect knowledge of all the states, i.e., we do not consider state estimation. This important subject is treated elsewhere (e.g., in Refs. BH and FPE). Suffice it to say here that one can feedback the estimated states with the same gains developed here to produce an efficient compensator.

6.2 Continuous Regulators

We consider the system

$$\dot{x} = Ax + Bu , \quad y = Cx , \tag{6.1}$$

and the quadratic performance index

$$J = \frac{1}{2} \int_0^\infty (y^T Qy + u^T Ru)dt , \tag{6.2}$$

where (A, B, C, Q, R) are constant matrices.

The EL equations for this system are

$$\begin{bmatrix} \dot{x} \\ \dot{\lambda} \end{bmatrix} = \begin{bmatrix} A & -BR^{-1}B^T \\ -C^T QC & -A^T \end{bmatrix} \begin{bmatrix} x \\ \lambda \end{bmatrix} , \tag{6.3}$$

$$u = -R^{-1}B^T\lambda , \tag{6.4}$$

with boundary conditions

$$x(0) = x_o , \quad \lambda(\infty) = 0 . \tag{6.5}$$

A *backward sweep solution* is obtained by assuming a solution of the form

$$\lambda(t) = S(t)x(t) . \tag{6.6}$$

Differentiating (6.6) and substituting for $\dot{x}$ and $\dot{\lambda}$ from (6.3) gives the following *Riccati equation* for S:

$$-\dot{S} = SA + A^T S + C^T QC - SBR^{-1}B^T S , \quad S(\infty) = 0 . \tag{6.7}$$

This set of nonlinear differential equations is stable for backward integration; hence S reaches a steady-state value S_{ss}. Now S_{ss} can, in principle, be found from the set of quadratic equations obtained by setting $\dot{S} = 0$ in (6.7). In practice, this is almost impossible for large systems due to the many sign ambiguities involved in solving sets of quadratic equations. Kalman, who first gave these relations around 1960, advocated integrating the Riccati equation backward to steady-state (Ref. KE). However, in many cases, this is quite time-consuming because very small integration steps are required to obtain accurate solutions. A more efficient solution of the steady-state Riccati equation is described below.

Once the steady-state Riccati matrix S is found, (6.4) becomes a *TI linear feedback law* using (6.6). •

An Eigenvector Solution for the Steady-State Riccati Matrix

An eigenvector solution of the steady-state Riccati equation was given in 1963 by Mac-Farlane (Ref. Ma). It requires finding the complex eigenvalues and eigenvectors of the EL equations, which, in 1963, did not seem an attractive alternative to integrating the Riccati equation backward to steady-state. However, in 1970, Wilkinson et al. (Ref. WMP) gave an efficient and accurate algorithm for finding complex eigensystems based on the QR algorithm of Francis (Ref. Fr). In 1971, Hall (Ref. Ha) used the Wilkinson algorithm with MacFarlane's eigenvector solution to produce a computer code he called OPTSYS, which gave rapid, accurate solutions of the steady-state Riccati equation. His code was the basis of subsequent codes used in professional software such as MatrixX, Control-C, and MATLAB, which made further improvements in speed and accuracy by using Schur decomposition instead of eigenvector decomposition.

Symmetry of the Eigenvalues of the EL Equations

If x is an n-vector, the EL equations (6.3) constitute a homogeneous linear TI system of order $2n$. The general solution can be expressed in terms of its eigensystem:

$$x = X \exp(st) , \quad \lambda = \Lambda \exp(st) . \tag{6.8}$$

where the possible values of s, X, and Λ are found from

$$\begin{bmatrix} sI - A & BR^{-1}B^T \\ C^T QC & sI + A^T \end{bmatrix} \begin{bmatrix} X \\ \Lambda \end{bmatrix} = 0 . \tag{6.9}$$

If (6.9) is transposed and the order of X and Λ is reversed, it may be written as

$$\begin{bmatrix} -\Lambda^T & X^T \end{bmatrix} \begin{bmatrix} -sI - A & BR^{-1}B^T \\ C^T QC & -sI + A^T \end{bmatrix} = 0 . \tag{6.10}$$

The characteristic equation is obtained by setting the determinant of the coefficient matrix to zero in either (6.9) or (6.10). However, (6.10) differs from (6.9) only in the replacement of s by $-s$; thus if s_i is an eigenvalue of the EL equations, then $-s_i$ is also an eigenvalue. Since complex eigenvalues occur in conjugate pairs, this means that the *poles in the complex s-plane are located symmetrically about both the real and imaginary axes.* The poles in the left-half s-plane are the closed-loop poles of the regulated system. The poles in the right-half s-plane are the poles of the adjoint system.

The Steady-State Riccati Matrix in Terms of the Eigenvectors of the EL Equations

The general solution of the EL equations can be written in terms of the eigenvalues and the eigenvectors as

$$\begin{bmatrix} x \\ \lambda \end{bmatrix} = \begin{bmatrix} X_- & X_+ \\ \Lambda_- & \Lambda_+ \end{bmatrix} \begin{bmatrix} \alpha_-(t) & 0 \\ 0 & \alpha_+(t) \end{bmatrix} \begin{bmatrix} C_- \\ C_+ \end{bmatrix} , \tag{6.11}$$

where

$$\begin{aligned} \alpha_-(t) &= \text{diag}[\exp(-s_1 t), \cdots, \exp(-s_n t)] , \\ \alpha_+(t) &= \text{diag}[\exp(+s_1 t), \cdots, \exp(+s_n t)] , \\ s_i &= i^{th} \text{ eigenvalue with positive real part} , \\ C_- \text{ and } C_+ &\text{ are arbitrary constant } n\text{-vectors} , \end{aligned}$$

and the ith columns of X_- and Λ_- are the eigenvectors corresponding to the eigenvalues $-s_i$, while the ith columns of X_+ and Λ_+ are the eigenvectors corresponding to the eigenvalues $+s_i$.

Going backward in time (as we do with the Riccati equation for S), the eigenvalues with negative real parts will dominate so that

$$x(t) \rightarrow X_-\alpha_-(t)C_- , \quad \lambda \rightarrow \Lambda_-\alpha_-(t)C_- . \tag{6.12}$$

It follows that

$$\lambda(t) \rightarrow \Lambda_-[X_-]^{-1}x(t) . \tag{6.13}$$

Comparing this with (6.6), we see that the steady-state Riccati matrix is

$$S_{ss} = \Lambda_-[X_-]^{-1} . \tag{6.14}$$

This is the MacFarlane eigenvector solution for the steady-state Riccati matrix; it involves the two n by n partitions of the eigenvector matrix of the EL equations that correspond to the eigenvalues with negative real parts.

The Steady-State Feedback Gain Matrix

The steady-state feedback gain matrix is given by (6.14) in (6.6) and (6.4):

$$K_{ss} = R^{-1}B^T S_{ss} , \tag{6.15}$$

where

$$u(t) = -K_{ss}x(t) . \tag{6.16}$$

Symmetric Root Locus (SRL) for SISO, SIMO, and MISO Systems

Taking the Laplace transform of the EL equations (6.1), (6.3), and (6.4), we obtain

$$y(s) = Cx(s) , \tag{6.17}$$

$$x(s) = (sI - A)^{-1}Bu(s) , \tag{6.18}$$

$$\lambda(s) = -(sI + A^T)^{-1}C^T QCx(s) , \tag{6.19}$$

$$Ru(s) = -B^T\lambda(s) . \tag{6.20}$$

Eliminating $x(s)$ and $\lambda(s)$ from these equations, we obtain

$$[R + Y^T(-s)QY(s)]u(s) = 0 , \tag{6.21}$$

where

$$Y(s) = C(sI - A)^{-1}B , \tag{6.22}$$

which is the open-loop transfer function matrix from $u(s)$ to $y(s)$, i.e.,

$$y(s) = Y(s)u(s) .\qquad(6.23)$$

For a SIMO (single-input/multiple-output) system, the characteristic equation of (6.21) can be written in the Evans root locus form

$$-R = Y^T(-s)QY(s) .\qquad(6.24)$$

Thus a locus of the closed-loop roots vs. $1/R$ for fixed Q can be made by starting with the poles of the open-loop transfer functions plus the reflection of these poles across the imaginary axis.

Eliminating $u(s)$ and $\lambda(s)$ from (6.17) to (6.20), we obtain

$$[Q^{-1} + Y(s)R^{-1}Y^T(-s)]Qy(s) = 0 ,\qquad(6.25)$$

For a MISO (multiple-input/single-output) system, the characteristic equation of (6.25) can be written in the Evans root locus form

$$-Q^{-1} = Y(s)R^{-1}Y^T(-s) .\qquad(6.26)$$

Thus a locus of the closed-loop roots vs. Q for fixed R can be made by starting with the poles of the open-loop transfer functions plus the reflection of these poles across the imaginary axis.

For SISO systems, the zeros of the SRL are the zeros of the open-loop transfer function and reflections of these zeros across the imaginary axis. For SIMO and MISO systems the SRL zeros are *compromise zeros* obtained by weighting the several transfer function numerators with the values of Q or R^{-1}.

Example 6.2.1—SRL for a SISO System

Consider the SISO system

$$\frac{y(s)}{u(s)} = \frac{s}{s^2 + 1} .$$

The SRCE is

$$\frac{R}{Q} = \frac{s^2}{(s^2 + 1)^2} ,$$

and the SRL is shown in Figure 6.1 (note it is a 0 deg locus).

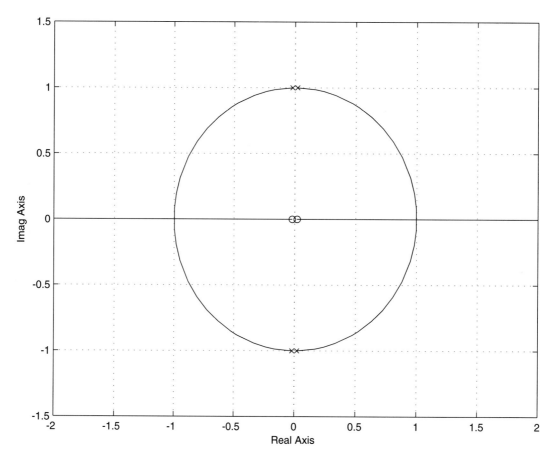

FIGURE 6.1   SRL vs. Q/R for the SISO Example

Example 6.2.2—SRL for a SIMO System

Consider the SIMO system

$$\frac{y_1(s)}{u(s)} = \frac{1}{s^2 + 1} \ , \quad \frac{y_2(s)}{u(s)} = \frac{s}{s^2 + 1} \ .$$

The SRCE is

$$\frac{R}{Q_2} = \frac{s^2 - Q_1/Q_2}{(s^2 + 1)^2} \ ,$$

and the SRL vs. Q_2/R is shown in Figure 6.2 (note it is also a 0 deg locus).

Example 6.2.3—SRL for a MISO System

Consider the MISO system

$$y(s) = \frac{u_1(s) + su_2(s)}{s^2 + 1} \ .$$

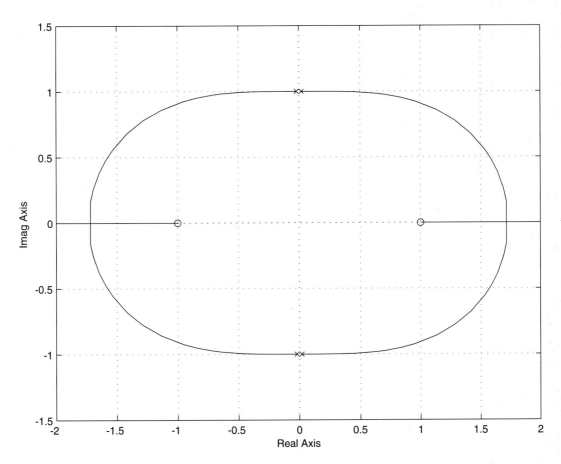

FIGURE 6.2   SRL vs. Q_2/R for the SIMO Example and vs. Q/R_2 for the MISO Example; $Q_1/Q_2 = 1$ or $R_2/R_1 = 1$

The SRCE is

$$\frac{R_2}{Q} = \frac{s^2 - R_2/R_1}{(s^2 + 1)^2} \, ,$$

and the SRL vs. Q/R_2 is the same as the one in Figure 6.2 if we replace Q_1/Q_2 by R_2/R_1.

Use of Professional Codes for LQ Regulator (LQR) Synthesis

There are several professional software packages available that provide reliable, fast, and accurate solutions for the LQR problem (e.g., MATLAB, MATRIX-X, and CONTROL-C). Anyone seriously interested in control design should have one of these packages, which have hundreds of other commands for control analysis and synthesis as well as excellent graphics.

Limitation of Bandwidth by Right-Half-Plane (RHP) Zeros

For SISO systems, the SRL vs. Q/R goes from the open-loop poles (and their reflections across the imaginary axis) either to infinity in a Butterworth pattern or to the zeros of the open-loop transfer function (and their reflections across the imaginary axis).

The closed-loop modes corresponding to the poles that tend toward infinity become very fast. The modes corresponding to the poles that tend toward left-half-plane (LHP) zeros of the open-loop transfer function have very small residues (almost pole-zero cancellations) so these modes hardly appear in the closed-loop response. Thus, systems with LHP zeros (or no zeros) can always be made to respond faster with more control authority.

However, the modes corresponding to the poles that tend toward LHP zeros that are reflections in the imaginary axis of RHP zeros of the open-loop transfer function do not have small residues and hence they dominate the closed-loop response (Ref. KS). *Thus RHP zeros place an upper limit on how fast one can make the closed-loop response even with large control authority.*

Example 6.2.4—SRL for a System with a RHP Zero; Controlling Tip Deflection of a Flexible Robot Arm

A flexible robot arm is modeled as two links of mass m and length ℓ with a tip load of mass m (see Problem 5.2.11 and Figure 5.11). The links are connected by a pin and a torsion spring of spring constant k. The equations of motion are

$$\begin{bmatrix} \dot{\theta}_1 \\ \dot{q}_1 \\ \dot{\theta}_2 \\ \dot{q}_2 \end{bmatrix} = \begin{bmatrix} 0 & 1 & 0 & 0 \\ -\epsilon & 0 & \epsilon & 0 \\ 0 & 0 & 0 & 1 \\ 1-\epsilon & 0 & -1+\epsilon & 0 \end{bmatrix} \begin{bmatrix} \theta_1 \\ q_1 \\ \theta_2 \\ q_2 \end{bmatrix} + \begin{bmatrix} 0 \\ b \\ 0 \\ b-\epsilon \end{bmatrix} T \ ,$$

where $\epsilon = .4251$, $b = .2003$, time is in units of $1/\omega$, $\omega \stackrel{\Delta}{=} 2.782(k/m\ell^2)^{1/2}$, q_i are in units of ω, and the torque T is in units of k.

The tip deflection is $y = \theta_1 + \theta_2$, where y is in units of ℓ. The transfer function from T to y is

$$\frac{y(s)}{T(s)} = \frac{.03918}{s^2} - \frac{.06368}{s^2+1} \equiv \frac{-.0245[s^2-(1.265)^2]}{s^2(s^2+1)} \ ,$$

which has a RHP zero.

We consider the quadratic performance index

$$J = \int_0^\infty (Qy^2 + T^2)dt \ .$$

The SRL is shown in Figure 6.3. The rigid-body poles tend to the two zeros at $s = -1.246$ as $Q \to \infty$; one of these zeros is a reflected zero, so this limits the bandwidth possible by

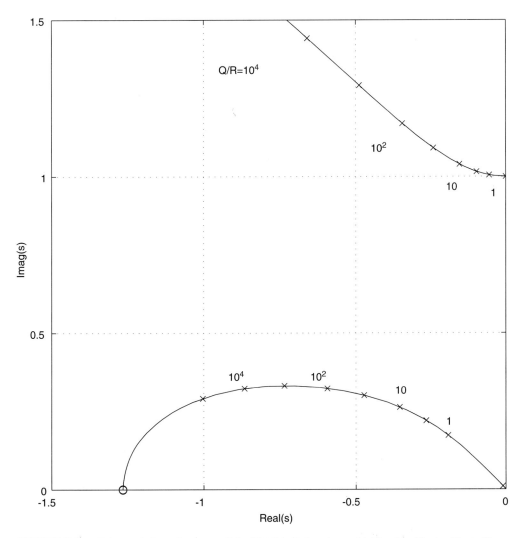

FIGURE 6.3 Symmetric Root Locus vs. Q for Flexible Robot Arm with Shoulder Torque Controlling Tip Deflection

feedback control. As $Q \to \infty$, two of the closed-loop poles tend to the zeros, but only one is canceled. The response to a unit step command for y for $Q = 100$ is shown on Figure 6.4; note the outer link turns in the "wrong" direction at first (and so does the tip), but then comes rapidly into place and stops.

In the actual flexible arm there are higher frequency modes but the behavior shown here is approximately correct. For a step command in y, the system looks as though a wave was being propagated down the arm from the shoulder torquer. Using time-varying gains the maneuver is completed in roughly half the time (Figure 5.11).

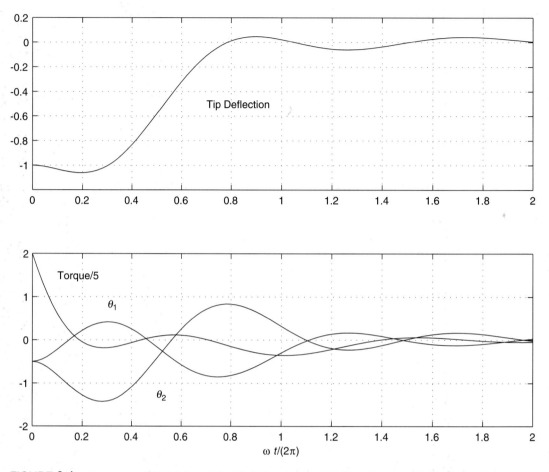

FIGURE 6.4   Response of Closed-Loop Flexible Robot Arm to Unit Step Command in Tip Deflection

SRL for MIMO Systems; Transmission Zeros (TZs)

SRL for MIMO systems are *not* of the Evans type, because the elements of the weighting matrices Q and R do not enter the SRCE linearly. As a result, MIMO SRL look strange to anyone used to Evans root loci.

Since there are at least four transfer functions in a MIMO system, there are at least four sets of zeros. In general, the SRL does *not* tend to any of these zeros as weights on the controls tend to zero ("cheap control"). Instead they tend to the TZs of the system.

A TZ is a closed-loop eigenvalue that corresponds to a mode of motion of the system where the outputs are zero. To find the state-feedback gains that produce such modes, we use the fact that if the output is zero, the time derivatives of the outputs must also be zero. Hence we differentiate the outputs one or more times and substitute for the state derivatives from the equations of motion until the resulting expressions contain the controls. (If one of the controls appears directly in one of the outputs, that output need not be differentiated.)

These equations are solved for the controls in terms of the states, giving a linear SFB law that keeps the outputs zero. When this law is substituted back into the EOM (with the outputs and lower order derivatives that did not contain the controls set equal to zero) we have a lower order system whose eigenvalues are the TZs.

Generalized Eigensystem Approach to Finding TZs

An equivalent, but more systematic, approach to finding TZs is to compute the eigenvalues of the dynamic system with constraints that all of the outputs be zero. If the system is

$$\dot{x} = Ax + Bu , \quad y = Cx + Du , \tag{6.27}$$

then let $x = X \exp(st)$, $u = U \exp(st)$, and $y = 0$, so that

$$\begin{bmatrix} A & B \\ C & D \end{bmatrix} \begin{bmatrix} X \\ U \end{bmatrix} = s \cdot \begin{bmatrix} I & 0 \\ 0 & 0 \end{bmatrix} \begin{bmatrix} X \\ U \end{bmatrix} . \tag{6.28}$$

This is a *generalized eigenvalue problem (GEP)* of the form

$$Fz = sGz . \tag{6.29}$$

For square systems, i.e., systems where dim(u) = dim(y), the eigenvalues can be found by finding the roots of the characteristic equation

$$\det(sG - F) = 0 , \tag{6.30}$$

by directly expanding the determinant.

An efficient algorithm for solving the GEP was given in 1973 (Ref. MS) called the "QZ algorithm". It is used in MatrixX and MATLAB commands to find TZs of a system [which can be rectangular, i.e., dim(u) not equal to dim(y)].

Example 6.2.5 — SRL for a MIMO System; Lateral Motions of an A/C

Consider the lateral motions of a small general aviation A/C (the Navion). The state vector is $x = [v \ r \ p \ \phi]^T$, where v = sideslip velocity, r = yaw angular velocity, p = roll angular velocity, and ϕ = roll angle. The control vector is $u = [\delta a \ \delta r]$, which are aileron and rudder deflections. We consider the outputs $y = [a_y \ \phi]^T = Cx + Du$, where a_y = the lateral specific force (which should be zero for a coordinated turn).

The system matrices are (units ft, sec, crad):

$$A = \begin{bmatrix} -0.254 & -1.76 & 0 & .322 \\ 2.55 & -.76 & -.35 & 0 \\ -9.08 & 2.19 & -8.40 & 0 \\ 0 & 0 & 1 & 0 \end{bmatrix} , \quad B = \begin{bmatrix} 0 & .1246 \\ -.222 & -4.60 \\ 29.0 & 2.55 \\ 0 & 0 \end{bmatrix} ,$$

$$C = \begin{bmatrix} -.254 & 0 & 0 & 0 \\ 0 & 0 & 0 & 1 \end{bmatrix} , \quad D = \begin{bmatrix} 0 & .1246 \\ 0 & 0 \end{bmatrix} .$$

The expression for $a_y = 0$ can be solved for δr:

$$\delta r = \frac{.254v}{.1246} \equiv 2.0835v .$$

Putting $\phi = 0 \Rightarrow \dot\phi \equiv p = 0 \Rightarrow \dot p = 0$. From the expression for $\dot p$, using $p = \phi = 0$, we have

$$29\delta a = -2.55\delta r + 9.08v - 2.19r .$$

Substituting the expression for δr in terms of v, we have

$$\delta a = .1339v - .0755r .$$

Using these "SFB laws" for δr and δa and $p = \phi = 0$, we have a second-order system for (v, r):

$$\begin{bmatrix} \dot v \\ \dot r \end{bmatrix} \begin{bmatrix} 0 & -1.76 \\ -6.8569 & -.7432 \end{bmatrix} \begin{bmatrix} v \\ r \end{bmatrix} .$$

The eigenvalues of this system are the TZs:

$$s = 3.1221 \quad\text{and}\quad s = -3.8654 .$$

The corresponding eigenvectors are

$$\begin{bmatrix} -.5637 & 0.4553 \\ 1.0000 & 1.0000 \end{bmatrix} .$$

Multiplying this matrix by the feedback gain matrix above gives the control eigenvectors; when normalized they are

$$\begin{bmatrix} .1314 & -.0157 \\ 1.0000 & 1.0000 \end{bmatrix} .$$

Since this example system has a RHP TZ at $s = 3.1221$, the LQR bandwidth will be limited to 3.1221. Figure 6.5 shows a SRL vs. the parameter η, where the performance index is

$$J = \int_0^\infty \{\eta[(5a_y)^2 + (\phi/10)^2] + \delta a^2 + \delta r^2\}dt .$$

Note the stabilized spiral mode eigenvalue goes from $-.0088$ to -3.1221 as η goes from 0 to ∞, and this mode will dominate the response for $\eta \gg 1$.

Figure 6.6 shows the response of the controlled A/C to a command for bank angle $\phi = 1$ crad and $a_y = 0$, using the gains corresponding to $\eta = 100$ in the QPI above. The max magnitude of a_y occurs at $t = 0$ and is less then .02 ft/sec^2, so the turn is well coordinated. The RHP zero effect is hardly noticeable here; however, if one looks closely, both ϕ and δa start out slightly negative, i.e., in the "wrong" direction.

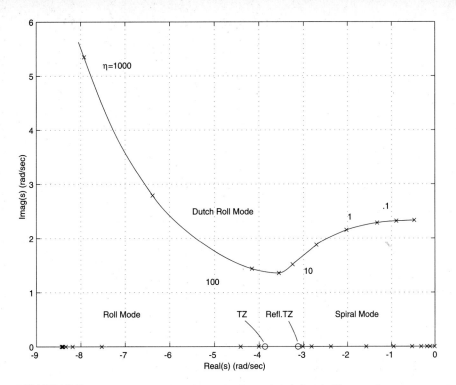

FIGURE 6.5 SRL for Lateral Motions of a General Aviation A/C vs. η where $J = \int_0^\infty \{\eta[(5a_y)^2 + (\phi/10)^2] + \delta a^2 + \delta r^2\}dt$.

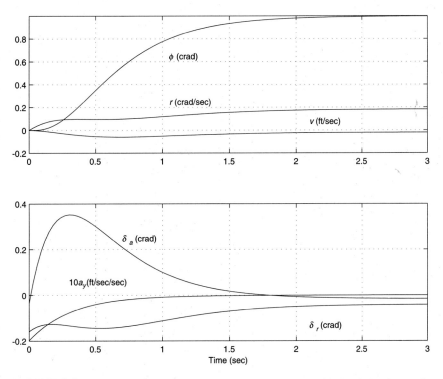

FIGURE 6.6 Closed-Loop Response of A/C to Step Command in Bank Angle ϕ and Lateral Specific Force $a_y = 0$

Coupling Numerator Interpretation of TZs

For square systems, i.e., systems with m inputs and m outputs, the TZ are the roots of the *coupling numerator* (Ref. MAG). Let

$$T(s) = \frac{N(s)}{d(s)} , \tag{6.31}$$

where $N(s)$ is an $m \times m$ matrix of polynomials, and $d(s)$ is the denominator polynomial of the transfer function matrix $T(s)$. Then the *coupling numerator is defined as*

$$N_c(s) = \frac{\det[N(s)]}{[d(s)]^{m-1}} . \tag{6.32}$$

The TZs are the roots of

$$N_c(s) = 0 . \tag{6.33}$$

For the MIMO Example 6.2.5 above the $N(s)$ matrix (units ft, sec, crad) is

$$\begin{bmatrix} -.0992s^2 - 7.7429s - 1.7628 & .1246s^4 + 1.1413s^3 - .6062s^2 - 12.122s + .7182 \\ 29s^2 + 21.990s + 132.08 & 2.55s^2 - 8.619s - 64.298 \end{bmatrix} , \tag{6.34}$$

and the factored denominator polynomial is

$$d(s) = (s + 8.4327)(s + .0088)[(s + .4862)^2 + (2.3335)^2] . \tag{6.35}$$

The determinant of $N(s)$ is a sixth-order polynomial. Its roots are the four roots of $d(s)$ plus two others that are the TZs

$$s = 3.1221 \quad \text{and} \quad s = -3.8654 . \tag{6.36}$$

Problems

6.2.1 QPI with Cross-Product Terms: Consider the more general QPI that has cross-product terms

$$J = \frac{1}{2} \int_0^\infty [x^T \quad u^T] \begin{bmatrix} Q & N \\ N^T & R \end{bmatrix} \begin{bmatrix} x \\ u \end{bmatrix} dt .$$

Show that the EL equations in this case become

$$\begin{bmatrix} \dot{x} \\ \dot{\lambda} \end{bmatrix} = \begin{bmatrix} A - BR^{-1}N^T & -BR^{-1}B^T \\ -Q + NR^{-1}N^T & -(A - BR^{-1}N^T)^T \end{bmatrix} \begin{bmatrix} x \\ \lambda \end{bmatrix} ,$$

$$u = -R^{-1}[B^T\lambda + N^Tx] .$$

6.2.2 Double-Integrator Plant: Consider the problem of finding $u(t)$ to minimize

$$J = \frac{1}{2} \int_0^{t_f} (Qy^2 + u^2)dt$$

subject to $\dot{y} = v$, $\dot{v} = u$ as $Q^{1/4}t_f \to \infty$.

(a) Show that the backward Riccati equation has a steady-state (SS) solution, giving the constant-gain feedback law

$$u = -Q^{1/4}(Q^{1/4}y + \sqrt{2}v) .$$

(b) Show that the closed-loop poles are at

$$s = \frac{Q^{1/4}(-1 \pm j)}{\sqrt{2}} .$$

(c) Sketch the SRL vs. Q, and show that it agrees with (b).

6.2.3 Undamped Oscillator Using the SS Riccati Equation: Consider the undamped oscillator

$$\begin{bmatrix} \dot{x}_1 \\ \dot{x}_2 \end{bmatrix} = \begin{bmatrix} 0 & 1 \\ -1 & 0 \end{bmatrix} \begin{bmatrix} x_1 \\ x_2 \end{bmatrix} + \begin{bmatrix} 0 \\ 1 \end{bmatrix} u ,$$

with performance index

$$J = \int_0^\infty (Qx_1^2 + u^2)dt .$$

(a) Show that the elements of the steady-state Riccati matrix are given by the solution of the simultaneous quadratic equations

$$0 = 2S_{12} + S_{12}^2 - Q , \quad 0 = -S_{11} + S_{22} + S_{12}S_{22} , \quad 0 = -2S_{12} + S_{22}^2 .$$

(b) Since S must be positive-definite, show that the solution of (a) is

$$S_{12} = (1 + Q)^{1/2} - 1 , \quad S_{22} = (2S_{12})^{1/2} , \quad S_{11} = S_{22}(1 + Q)^{1/2} .$$

(c) Show that the steady-state gain matrix is

$$K = \begin{bmatrix} S_{12} & S_{22} \end{bmatrix} .$$

6.2.4 Undamped Oscillator Using the EL Eigensystem: Consider the same system and performance index as in Problem 6.2.3.

(a) Show that the characteristic equation of the EL equations in Evans' form is

$$-Q = (s^2 + 1)^2 ,$$

which has the four roots

$$s = \pm \{[(\beta - 1)/2]^{1/2} \pm j[(\beta + 1)/2]^{1/2}]\} ,$$

with

$$\beta \triangleq (1 + Q)^{1/2} .$$

(b) For $Q = 1$, show that the eigenvectors of the EL equations corresponding to the eigenvalues with negative real parts may be written as (normalizing the largest element in each eigenvector to unity)

$$\begin{bmatrix} X_- \\ P_- \end{bmatrix} = \begin{bmatrix} .7769 - .3218j & .7769 + .3218j \\ j & -j \\ 1 & 1 \\ .3218 + .7769j & .3218 - .7769j \end{bmatrix} .$$

(c) Show that the SS Riccati matrix S is

$$S = P_- X_-^{-1} = \begin{bmatrix} 1.2872 & .4142 \\ .4142 & .9102 \end{bmatrix} ,$$

and the SS gain matrix is

$$K = \begin{bmatrix} .4142 & .9102 \end{bmatrix} .$$

Check these results with the results in Problem 6.2.3(b) and (c) for $a = 1$.

6.2.5 Inverted Pendulum: Consider the inverted pendulum

$$\begin{bmatrix} \dot{x}_1 \\ \dot{x}_2 \end{bmatrix} = \begin{bmatrix} 0 & 1 \\ 1 & 0 \end{bmatrix} \begin{bmatrix} x_1 \\ x_2 \end{bmatrix} + \begin{bmatrix} 0 \\ 1 \end{bmatrix} u ,$$

with PI

$$J = \int_0^\infty (Qx_1^2 + u^2)dt .$$

For $Q = 1$, find the LQR gains from the SS Riccati equation and also from the eigensystem of the EL equations.

6.2.6 Helicopter in Hover: The forward-velocity/pitch motions of a medium-size helicopter (the OH-6A) near hover are modeled by

$$
\begin{bmatrix} \dot{u} \\ \dot{q} \\ \dot{\theta} \end{bmatrix} = \begin{bmatrix} -.026 & .013 & -.322 \\ 1.260 & -1.765 & 0 \\ 0 & 1 & 0 \end{bmatrix} \begin{bmatrix} u \\ q \\ \theta \end{bmatrix} + \begin{bmatrix} .086 \\ -7.41 \\ 0 \end{bmatrix} \delta_{cy} \ ,
$$

where u = forward velocity (ft/sec), q = pitch rate (crad/sec), θ = pitch angle (crad), δ_{cy} = longitudinal cyclic pitch (deci-inches).

(a) Sketch the locus of closed-loop poles vs. Q for the velocity-hold performance index

$$
J = \int_0^\infty (Qu^2 + \delta_{cy}^2)dt \ .
$$

(b) Sketch the locus of closed-loop poles vs. Q for the position-hold performance index

$$
J = \int_0^\infty (Qx^2 + \delta_{cy}^2)dt \ .
$$

where $\dot{x} = u$.

6.2.7 Station-Keeping of a S/C near the Earth-Moon L_1 Point: Synthesize a regulator to stabilize the S/C of Problem 5.2.13 at the L_1 point.

(a) Sketch the locus of closed-loop poles that would be obtained with LQ synthesis using the PI

$$
J = \int_0^\infty (Ay^2 + u^2)dt \ .
$$

(b) Using MATLAB (or any other LQR synthesis code), determine the SS feedback gains to minimize J, where

$$
J = \int_0^\infty (1000y^2 + u^2)dt \ ,
$$

(c) Plot the transient response $[x(t), y(t), u(t)]$ of the closed-loop system to a unit initial velocity perpendicular to the Earth-Moon line $[\dot{y}(0) = 1]$.

6.2.8 Station-Keeping of a Solar Sail S/C near the Sun-Earth L_2 Point: Synthesize an LQR to stabilize the solar sail S/C of Problem 5.2.14 at the equilibrium point near the Sun-Earth L_2 point.

(a) Plot the locus of closed-loop poles (versus Q) that would be obtained using an LQR with the QPI

$$
J = \int_0^\infty (Qy^2 + \theta^2)dt \ .
$$

(b) Show that the fastest possible response is exponential with a time constant of about 16 days.

(c) Determine the feedback gains for $Q = 100$ and plot the closed-loop response to an initial lateral offset $y(0) = .01$ distance unit.

6.2.9 Cart with a Pendulum: Consider a SIMO regulator version of the terminal control Problem 5.2.9, i.e., add an integral-quadratic penalty on y and θ, and let $t_f \to \infty$.

(a) Put equal weights on normalized y and θ, i.e., put $Q_y = Q_\theta \triangleq Q$. Plot the SRL vs. Q/R, and locate the compromise zeros.

(b) For $Q/R = 1$ plot the response of the closed-loop system to initial conditions $y = -1, \dot{y} = 0, \theta = 0, \dot{\theta} = 0$. Compare the results with Figures 5.13 and 5.14.

6.2.10 Cart with an Inverted Pendulum: Consider a continuous SIMO regulator version of the terminal control Problem 5.4.10, i.e., add an integral-quadratic penalty on y and θ, and let $t_f \to \infty$.

(a) Put equal weights on normalized y and θ, i.e., put $Q_y = Q_\theta \triangleq Q$. Plot the SRL vs. Q/R, and locate the compromise zeros.

(b) For $Q/R = 1$ plot the response of the closed-loop system to initial conditions $y = -1, \dot{y} = 0, \theta = 0, \dot{\theta} = 0$. Compare the results with Figures 5.10 and 5.11.

6.2.11 A/C Lateral Control with Rudder Only (RHP Zero):   Consider a SISO version of the MIMO Example 6.2.5 in which the output is roll angle ϕ and the only control is rudder δr. This type of control is sometimes used for pilotless A/C since turn coordination is not as important there and ailerons do not have to be supplied. The transfer function from δr to ϕ is

$$\frac{\phi(s)}{\delta r(s)} = \frac{2.55(s + 3.61)(s - 6.99)}{(s + 8.43)[(s + .49)^2 + (2.33)^2](s + .009)} \; .$$

(a) Sketch a SRL vs. Q/R, where the QPI is

$$J = \int_0^\infty [Q\phi^2 + R(\delta r)^2]dt \; .$$

(b) Show that for $Q/R = 4$ the closed-loop poles are at $s = [-8.428, \; -1.183 \pm 2.812i, \; -1.639]$ and the feedback gain vector is $k = [.615 \; -.752 \; -.202 \; -1.886]$, where the state vector is $x = [v \, r \, p \, \phi]^T$.

(c) For $Q/R = 4$ calculate and plot the ϕ and δr responses to a step command in ϕ (entry into a steady turn).

6.2.12 Stabilizing an Unstable S/C Using a CMG: A S/C in circular orbit has its long axis in the flight direction (see Figure 6.7), which is an unstable attitude in pitch (angle θ) due to the gravity gradient torque. A control moment gyro (CMG) with spin angular momentum h is attached to the S/C with its spin axis nominally in the flight direction; it is free to rotate about the nominally vertical z-axis (angle ϕ) and a control torque T can be applied to it about this axis from the S/C. The normalized EOM for small θ and ϕ can be written as

$$\ddot{\theta} = \mu^2\theta - \epsilon\dot{\phi} \; , \quad \ddot{\phi} = \frac{1}{\epsilon}\dot{\theta} + T \; ,$$

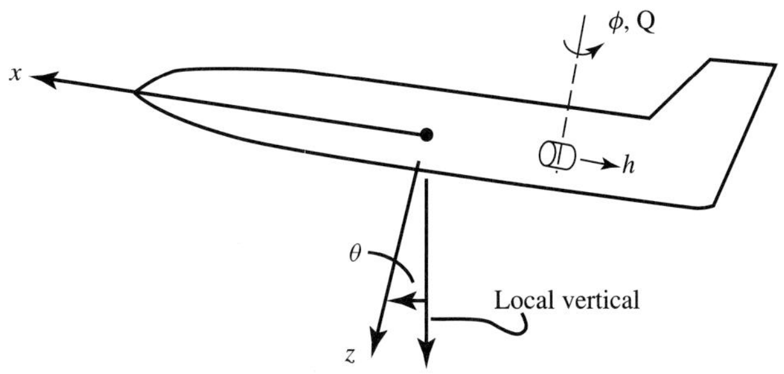

FIGURE 6.7 Spacecraft with a Pitch Control Moment Gyro

$\omega = h/\sqrt{I_y J}$ = nutation frequency, $\mu = n\sqrt{3(I_z - I_x)/I_y}$ = libration frequency, $\epsilon = \sqrt{J/I_y}$, (I_x, I_y, I_z) = principal moments of inertia of S/C + CMG about the (roll, pitch, yaw) axes, J = moment of inertia of CMG about vertical axis, and n = orbit rate. Units are T in $J\omega^2$, t in $1/\omega$, h in $J\omega$, μ in ω.

(a) Sketch the SRL vs. Q for the controlled system that minimizes

$$\int_0^\infty (Q\phi^2 + T^2)dt \ ,$$

for the case where $\mu = .01, \epsilon = .01$.

(b) What is the max possible bandwidth of this controlled system, and why is there a maximum?

(c) Select a value of Q from the root locus in (a) and calculate the corresponding feedback gains and closed-loop eigenvalues.

(d) Plot the θ, ϕ, and T responses to an initial condition of $\theta = 1$ deg, $\dot\theta = \phi = \dot\phi = 0$.

6.2.13 S-Turn of a 747 A/C (Ref. HB): This is a constant gain version of Problem 5.2.20 (where time-varying gains were used). The maneuver takes longer to perform but is again graceful (small, smooth control deflections) and well-coordinated (lateral specific force a_y very small).

Using the QPI $J = \int_0^\infty (y^T Qy + u^T Ru)dt$ with

$$Q = \begin{bmatrix} 1 & 0 \\ 0 & 64 \end{bmatrix}, \quad R = \begin{bmatrix} 1 & 0 \\ 0 & 1 \end{bmatrix},$$

and the initial states all zero except $y(0) = -10$ ft, verify Figures 6.8 and 6.9 for the S-turn maneuver. Note the QPI contains cross-product terms in the states and the controls and that $D \neq 0$.

6.2.14 Helicopter Acceleration from Hover (Four Controls): This is a constant gain version of Problem 5.2.21 (where time-varying gains were used). The maneuver takes longer to perform but is again almost completely decoupled (small responses in v, w, r for a command in u).

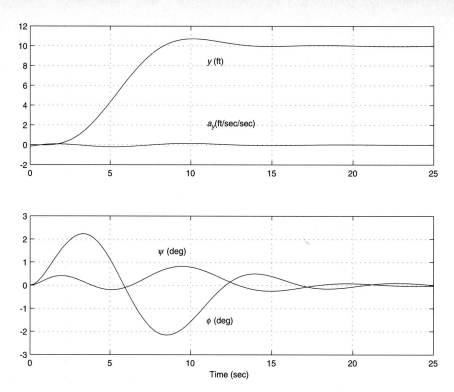

FIGURE 6.8   747 S-Turn; Lateral Distance/Specific Force and Roll/Yaw Angles

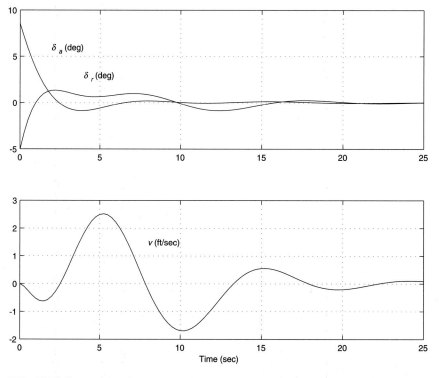

FIGURE 6.9   747 S-Turn; Aileron/Rudder Angles and Sideslip Velocity

Using the performance index $J = \int_0^\infty (y^T Q y + u^T R u) dt$ with $y = [u \ w \ v \ r]^T$ and $u = [\delta c \ \delta e \ \delta a \ \delta r]^T$, use LQR to determine the feedback gains. Then, using STEPCMD, plot the response to a command $u_c = 10$ ft/sec and compare the results with the time-varying gains case shown in Figures 5.25 and 5.26.

6.2.15 Helicopter Deceleration to Hover (Four Controls): Interchange the initial and final conditions in the previous problem so that the helicopter decelerates to hover from an equilibrium cruise condition at an altitude of 10 ft, a horizontal velocity of 10 ft/sec, and a vertical descent rate of 1 ft/sec.

6.3 Discrete Regulators

Reciprocal Eigenvalues of the EL Equations

We consider the TI discrete system

$$x(i + 1) = \Phi x(i) + \Gamma u(i) , \quad y(i) = C x(i) , \quad (6.37)$$

with the QPI

$$J = \frac{1}{2} \sum_{i=1}^\infty y(i)^T Q_d y(i) + \frac{1}{2} \sum_{i=1}^\infty u^T(i) R_d u(i) , \quad (6.38)$$

where $(\Phi, \Gamma, C, Q_d, R_d)$ are constant matrices.

The discrete EL equations are (5.56) with $N_d = 0$ and $C^T Q_d C$ replacing Q_d. The eigensystem can be found by assuming solutions of the form $x(i) = x(z)z^i$, $u(i) = u(z)z^i$, and $\lambda(i) = \lambda(z)z^i$. This yields

$$x(z) = (zI - \Phi)^{-1}\Gamma u(z) , \quad \lambda(z) = (I - z\Phi^T)^{-1}C^T Q_d C x(z) , \quad u(z) = -R_d^{-1}\Gamma^T z\lambda(z) . \quad (6.39)$$

Eliminating $u(z)$ and $\lambda(z)$ gives

$$\left[Q_d^{-1} + Z(z)R_d^{-1}Z^T\left(\frac{1}{z}\right) \right] Q_d y(z) = 0 , \quad (6.40)$$

where

$$y(z) = Z(z)u(z) , \quad Z(z) = C(zI - \Phi)^{-1}\Gamma . \quad (6.41)$$

Alternatively, eliminating $\lambda(z)$ and $x(z)$ gives

$$R_d^{-1}\left[R_d + Z^T\left(\frac{1}{z}\right) Q_d Z(z) \right] u(z) = 0 . \quad (6.42)$$

The eigenvalues may be obtained by setting the determinant of the coefficient matrices to zero in either (6.40) or (6.42). Clearly, if z_1 is an eigenvalue, then $1/z_1$ is also an eigenvalue; this implies that the eigenvalues *are reciprocal with respect to the unit circle in the complex z-plane*. The eigenvalues with magnitudes less than one are the eigenvalues of the steady-state regulator.

Reciprocal Root Locus (RRL)

If both y and u are scalar (SISO system), then either (6.40) or (6.42) immediately yields the characteristic equation in Evans' form:

$$-\frac{R_d}{Q_d} = Z(z)Z\left(\frac{1}{z}\right) . \tag{6.43}$$

We shall call a root locus vs. Q_d/R_d a RRL.

If y is a scalar but u is a vector (MISO system), then Q_d is a scalar, so (6.40) yields the characteristic equation

$$-1/Q_d = Z(z)R_d^{-1}Z^T\left(\frac{1}{z}\right) . \tag{6.44}$$

By fixing R_d we can plot a RRL vs. Q_d.

If u is a scalar but y is a vector (SIMO system), then R_d is a scalar so (6.42) yields the characteristic equation

$$-R_d = Z^T\left(\frac{1}{z}\right)Q_d Z(z) . \tag{6.45}$$

By fixing Q_d we can plot a RRL vs. R_d.

Example 6.3.1—RRL for an Undamped Oscillator

Consider an undamped oscillator with zero-order hold control $u(i)$. The discrete EOM are then

$$\begin{bmatrix} x_1(i+1) \\ x_2(i+1) \end{bmatrix} = \begin{bmatrix} \cos T & \sin T \\ -\sin T & \cos T \end{bmatrix} \begin{bmatrix} x_1(i) \\ x_2(i) \end{bmatrix} + \begin{bmatrix} 1 - \cos T \\ \sin T \end{bmatrix} u(i) ,$$

where time is in units of $1/\omega$, ω = natural frequency, and T = normalized sample time.

With $y = x_1$ so that $C = [1\ 0]$, $Z(z)$ is given by (6.41) with Φ, Γ from above:

$$Z(z) = \frac{[1 - \cos T](z + 1)}{[z - \exp(jT)][z - \exp(-jT)]} .$$

Hence the SISO reciprocal root characteristic equation (RRCE) (6.40) or (6.42) is

$$-\frac{R_d}{Q_d} = \frac{[1 - \cos T]^2 z(z + 1)^2}{[z - \exp(jT)]^2[z - \exp(-jT)]^2} .$$

An RRL vs. Q_d/R_d is shown in Figure 6.10 for $T = \pi/16$. Notice the zero at $z = 0$ and the double zero at $z = -1$; the root locus outside is a reflection in the unit circle of the root locus inside.

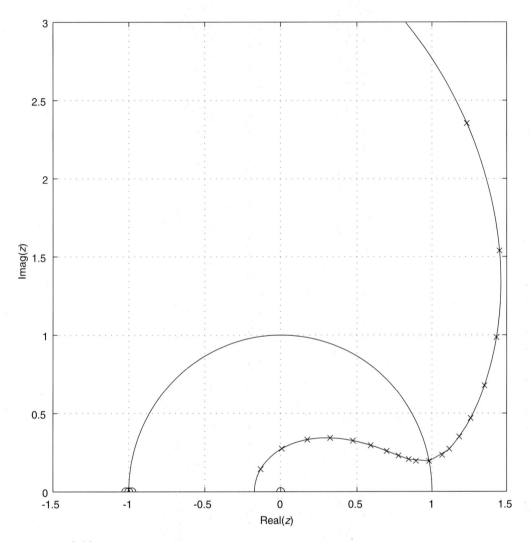

FIGURE 6.10   RRL vs. Q_d/R_d for Undamped Oscillator with ZOH, $T = \pi/16$

Eigenvector Solution of the SS Riccati Equation

An eigenvector decomposition algorithm was given by Vaughan in 1970 (Ref. Va). Letting $x(i) = Xz^i$, $\lambda(i) = \Lambda z^i$, the EL equations may be put into the form of a GEP:

$$(F - zG)\begin{bmatrix} X \\ \Lambda \end{bmatrix} = 0 , \tag{6.46}$$

where

$$F \triangleq \begin{bmatrix} \Phi & 0 \\ C^T Q_d C & -I \end{bmatrix} , \quad G \triangleq \begin{bmatrix} I & \Gamma R_d^{-1} \Gamma^T \\ 0 & -\Phi^T \end{bmatrix} . \tag{6.47}$$

For distinct eigenvalues the general solution can be written in terms of eigenvectors and eigenvalues exactly as in the continuous case (6.11), with $(\cdot)_-$ representing the eigenvalues inside the unit circle, $(\cdot)_+$ representing the eigenvalues outside the unit circle. Thus the SS Riccati matrix is given by (6.14), the ratio of the two stable $n \times n$ eigenvector matrices.

Codes for Solving the Steady-State Riccati Equation

The simplest method to program is sequencing the backward Riccati equation to steady-state. However, as in the continuous case, this algorithm is slow and not very accurate.

The Vaughan eigenvector decomposition algorithm (above) combined with the iterative QR algorithm of Francis and Wilkinson for finding the eigenvalues of a matrix, was given in (Ref. Ha1). Professional compiled codes are now available (for example) in MATLAB, MatrixX, and Control-C, that use Schur decomposition (that works even for repeated eigenvalues). The MATLAB command is DLQR for discrete LQR (see example below).

Example 6.3.2—Discrete LQR Control of an Undamped Oscillator

Consider the undamped oscillator treated in Example 6.3.1. The use of the MATLAB command DLQR is shown below with $Q_d = 9$, $R_d = 1$ to find the feedback gains "k". The command DLSIM is then used to calculate the response to the initial conditions $x(0) = [-1 \ 0]^T$.

```
% Script e06_3_2.m; response of controlled undamped oscillator
% to x(0)=[-1 0]';
%
a=pi/16; Ph=[cos(a) sin(a);-sin(a) cos(a)];
Ga=[1-cos(a);sin(a)]; C=[1 0]; Qd=9; Rd=1;
k=dlqr(Ph,Ga,Qd*C'*C,Rd);
xo=[-1 0]; w=zeros(33,1); t=[0:1/32:1]';
[y,x]=dlsim(Ph-Ga*k,Ga,C,0,w,xo); u=-x*k';
%
figure(1); clf; subplot(211),plot(t,x); grid;
axis([0 1 -1 1]); text(.25,.7,'x2'); text(.25,-.3,'x1');
subplot(212), zohplot(t,u); grid; text(.35,.25,'u');
xlabel('om*t/(2*pi)');
```

The response is shown in Figure 6.11. Note the control first helps the spring by pushing $(u > 0)$, then pulls to slow the mass down as it approaches equilibrium at $x_1 = 0$.

Discretization of Continuous QPIs

If the discrete system is a zero-order-hold (ZOH) version of a continuous system, the continuous QPI should also be discretized. The QPI

$$J = \int_0^\infty (x^T Q x + u^T R u)dt \ , \tag{6.48}$$

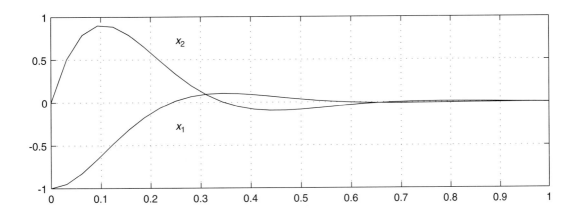

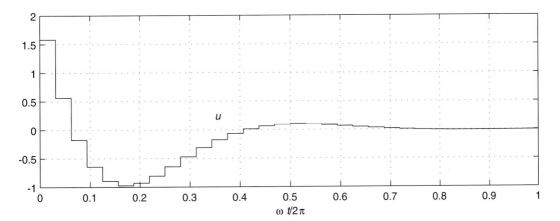

FIGURE 6.11   Response of Undamped Oscillator for $Q_d/R_d = 9$ with $T = \pi/16$

becomes

$$J = \sum_{i=0}^{\infty} \int_{i\Delta}^{(i+1)\Delta} [x(t)^T Q x(t) + u^T(i\Delta)Ru(i\Delta)]dt \ , \tag{6.49}$$

where

$$x(t) = \Phi(t)x(i\Delta) + \Gamma(t)u(i\Delta) \ , \quad i\Delta \le t \le (i+1)\Delta \ . \tag{6.50}$$

Substituting (6.50) into (6.49) and shifting to the usual discrete notation, gives

$$J = \sum_{i=0}^{\infty} \begin{bmatrix} x^T(i) & u^T(i) \end{bmatrix} \begin{bmatrix} Q_d & N_d \\ N_d^T & R_d \end{bmatrix} \begin{bmatrix} x(i) \\ u(i) \end{bmatrix} \ , \tag{6.51}$$

TABLE 6.1   MATLAB Code CVRT

```
function [Ad,Bd,Qd,Nd,Rd]=cvrt(A,B,Q,N,R,Ts)
% Converts continuous system and quadratic performance index to discrete
% system and equivalent quadratic performance index; Jcont=integral
% (x'Qx+2*x'Nu+u'Ru)dt, dx/dt=Ax+Bu; Jdisc=sum(x'Qdx+2x'Ndu+u'Rdu),
% x(i+1)=Ad*x(i)+Bd*u(i); Ts=sample time; algorithm from Ref.~VL.
%
[ns,nc]=size(B); z1=zeros(ns); z2=zeros(ns,nc); R2=sqrt(R);
S=[-A',eye(ns),z1,z2;z1,-A',Q,N/R2;z1,z1,A,B/R2; z2',z2',z2',zeros(nc)];
Sd=expm(S*Ts); k1=Sd([1:ns],[3*ns+1:3*ns+nc]);
B2=Sd([ns+1:2*ns],[2*ns+1:3*ns]); H2=Sd([ns+1:2*ns], [3*ns+1:3*ns+nc]);
Ad=Sd([2*ns+1:3*ns],[2*ns+1:3*ns]); B3=Sd([2*ns+1:3*ns], [3*ns+1:3*ns+nc]);
Bd=B3*R2; Qd1=Ad'*B2; Qd=(Qd1+Qd1')/2; Nd=Ad'*H2*R2; Rd1=B'*Ad'*k1;
Rd=Ts*R+R2'*(Rd1+Rd1')*R2;
```

where

$$Q_d = \int_0^\Delta \Phi^T(t)Q\Phi(t)dt \;, \quad N_d = \int_0^\Delta \Phi^T(t)Q\Gamma(t)dt \;, \quad R_d = \int_0^\Delta [R+\Gamma(t)^T Q\Gamma(t)]dt \;.$$

(6.52)

Note the $x^T(i)N_d u(i)$ cross-product term that occurs even when there is no cross-product term in the continuous QPI. This requires the use of the cross-product option in the MATLAB command DLQR.

A MATLAB code CVRT for doing this conversion is given in Table 6.1 for the more general continuous PI that contains a cross-product term $x^T N u$. It uses the MATLAB function EXPM to perform the integrations (Ref. VL). The MATLAB code LQRD also does this conversion and gives the discrete gain matrix and z-plane eigenvalues.

Example 6.3.3—Helicopter Near Hover

Consider control of the horizontal longitudinal velocity of a helicopter near hover. The state vector is $[\,u \; q \; \theta\,]^T$ and the control is δ_c, where u = forward velocity, q = pitch angular velocity, θ = pitch angle, δ = longitudinal cyclic stick motion. For the OH-6A, a small utility helicopter, at 2550 lb gross weight, the coefficient matrices in the state variable form $\dot{x} = Ax + B\delta_c$, are

$$A = \begin{bmatrix} -.0257 & .0130 & -.322 \\ 1.260 & -1.765 & 0 \\ 0 & 1 & 0 \end{bmatrix} \;, \quad B = \begin{bmatrix} .0860 \\ -7.41 \\ 0 \end{bmatrix} \;, \quad C = \begin{bmatrix} 1 & 0 & 0 \end{bmatrix} \;,$$

where the units are ft, sec, crad for $(u, \; q, \theta)$ and deci-inches for δ_c. The performance index is (6.2).

If we use a sample period of .1 sec, the discrete Φ, Γ, Q_d, N_d, R_d matrices are calculated below using the function file CVRT, followed by calculation of the closed-loop eigenvalues using DLQR. A RRL vs. Q/R (*not* Q_d/R_d) is shown in Figure 6.12.

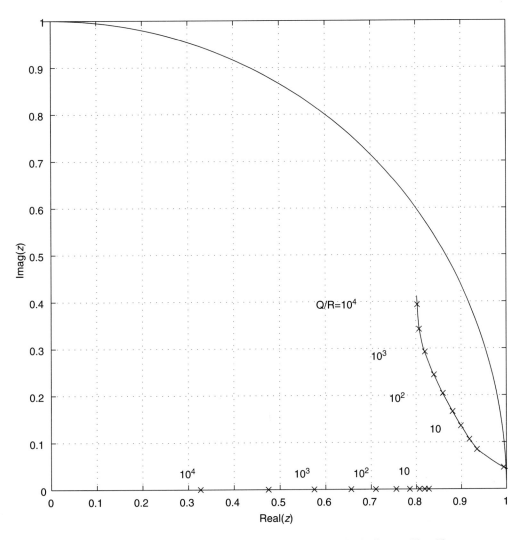

FIGURE 6.12   RRL vs. Q/R for Controlling Horizontal Velocity of a Helicopter Near Hover

```
% Script e06_3_3.m; RRL for OH-6a helicopter in hover.
%
A=[-.0257 .0130 -.322;1.260 -1.765 0;0 1 0];
B=[.0860 -7.41 0]'; C=[1 0 0];
N=[0 0 0]'; R=1; Ts=.1; Q1=[0 1 3 10 30 100 300 1000 3000 10000];
for i=1:10, Q=C'*Q1(i)*C;
 [Ad,Bd,Qd,Nd,Rd]=cvrt(A,B,Q,N,R,Ts);
 k=dlqr(Ad,Bd,Qd,Rd,Nd); ev(:,i)=eig(Ad-Bd*k);
end
```

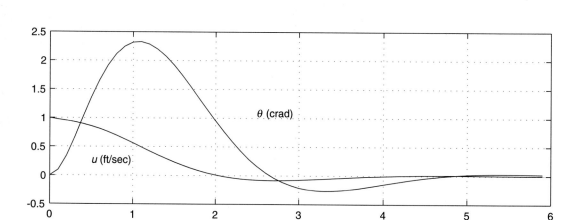

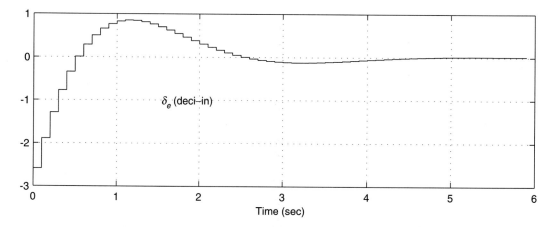

FIGURE 6.13   Response of OH6A Helicopter to Initial Conditions $u = 1$, $q = 0$, $\theta = 0$ for the Case $Q/R = 1$

```
for i=1:91, x(i)=cos((i-1)*pi/180);y(i)=sin((i-1)*pi/180);end
clf; plot(real(ev),imag(ev),'x'); grid; hold on; plot(x,y); hold off;
axis([0 1 0 1]); axis('square'); xlabel('Real(z)'); ylabel('Imag(z)');
```

Figure 6.13 shows a simulation of the response of the helicopter to initial conditions $u = 1$, $q = 0$, $\theta = 0$ for the case $Q/R = 1$. Note that the vehicle is tipped backward ($\theta > 0$) to provide a backward component of rotor force that slows it down to $u = 0$.

Problems

6.3.1 Discrete EL Equations for QPI with Cross-Product Terms: If a discrete system is a ZOH version of a continuous system, the continuous QPI should also be discretized (see the MATLAB code CVRT above or the MATLAB command LQRD). This gives cross-product terms in the discrete QPI even when they are absent in the continuous QPI, i.e.,

$$J = \frac{1}{2} \sum_{i=0}^{\infty} \left[x^T(i) \quad u^T(i) \right] \begin{bmatrix} Q_d & N_d \\ N_d^T & R_d \end{bmatrix} \begin{bmatrix} x(i) \\ u(i) \end{bmatrix} .$$

Show that the EL equations in this case become

$$\begin{bmatrix} x(i+1) \\ \lambda(i) \end{bmatrix} = \begin{bmatrix} \Phi - \Gamma R_d^{-1} N_d^T & -\Gamma R_d^{-1} \Gamma^T \\ Q_d - N_d R_d^{-1} N_d^T & [\Phi - \Gamma R_d^{-1} N_d^T]^T \end{bmatrix} \begin{bmatrix} x(i) \\ \lambda(i+1) \end{bmatrix},$$

$$u(i) = -R_d^{-1}[\Gamma^T \lambda(i+1) + N_d^T x(i)] .$$

6.3.2 Double-Integrator Plant: Consider a double-integrator plant with ZOH control having the continuous QPI (6.2) with $y = x_1 =$ displacement. If time is measured in units of $(Q/R)^{1/4}$ and x_1 in $\sqrt{R/Q}$, then we may place $R = 1$ and

$$Q = \begin{bmatrix} 1 & 0 \\ 0 & 0 \end{bmatrix} .$$

Denote the normalized sample time as Δ.

(a) If we approximate $Q_d \approx Q\Delta$, $N_d \approx 0$, $R_d \approx R\Delta$, show that the RRCE may be written in Evans' root locus form as

$$-\frac{4}{\Delta^4} = \frac{z(z+1)^2}{(z-1)^4} .$$

(b) Sketch the reciprocal root locus vs. $4/\Delta^4$ in the complex z-plane.
(c) Using (6.52) show that the equivalent discrete QPI has

$$Q_d = \begin{bmatrix} \Delta & \Delta^2/2 \\ \Delta^2/2 & \Delta^3/3 \end{bmatrix}, \quad N_d = \begin{bmatrix} \Delta^3/6 \\ \Delta^4/8 \end{bmatrix}, \quad R_d = \Delta + \Delta^5/20 .$$

For $\Delta = .25$, show that

$$\Phi = \begin{bmatrix} 1 & .25 \\ 0 & 1 \end{bmatrix}, \quad \Gamma = \begin{bmatrix} .0313 \\ .25 \end{bmatrix}, \quad \begin{bmatrix} Q_d & N_d \\ N_d^T & R_d \end{bmatrix} = \begin{bmatrix} .2500 & .0313 & .0026 \\ .0313 & .0052 & .0005 \\ .0026 & .0005 & .2500 \end{bmatrix} .$$

Use the MATLAB code LQRD or the code CVRT to check the result above.
(d) Using the MATLAB command LQRD or the code CVRT show that the closed-loop eigenvalues are $.8251 \pm .1476j$ and the feedback gain vector is $k = [.8381, \ 1.2947]$.
(e) Plot the response to an initial dispacement of -1 with zero initial velocity for 35 time steps using the approximation in (a) and then with the equivalent QPI of (c).

6.3.3 Undamped Oscillator: Do Example 6.3.1 with the output $y = x_2 =$ velocity instead of $x_1 =$ position.

6.3.4 Inverted Pendulum: Consider 6.2.5 with ZOH control. Use time in units of $1/\sigma$, where σ is the fall-down time constant.

(a) Show that the ZOH discrete model of the inverted pendulum with sample time T is

$$x(i+1) = \Phi x(i) + \Gamma u(i) ,$$

where

$$\Phi = \begin{bmatrix} \cosh T & \sinh T \\ \sinh T & \cosh T \end{bmatrix} , \quad \Gamma = \begin{bmatrix} 1 - \cosh T \\ \sinh T \end{bmatrix} .$$

6.3.5 Helicopter near Hover: Extend Example 6.3.3 to the control of position y, where $\dot{y} = u$. Make a RRL plot, choose a value of Q/R, and the plot the response to $y(0) = -10$ ft with the other initial states zero.

6.3.6 Cart with a Pendulum: Consider Problem 6.2.9 with ZOH control and normalized sample time of $\pi/16$. In addition to RRL plots, choose a value of Q/R and plot the responses to $y(0) = -1$ with the other initial states zero.

6.3.7 Cart with an Inverted Pendulum: Consider Problem 6.2.10 with ZOH control and normalized sample time of .1. In addition to RRL plots, choose a value of Q/R and plot the responses to $y(0) = -1$ with the other initial states zero.

6.3.8 Flexible Robot Arm: Consider the flexible robot arm Example 6.2.4 with ZOH control and normalized sample time of .05. In addition to RRL plots, choose a value of Q/R and plot the responses to $y(0) = -1$ with the other initial states zero.

6.3.9 Lateral Control of an Airplane: Consider the MIMO lateral A/C control Example 6.2.5 with ZOH control and sample time of .05 sec. In addition to a RRL plot like Figure 6.5, choose a value of Q/R and plot the response to recovery from a coordinated turn (initial conditions corresponding to steady-state with $a_y = 0$, $\phi = 10$ deg).

6.3.10 to 6.3.14:   Do discrete versions of Problems 6.2.11 to 6.2.15.

6.4 Chapter Summary

LQ regulators are special cases of the LQ problem treated in Chapter 5 where (A, B, C, D, Q, R) are all constant matrices, and $t_f - t \gg$ time constants of closed-loop system. In this case $S \to$ constant matrix $\Rightarrow u = -Kx$, where K is a constant gain matrix.

- Obtain constant S by (a) integrating (or sequencing) Riccati equation backward to steady-state, or (b) from eigenvectors of EL equations above (McFarlane-Potter algorithm; cf. Matrix-X, MATLAB, Ctrl-C, etc.).
- SRL. For scalar u and y:

$$-\frac{R}{Q} = Y(-s)Y(s) ,$$

where

$$y(s) = Y(s)u(s) , \quad Y(s) = D + C(sI - A)^{-1}B .$$

$Y(s)$ is the open-loop transfer function. The stable poles of the root locus are the closed-loop poles.

- RRL. For scalar u and y:

$$-\frac{R_d}{Q_d} = Z\left(\frac{1}{z}\right)Z(z) ,$$

where

$$y(z) = Z(z)u(z) , \quad Z(z) = D + C(zI - \Phi)^{-1}\Gamma .$$

$Z(z)$ is the open-loop transfer function. The stable poles of the root locus are the closed-loop poles.

CHAPTER 7

Dynamic Programming

7.1 Introduction

In Chapters 2 through 4, we determined feedforward optimal control histories $u = u(t)$ for nonlinear systems, or, more precisely, "one-sample" feedback control histories $u = u[t; x(t_0)]$, since they depended on the initial state $x(t_0)$.

In Chapters 5 and 6 for linear systems with quadratic criteria, we noted that the sample could be taken at any time along an optimal path so that we had, in effect, a closed-loop linear feedback control. In the next chapter, we extend this idea to nonlinear systems by using the *second variation* to develop *neighboring optimum feedback control* that feeds forward the nominal optimum control and feeds back the deviation from the nominal optimum path.

In this chapter, we develop Bellman's concept of *dynamic programming (DP)* (Refs. Be and Dr). Bellman derived a partial differential equation for the optimal value of the performance index (the optimal return function) as a function of the state; the optimal control is proportional to the gradient of the optimal return function. Hence a more descriptive name would be *nonlinear optimal feedback control*. While the DP concept is elegant, it is feasible for implementation only for very simple systems. However, *differential DP* is feasible for complex systems, and we shall show (in the next chapter) that this concept is identical to the second variation or neighboring optimum control solutions mentioned above, and is closely related to the LQ control concepts discussed in Chapters 5 and 6.

7.2 Extremal Fields

All points on an optimal path are possible initial points for that path; Bellman called this simple but powerful idea *the Principle of Optimality*.

If we calculate a *family of optimal paths* going backward from a terminal manifold $\psi(x, t) = 0$, and store them in memory, we could, in principle, interpolate the optimal control at any point in the state space, i.e., we would have an optimal feedback control law $u = u(x)$. For a few very simple systems, we can find analytical solutions, in the form of sets of implicit equations, for arbitrary initial conditions. If we set up a scheme to solve these equations quickly, we have a nonlinear feedback scheme for determining the optimal control as a function of the current state x and the current time t, without the requirement of extensive memory.

For realistic systems we would have to calculate and store in memory a *family of optimal paths* so that all the possible initial points are close to one of our stored optimal paths. In the calculus of variations literature, such a family is called a *field of extremals*.

In general, only one optimal path to the terminal manifold will pass through a given point (x, t), so that a unique optimal control vector $u^o(x, t)$ is associated with each point. In Chapter 8 we consider what happens in the unusual case where more than one extremal passes through a given point. Hence, we may write

$$u^o = u^o(x, t) \ . \tag{7.1}$$

i.e., the optimal control vector is a function of the present state x and the present time t.

In Figure 7.1, an extremal field is sketched for a system with only one state variable x, along with contours of constant values of u^o. In this case the terminal manifold $\psi(x, t) = 0$ is simply a curve.

Associated with starting from a point (x, t) and proceeding optimally to the terminal manifold, there is a unique optimal value of the performance index, J^o. We may therefore regard J^o as a function of the initial point, i.e.,

$$J^o = J^o(x, t) \ . \tag{7.2}$$

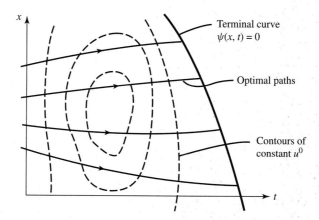

FIGURE 7.1   A Family of Optimal Paths and Contours of Constant Optimal Control u^o

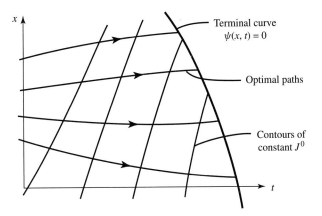

FIGURE 7.2 A Family of Optimal Paths and Contours of Constant Optimal Return Function J^o

Bellman called this is the *optimal return function.* For the one-dimensional case, contours of constant J^o may be sketched on the (x, t)-plane as in Figure 7.2.

The contours of constant J^o are like "wave fronts," and the optimal paths are like "rays." In general, the rays are *not* orthogonal to the wave fronts (in optics, they are orthogonal because $\vec{r}$ is proportional to $\vec{\lambda}$; see the Fermat problems of Chapters 2 to 4).

One aspect of the classical *Hamilton-Jacobi theory* is concerned with finding the partial differential equation satisfied by the optimal return function $J^o(x, t)$. Bellman generalized the Hamilton-Jacobi theory to include discrete and combinatorial systems, and he called this overall theory *dynamic programming* (Ref. Be).

Autonomous Systems

If the system, the performance index, and the constraints are not explicit functions of the time, and if the terminal time is unspecified, the optimal control u^o, the optimal return function J^o, and the time-to-go T^o are not explicit functions of time; i.e., we have

$$u^o = u^o(x) , \quad J^o = J^o(x) , \quad T^o = T^o(x) . \tag{7.3}$$

We will call such systems *autonomous systems.* For min time-time problems, the optimal return function is the same as the time-to-go.

Example 7.2.1—DP Solution for a Ship Moving in a Current $u_c = -Vy/h$

Consider min time paths to $x = y = 0$ for a ship moving through a region with current $u_c = -Vy/h$ (a Zermelo Problem), where V is the ship velocity with respect to the water; at $y = \pm h$ the current velocity equals the ship velocity. A family of min time paths (a field of extremals) may be calculated for this problem using the parametric equations developed from the analytic solution of the EL equations in Problem 3.3.2; here we have reversed the direction of the current and of the ship motion so that θ_o becomes θ_f:

$$T \triangleq \frac{V(t_f - t)}{h} = \tan \theta_f - \tan \theta , \tag{7.4}$$

$$\frac{y}{h} = \sec \theta_f - \sec \theta , \tag{7.5}$$

$$\frac{x}{h} = \frac{1}{2}\left[\sinh^{-1}(\tan \theta_f) - \sinh^{-1}(\tan \theta) - \frac{y}{h}\tan \theta + T \sec \theta_f \right] . \tag{7.6}$$

Equations (7.5) and (7.6) constitute a *nonlinear feedback law* since they are implicit equations for θ and θ_f as functions of $(x/h, \ y/h)$. Figure 7.3 shows the family of min time paths and contours of constant control angle θ. Figure 7.4 shows the family of min time paths and contours of constant time-to-go $T \triangleq V(t_f - t)/h$.

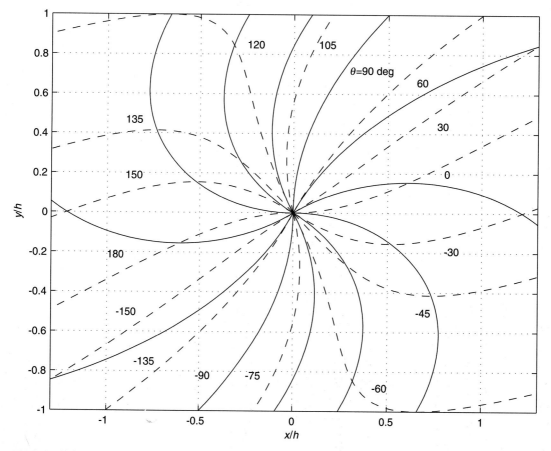

FIGURE 7.3   Min Time Paths to $x = y = 0$ with $u_c = -Vy/h$ and Contours of Constant Heading Angle

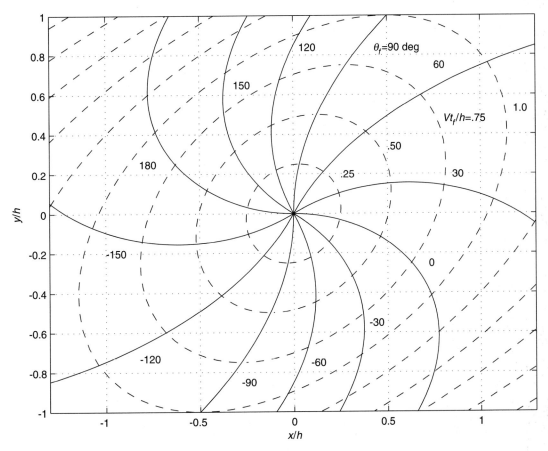

FIGURE 7.4   Min Time Paths to $x = y = 0$ with $u_c = -Vy/h$ and Contours of Constant Time-to-Go

7.3 The Continuous Minimum Principle

The Optimal Return Function

Consider the general control problem discussed in previous chapters, in the Mayer form, for an arbitrary initial point (x, t). The performance index is

$$J = \phi[x(t_f), t_f] \,, \tag{7.7}$$

and the system equations are

$$\dot{x} = f(x, u, t) \,, \tag{7.8}$$

with terminal boundary conditions

$$\psi[x(t_f), t_f] = 0 \,. \tag{7.9}$$

The *optimal return function* is defined as

$$J^o(x, t) = \min_{u(t)} \{\phi[x(t_f), t_f]\} \ , \tag{7.10}$$

with the boundary condition

$$J^o(x, t) = \phi(x, t) \ \text{ on } \ \psi(x, t) = 0 \ . \tag{7.11}$$

The Hamilton-Jacobi-Bellman (HJB) Equation

Assume that $J^o(x, t)$ is continuous and possesses continuous first derivatives. Start at a point $(x, \ t)$ and go *backward* in time by an amount Δt using some control u; this implies a change in state $\Delta x = -f(x, u)\Delta t$. It follows that

$$J^o[x - f(x, u, t)\Delta t, t - \Delta t] \geq J^o(x, t) \ . \tag{7.12}$$

If u is the optimal control, then the equality holds in (7.12). Expanding the left side of (7.12) in a Taylor series to first order in Δt gives

$$\min_u \left[J^o(x, t) - J_x^o f(x, u, t)\Delta t - J_t^o \Delta t \right] \approx J^o(x, t) \ . \tag{7.13}$$

Since J^o and J_t^o do not explicitly depend on u, (7.13) may be written, by passing to the limit as $\Delta t \to 0$, as

$$-J_t^o = \min_u \left[J_x^o f(x, u, t) \right] \ . \tag{7.14}$$

We shall call (7.14) the HJB equation; it is a first-order nonlinear partial differential equation and it must be solved with the boundary condition (7.11).

Connection to the Calculus of Variations (Ref. Dr)

In Chapter 2 we showed that the Lagrange multipliers $\lambda(t)$ are influence functions. Small changes in the initial state δx, holding the initial time constant, produce small changes in the performance index according to

$$dJ^o = \lambda^T(t)\delta x \ . \tag{7.15}$$

If we increase the initial time by an amount dt, the initial state becomes

$$dx = \delta x + \dot{x} dt \ , \tag{7.16}$$

Hence we may rewrite (7.15) as

$$dJ^o = \lambda^T(t)dx - H(t)dt \ , \tag{7.17}$$

where

$$H(x, u, t, \lambda) \stackrel{\Delta}{=} \lambda^T f(x, u, t) \ . \tag{7.18}$$

Thus

$$\lambda^T \equiv J_x^o, \quad H \equiv -J_t^o \ . \tag{7.19}$$

Using (7.19), Equation (7.14) may be written as

$$-J_t^o = H^o\left(x, t, J_x^o\right) \ , \tag{7.20}$$

where

$$H^o\left(x, J_x^o, t\right) \overset{\Delta}{=} \min_u H\left(x, u, t, J_x^o\right) \ . \tag{7.21}$$

The pair of equations (7.20) and (7.21) is another statement of the HJB equation. It shows that u^o *is the value of u that globally minimizes the Hamiltonian* $H(x, u, t, J_x^o)$, holding x, J_x^o, and t constant. We shall call this the *Minimum Principle*. In classical mechanics, H is defined with the opposite sign, so that to minimize J one must maximize H. This convention was adopted by Pontryagin in his famous statement of this principle, so it is also called *Pontryagin's Maximum Principle* (Ref. PBGM).

One of the most effective ways to solve the nonlinear partial differential equation (7.21) is by the *method of characteristics* (see, for example, Ref. CH) which amounts to finding a field of extremals as in Section 7.1 using the calculus of variations.

The great drawback of DP is, as Bellman himself called it, the "curse of dimensionality." Even recording the solution to a moderately complicated problem involves an enormous amount of storage. If we want optimal paths in the neighborhood of only one optimal path, it is tedious and uses unnecessary amounts of memory to find a whole field of extremals; if we need feedback, a perturbation feedback scheme is usually quite adequate. We develop such a scheme in the next chapter using the *second variation* and call it a "neighboring extremal perturbation feedback" scheme. It requires considerably less memory storage than DP, and is almost identical to the time-varying LQ feedback scheme developed in Chapter 5, with the addition of a feedforward control.

The Legendre-Clebsch Condition

If there are no bounds on x and u, it follows from our differentiability assumptions that u^o must be such that

$$H_u \overset{\Delta}{=} J_x^o f_u = 0 \ , \tag{7.22}$$

$$H_{uu} \overset{\Delta}{=} J_x^o f_{uu} \geq 0 \tag{7.23}$$

for all $t \leq t_f$, i.e., every component of the vector H_u must vanish and H_{uu} must be a nonnegative-definite matrix. Equation (7.23) is known as the *Legendre-Clebsch condition* in the calculus of variations. Equations (7.22) and (7.23) are local versions of the Minimum Principle for differentiable systems without control bounds.

The EL Equations from the HJB Equation

Consider a particular optimal path and its associated optimal control function. Then we have

$$\dot{\lambda}^T \equiv \dot{J}_x^o \equiv J_{xt}^o + J_{xx}^o \dot{x} \ . \tag{7.24}$$

Partial differentiation of (7.14) with respect to x, considering $u^o = u^o(x, t)$, gives

$$J_{xt}^o + J_{xx}^o \dot{x} + J_x^o \left(f_x + f_u u_x^o \right) = 0 . \tag{7.25}$$

Now the coefficient of u_x^o in (7.25) vanishes on an optimal path according to (7.22). Using (7.19) in (7.24), we obtain

$$\dot{\lambda}^T = -\lambda^T f_x , \tag{7.26}$$

which, along with (7.22), are the EL equations.

Furthermore, the fact that J^o is equal to ϕ on $\psi = 0$ implies that

$$dJ^o = J_x^o dx + J_t^o dt \equiv \phi_x dx + \phi_t dt , \tag{7.27}$$

where (dx, dt) are related by the constraint that

$$d\psi = \psi_x dx + \psi_t dt = 0 . $$

From Chapter 1, this implies that there exists a constant vector v such that

$$J_x^o - \phi_x = v^T \psi_x , \quad J_t^o - \phi_t = v^T \psi_t . $$

Using (7.19), these relations may be written as

$$\lambda^T(t_f) = \phi_x + v^T \psi_x , \quad -H = \phi_t + v^T \psi_t , \tag{7.28}$$

which are the terminal boundary conditions for the EL equations with open end time t_f.

Equivalence of the Sweep Method and DP

The HJB equation, applied to LQ optimal control problems leads directly to the matrix Riccati equation. For example, consider the continuous LQ terminal controller problem of Section 5.2. The HJB equation is (reverting to the Bolza form of H)

$$-J_t^o = \min_u \left[J_x^o (Ax + Bu) + \tfrac{1}{2}(x^T Qx + u^T Ru) \right] , \tag{7.29}$$

with the terminal boundary condition

$$J^o(x, t_f) = \tfrac{1}{2} x^T S_f x . \tag{7.30}$$

The minimization of the Hamiltonian over u leads to $u = -R^{-1} B^T (J_x^o)^T$. Substituting this into (7.20) gives

$$-J_t^o = J_x^o Ax + \tfrac{1}{2}[x^T Qx - J_x^o BR^{-1} B^T (J_x^o)^T] . \tag{7.31}$$

Motivated by (7.30), we try a product solution of the form

$$J^o = \tfrac{1}{2} x^T S(t) x ,$$ (7.32)

Substituting (7.32) into (7.29), we have

$$0 = \tfrac{1}{2} x^T [\dot{S} + SA + A^T S - SBR^{-1}B^T S + Q] x ,$$ (7.33)

which, since it must hold for all x, implies the Riccati equation (5.31).

Thus *the sweep method and DP are one and the same thing for LQ problems.* Equation (7.32) agrees with (5.34), namely, $\tfrac{1}{2} x^T S(t) x$ is the optimal return function, i.e., the min value of J starting at time t with state vector x.

LQ terminal control problems with zero terminal error are treated in Problem 7.3.8.

Problems

7.3.1 VDP for Min Time to a Vertical Line Using Gravity: Consider the VDP problem of finding $\gamma(t)$ to minimize the time to go to $x = 0$ from arbitrary (V, x) under the action of gravitational force only (see Figure 7.5), where

$$\dot{x} = -V \cos \gamma , \quad \dot{V} = g \sin \gamma ,$$

(a) Show that optimal γ and min time-to-go T are solutions of the implicit equations

$$T = \frac{\gamma}{\cos \gamma} , \quad x = \tan \gamma + \frac{\gamma}{\cos^2 \gamma} ,$$

where T is in units of V/g and x is in units of $V^2/(2g)$. Note that this is a non-linear feedback law where the optimal control and the time-to-go depend on only one dimensionless variable $2gx/V^2$. Plot gT/V and γ vs. $2gx/V^2$.

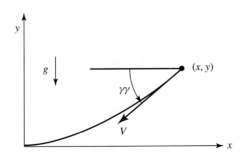

FIGURE 7.5   Nomenclature for Min Time Paths to $x = 0$ Using Gravity

(b) y is an ignorable coordinate in this problem. However, using $\dot{y} = -v \sin \gamma$ show that $y(t_f)$ is given by

$$\frac{y(t_f) - y}{x} = \frac{\sin^2 \gamma}{\gamma + \sin \gamma \cos \gamma} ,$$

and plot y/x vs. $2gx/V^2$.

7.3.2 VDP for Min Time to a Point Using Gravity: Consider Problem 7.3.1 with terminal conditions $x = y = 0$. The equations of motion are

$$\dot{V} = g \sin \gamma , \quad \dot{x} = -V \cos \gamma , \quad x(t_f) = 0 , \quad \dot{y} = -V \sin \gamma , \quad y(t_f) = 0 .$$

(a) Show that conservation of potential and kinetic energy gives

$$\frac{V^2}{2} + gy = E = \text{constant} ,$$

so that the EOM may be written as a Fermat problem with the magnitude of the velocity $= \sqrt{1 - y}$:

$$\dot{x} = -2\sqrt{1 - y} \cos \gamma , \quad \dot{y} = -2\sqrt{1 - y} \sin \gamma ,$$

where (x, y) are in units of E/g and time is in units of $\sqrt{2E}/g$.

(b) Show that the nonlinear feedback law is obtained by inverting:

$$x = \frac{\gamma - \gamma_f + (\sin 2\gamma - \sin 2\gamma_f)/2}{\cos^2 \gamma_f} , \quad y = 1 - \frac{\cos^2 \gamma}{\cos^2 \gamma_f} ,$$

to find $\gamma = \gamma(x, y)$, $\gamma_f = \gamma_f(x, y)$.

(c) Show that the time-to-go is

$$T = \frac{\gamma - \gamma_f}{\cos \gamma_f} .$$

(d) Plot contours of constant γ and constant γ_f on an x vs. y chart. Note that the latter contours are the optimal paths.

(e) Plot contours of constant T on an x vs. y chart; note that they are orthogonal to the optimal paths.

(f) Set up a MATLAB code using the command FSOLVE to compute the min time-to-go and the corresponding γ given g and (x, y, v). Check it out for normalized $x = .8$, $y = .4$.

7.3.3 VDP for Min Distance to a Point on a Sphere: In Problem 3.3.3, the calculus of variations solution was found for min distance between two points on a sphere. The equation for the optimal path (a great circle) and the heading angle β can be written as

$$\tan \theta = \tan \theta_m \sin(\phi - \alpha) , \quad \cos \beta = \frac{\cos \theta_m}{\cos \theta} ,$$

where $(\theta, \phi) = $ (latitude, longitude), and (θ_m, α) are unknown constants. Note that θ_m is the max latitude on the path (or extension of the path), while α is the longitude where the path (or extension of the path) crosses the equator.

(a) Show that α and θ_m can be expressed in terms of the current position and the desired final position (θ_f, ϕ_f) as

$$\alpha = \tan^{-1}\left(\frac{\tan\theta_f \sin\phi - \tan\theta \sin\phi_f}{\tan\theta_f \cos\phi - \tan\theta \cos\phi_f}\right), \quad \theta_m = \tan^{-1}\left[\frac{\tan\theta}{\sin(\phi - \alpha)}\right],$$

which thus yields an *explicit DP solution* for the current heading β in terms of the current state:

$$\beta = \cos^{-1}\left(\frac{\cos\theta_m}{\cos\theta}\right).$$

(b) Plot a family of min distance paths to New York along with contours of constant β on a latitude vs. longitude chart. The latitude and longitude of New York are 40.7 deg N and 73.8 deg E respectively.
(c) Plot contours of constant distance-to-go to New York on a latitude vs. longitude chart.
(d) Set up a MATLAB code to compute the optimal heading angle β given current and desired positions. Check it out for the Tokyo-to-New York great circle route considered in Problems 2.3.7, 3.n.3, 4.n.3, n=1,2,3,4.

7.3.4 VDP for Min Time to $x = y = 0$ with $V = V_o(1 + y/h)$: By reversing the direction of motion show that Figure 3.7 may be interpreted as a chart of min times-to-go to $x = y = 0$ (cf. Problem 3.3.4) and plot the corresponding feedback chart that has contours of constant control (the velocity-direction-angle θ).

7.3.5 TDP for Min Time Intercept at $x = r, y = 0$: This is a DP version of Problem 4.3.5 (which see). There is no lack of generality in letting $y_f = 0$ since we are free to choose the orientation of the reference axes. Thus $x_f \equiv r =$ distance to go and the x-axis is the current LOS (see Figure 7.6). Replace the thrust direction angle θ by $-\theta$ and r by $r/2$.

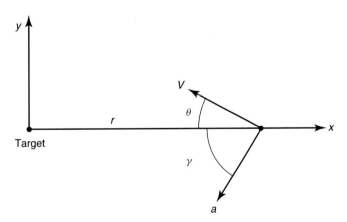

FIGURE 7.6   Nomenclature for Min Time Intercept and Rendezvous Paths to $r = 0$

(a) Making the changes indicated show that the nonlinear feedback law (DP solution) may be put into the implicit form:

$$0 = T^4 - 4T^2 + (4r \cos \gamma)T - r^2 \, ,$$

$$\tan \theta = \frac{T \sin \gamma}{r/2 - T \cos \gamma} \, .$$

The normalized variables r and γ determine the normalized min time-to-go T and the thrust direction angle θ from these relations.

(b) Plot contours of constant θ and T on a chart of $2ar/V^2$ vs. γ and compare your results with Figures 7.7 and 7.8. Note the "barrier" that divides direct paths from those that go beyond the target and come back (cf. Problem 4.3.5 and Refs. Sm, Br2).

7.3.6 TDP for Min Time to $v_f = 0$ and Specified u_f, y_f: This is a nonlinear feedback (DP) version of Problem 4.3.6. Obtain the feedback law in the form

$$\theta = \theta \left[\frac{a(y_f - y)}{(u_f - u)^2}, \frac{v}{u_f - u} \right] \, ,$$

where the nomenclature is shown in Figure 7.11; $u_f = $ final (orbital) velocity, $y_f = $ final (orbital) altitude, and (u, v, y) are the current state variables. Thus θ depends on only two normalized variables.

7.3.7 TDP for Min Time Rendezvous at $x = r$, $y = 0$:   Same as Problem 7.3.5 but with the added requirement that final $u = v = 0$ when the pursuer reaches $x = r$, $y = 0$. Verify the results shown in Figures 7.9 and 7.10 (cf. Problem 4.3.7 and Ref. Br1).

7.3.8 TDP for Min Time Intercept with Gravity: Do a DP version of Problem 4.3.8 (which see).

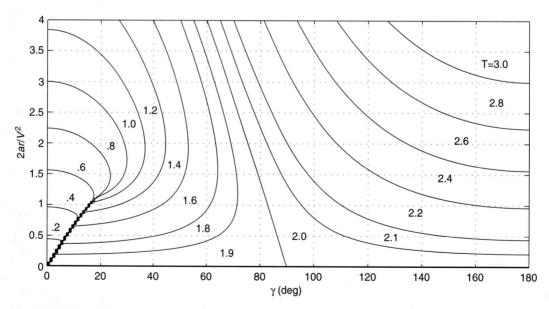

FIGURE 7.7   Feedback Chart of Thrust Direction Angle β for Min Time Intercept Paths to $r = 0$

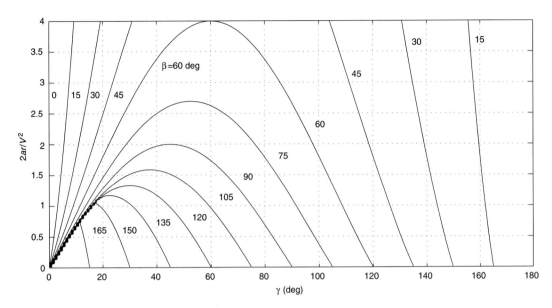

FIGURE 7.8   Time-to-Go Chart for Min Time Intercept Paths to $r = 0$

7.3.9 TDP for Min Time with Gravity to $v_f = 0$ and Specified u_f, y_f: Do a nonlinear feedback (DP) version of Problem 4.3.9, orbit injection with a, g constant. Obtain the feedback law in the form

$$\theta = \theta\left[\frac{a(y_f - y)}{(u_f - u)^2}, \ \frac{v}{u_f - u}; \ \frac{g}{a}\right] ,$$

where the nomenclature is shown in Figure 7.11; u_f = final (orbital) velocity, y_f = final (orbital) altitude, and (u, v, y) are the current state variables. Thus θ depends on only two normalized variables and the parameter g/a.

7.3.10 TDP for Min Time with Gravity to $v_f = 0$ and Specified u_f, y_f, x_f: Do a nonlinear feedback (DP) version of Problem 4.3.10, orbit injection with a specified downrange insertion point and a, g constant. Obtain the feedback law in the form

$$\theta = \theta\left[\frac{a(y_f - y)}{(u_f - u)^2}, \ \frac{a(x_f - x)}{(u_f - u)^2}, \ \frac{v}{u_f - u}, \ \frac{g}{a}\right] ,$$

where the nomenclature is shown in Figure 7.11; u_f = final (orbital) velocity, y_f = final (orbital) altitude, x_f = downrange orbit insertion distance, and (u, v, y, x) are the current state variables. Thus θ depends on only three normalized variables and the parameter g/a.

7.3.11 LQ Hard Terminal Control Using the HJB Equation: Referring to Section 5.4 which treated "hard" terminal constraints $\psi \overset{\Delta}{=} M_f x(t_f) = 0$, the form of the solution to the HJB equation is

$$J^o = \frac{1}{2}\begin{bmatrix} x^T & v^T \end{bmatrix}\begin{bmatrix} S(t) & M^T(t) \\ M(t) & -Q(t) \end{bmatrix}\begin{bmatrix} x \\ v \end{bmatrix} .$$

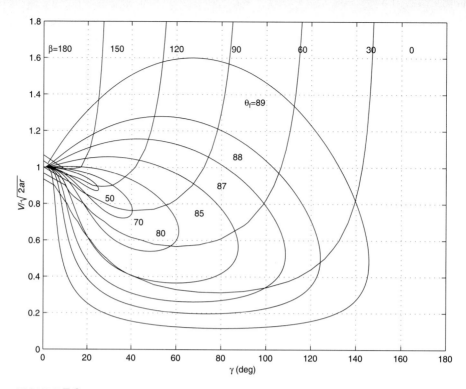

FIGURE 7.9 Feedback Chart for Minimum Time Rendezvous Paths

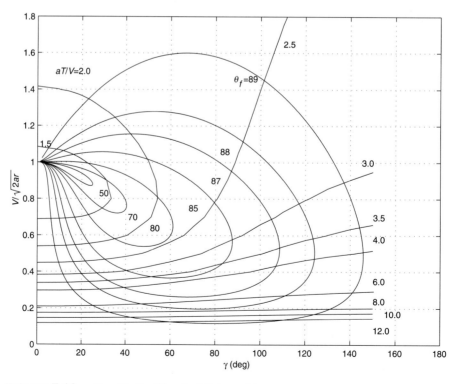

FIGURE 7.10 Time-To-Go Chart for Minimum Time Rendezvous Paths

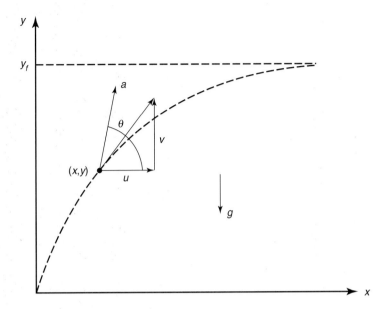

FIGURE 7.11   Nomenclature for Minimum Time Orbit Injection

(a) Show that this leads to the matrix differential equations (5.31) and (5.83) with boundary conditions (5.82).

(b) Show that $\hat{\psi} \triangleq M(t)x$ is the predicted value of $\psi \triangleq M_f x(t_f)$ neglecting the terminal constraints, i.e., using the control $u = -R^{-1}B^T Sx$, so that the optimal control law (5.100) could be written as

$$u = -R^{-1}B^T(Sx + M^T Q^{-1}\hat{\psi}) .$$

7.4 The Combinatorial Minimum Principle

Dynamic programming becomes a feasible computational method for discrete step optimal control problems *if the states and the controls are coarsely quantized*, i.e., if there are only a small number of possible values for each state and control variable.

Example 7.4.1—DP Solution of a Combinatorial Problem with a Three by Three Grid

Consider a simple example: we wish to find the path from A to B in Figure 7.12 traveling only to the right, such that the sum of the numbers on the segments along this path is a minimum. If we consider these numbers to be travel times, then we are looking for the min time path from A to B. The "control" is the decision to go up-right or down-right at each node (only two possible values); the "state" is the vertical location in the grid that has between 1 and 4 possible values.

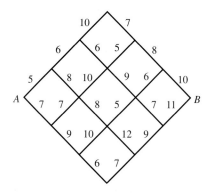

FIGURE 7.12   A Min Time
Combinatorial Problem; Numbers are
Times to Travel Legs of the Grid

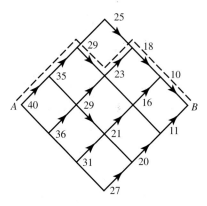

FIGURE 7.13   DP Solution to the
Problem of Figure 7.12

It would be tedious to try all 20 of the possible routes from A to B traveling only to the right. Instead of starting at A and trying different routes to B, we work *backward* from B to find the min time routes to B from each of the 15 nodes on the grid as indicated in Figure 7.13.

The first step backward can be taken either up or down. In Figure 7.14 we have placed the time figures (10 and 11) near these two nodes, and we have indicated by a small arrow the direction from these nodes to the end point B.

Now let us examine the min time from the node marked with an "x" in Figure 7.14 to the terminal point B. Two paths are possible: one through the '10' node, which would take 16 units of time (6 + 10); and the other through the "11" node, which would take 18 units of time (7 + 11). Clearly, the faster path passes through the "10" node so we replace the "x" by a "16" and draw an arrow pointing from that node to the "10" node.

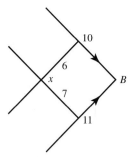

FIGURE 7.14   First
Step in DP Solution to
the Problem of
Figure 7.12

This same procedure is repeated over and over for new nodes lying to the left of nodes that have already been labeled by their shortest times to B. Finally, we have the min time values for every node, shown in Figure 7.13, and the optimal direction to follow in leaving each node.

The shortest-time path can now be traced from A to B by moving always in the direction indicated by the arrows; this route is shown on Figure 7.13 by a dashed line and takes 40 units of time. We also have a useful by-product, namely the min time routes from each of the nodes to B.

We had to find only 15 numbers using this algorithm instead of computing the time for each of the 20 possible paths. The savings in computation are even more impressive as the number of segments on a side increases:

Segments on a side	3	4	5	6	7	n
Possible routes	20	70	252	724	2632	$(2n)!/n!n!$
Computations	15	24	35	48	63	$(n+1)^2 - 1$

Problems

7.4.1 Min Distance in a Plane: Consider the simple problem of finding the min time path traveling at unit velocity (equivalent to the min distance path) to the origin in (x, y) space:

$$\min_{\theta(t)} J = t_f$$

subject to

$$\dot{x} = -\cos\theta , \quad \dot{y} = -\sin\theta .$$

Solve this problem by the DP method, and identify the optimal return function and the feedback control function.

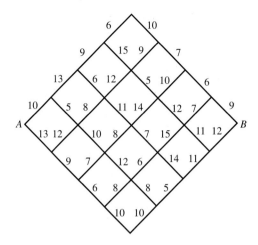

FIGURE 7.15   Grid for 7.4.2

7.4.2 Min Time in a Five-by-Five Grid: In the five-by-five grid shown in Figure 7.15, find the min time path from A to B moving only to the right. There are 70 possible routes, but you need to calculate only 24 numbers.

7.4.3 Max Time in a Five-by-Five Grid: Find the max time path from A to B for the grid of Figure 7.13, still moving only to the right.

7.4.4 Min Time Air Routes: Many airlines now use combinatorial DP to find min time flight paths, taking into account the variable temperatures and the strong winds that usually occur at jet cruising altitudes and the restrictions on possible routes imposed by air traffic control. Savings on the order of 15 minutes to half an hour on a nominal 7-hour flight are obtained this way. A grid of "checkpoints" is selected and all routes must consist of straight-line segments between checkpoints. A simplified version of such a grid is shown in Figure 7.16. Imagine, for example, that point A is New York and point M is London. Checkpoints B through L are points above the ocean, located by giving their longitude and latitude. Using wind and temperature data collected by weather ships, the flight planner computes the time to fly each segment. In practice, there are more checkpoints (on the order of 15) than are shown in Figure 7.16, so a computer solution is essential. For the times (in minutes) given on the figure, use the DP algorithm to find the min time route.

7.4.5 Min Time in an Irregular Grid: The DP algorithm developed in the example can also be used on irregular patterns with more than two choices at some nodes, as in Figure 7.17. Find the min time path from node 12 to node 1, moving only to the right.

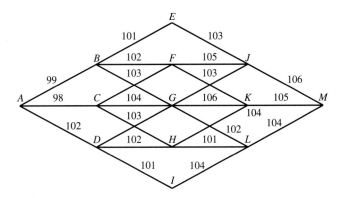

FIGURE 7.16   Grid for Problem 7.4.4

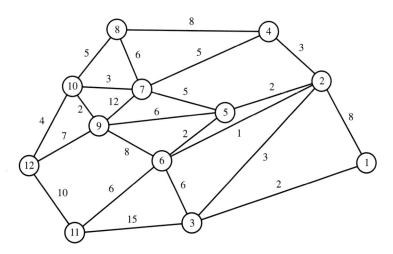

FIGURE 7.17   Grid for Problem 7.4.5

7.4.6 Min Time in a Grid with Terminal Costs: Consider the routing problem illustrated in Figure 7.18. We wish to proceed along a path from level 0 to level 4, with the cost of each leg of the path given by the number associated with that particular segment. The total cost is measured by the sum of the cost of each segment and the terminal cost associated with each position of level 4.

(a) Find the min cost path from point A to level 4.
(b) Find the min cost path from point A to point B.
(c) Find the min cost path from level 0 to level 4.
(d) Find the min cost path from point A to the surface S, assuming 0 terminal costs on S.

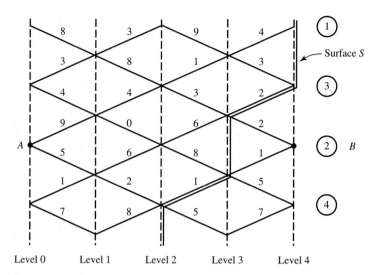

FIGURE 7.18   Grid for Problem 7.4.6

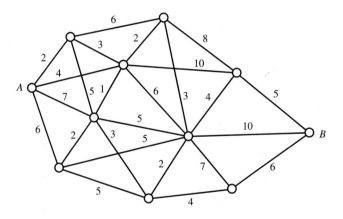

FIGURE 7.19   Grid for Problem 7.4.7

(e) (Suggested by S. E. Dreyfus) Solve (a) with the added consideration that, each time we change directions, one additional unit of cost is added to the cost of the segment following. What is different about this problem?

7.4.7 Min Time on an Irregular Grid: Find the min time and the min time path from point A to point B, moving only to the right. The time to travel between points is shown on each segment of Figure 7.19.

7.4.8 Min Time on a Grid with Initial and Final Costs: In Figure 7.20 find the min cost and the corresponding path (or paths) from level 0 to level 3 moving only to the right. The

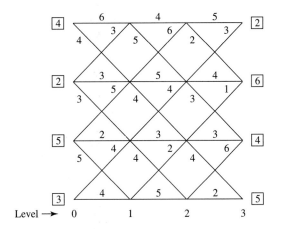

FIGURE 7.20 Grid for Problem 7.4.8

cost of traveling each segment is shown on the segment. There are also initial and terminal costs which are shown in boxes.

7.5 Chapter Summary

DP treats the same dynamic optimization problem as the one in Chapter 4, but finds $u(x, t)$ instead of $u(x_o, t_o, t)$, i.e., a feedback solution for all possible "initial" states at any time t, instead of only one specified initial state x_o at $t = t_o$.

The basic relation is the HJB Equation:

$$J_t + H^o(x, t, J_x^o) = 0 , \quad \text{(a partial differential equation)}$$
$$J^o(x, t) = \phi(x, t) \quad \text{on} \quad \psi(x, t) = 0 , \quad \text{(boundary condition)}$$

where

$$H = J_x^o f(x, u, t) \quad \Rightarrow \lambda^T \equiv J_x^o ,$$
$$H^o = \min_u H(x, u, t, J_x^o) ,$$
$$u(x, t) = \arg\left[\min_u H(x, u, t, J_x^o)\right] ,$$
$$J^o(x, t) = \text{optimal return function} .$$

The DP concept is elegant and powerful. However, DP numerical algorithms are only feasible for low-order discrete-step problems with coarsely quantized state and control variables.

CHAPTER 8

Neighboring Optimum
Control

8.1 Introduction

For many nonlinear control problems a feedback (DP) solution is needed only in a neighborhood of a nominal optimum path in the state space. This can be provided by storing the nominal optimum path and the gradient of the optimum control with respect to the states on the nominal optimum path, i.e., a time-varying feedback gain on state deviation. Neighboring optimum paths can then be generated in real time by feedforward of the nominal optimum control plus linear feedback of the deviation of the state from the nominal optimum state.

DP requires storing the optimal control $u^o(x)$ over an *n-dimensional* grid in the neighborhood of the nominal path. For open-end time problems the time-to-go $T^o(x)$ must also be stored. Storing a function of one variable can be done in a few lines; two variables requires a few pages; three variables a book; four variables a shelf of books; five variables a bookcase; six variables a room of bookcases; seven variables a library, etc. Bellman called this the "curse of dimensionality."

By contrast, neighboring optimum feedback control is *one-dimensional* independent of the dimension of x. It requires storing only $u^o_N(T)$ and $x^o_N(T)$, the nominal optimum control and state histories as a function of $T =$ time-to-go, plus the gain matrix u^o_x. For open-end time problems $T^o_x =$ the gradient of the time-to-go with respect to the states must also be stored.

Determining u^o_x and T^o_x requires the solution of the *Accessory Minimum Problem (AMP)* of the calculus of variations, which is the basis for investigating the *second variation,* i.e., expansion of the performance index to second order, and linearization of the plant

315

equations, about the nominal optimum path. The AMP is a time-varying LQ terminal control problem with zero terminal errors (see Section 5.4). The weighting matrices in the quadratic performance index are the second derivatives of the augmented performance index $\Phi_{xx} \overset{\Delta}{=} \phi_{xx} + \nu^T \psi_{xx}$ and of the Hamiltonian with respect to $y = [x; u]$ (see below). If Φ_{xx} is not positive definite, *focal points* will occur on the optimal paths for large T (Section 8.6). If H_{yy} is not positive definite, *conjugate points* will occur on the extremal paths for large enough T (Section 8.7).

The solution of the AMP also provides *sufficient conditions for determining the local minimality* of the path. Thus, the AMP plays the same role in the calculus of variations that the second derivative test plays in parameter optimization (see Section 1.4).

An NR algorithm for computing the nominal optimum path and the feedback gains together can be obtained by an extension of the AMP. In this algorithm H_u is a forcing function that is driven to zero in successive iterations. While this algorithm is computationally intensive compared to the first-order gradient schemes discussed in Chapters 2 to 4, it finds the solution with fewer iterations like shooting algorithms. Unlike shooting algorithms the integration is stable in both directions; the trajectory is improved by changing the control throughout the path rather than by changing only the boundary conditions. However, the NR algorithm must be supplied with the second derivatives of the terminal cost and the system equations (also the case for the AMP problem), which is not the case for shooting algorithms.

Outline of Neighboring Optimum Control (NOC)

This scheme was developed in Refs. SB2, DM, JM, WB, and BWC. It involves calculation and storage of the nominal optimum control and state vectors $u_N^o(T)$ and $x_N^o(T)$, where $T = t_f - t =$ time-to-go. At the same time, the gradient of the control vector $u_x^o(T)$ and, for open-end time problems, the gradient of the time-to-go T_x^o are calculated and stored. These latter quantities are found by solving the AMP, which requires backward solution of a matrix Riccati equation over the nominal optimum path. For min time problems, $T_x^o(T) = \lambda(T)$ since $J = T$.

Then during the operation or simulation of the system, T^o and $u^o(T)$ are determined as follows:

(a) Obtain an estimate of the current state x.
(b) For fixed-end time problems, $T = t_f - t$, where $t_f =$ specified final time, and $t =$ the current time. For open-end time problems, find T by interpolation in a table of $x_N^o(T)$ and $T_x^o(T)$ so that the point on the nominal path has the same time-to-go as the current point, i.e., find T so that

$$0 = T_x^o(T)[x - x_N^o(T)].$$

This procedure implicitly determines T, hence also $x_N^o(T)$, and is called a "transversal comparison" in the calculus of variations.
(c) With the value of T obtained in (b), interpolate in the stored table to find $u_N^o(T)$ and $u_x^o(T)$.
(d) Calculate the neighboring optimal control $u^o(T) = u_N^o(T) + u_x^o(T)[x - x_N^o(T)]$.

8.2 The AMP for Continuous Systems

Consider the Mayer problem of Section 3.3, where final time t_f is specified. *First-order necessary conditions* are (3.39) to (3.41) and the EL equations are (3.45) to (3.47).

Second-order necessary conditions may be derived by considering perturbations of the first-order necessary conditions or, equivalently, expanding the performance index to second order and linearizing the plant equations about the nominal optimum path. This leads to the following AMP:

$$\min_{\delta u(t)}(J - J_N^o) \approx \frac{1}{2}\delta x^T(t_f)\Phi_{xx}\delta x(t_f) + \frac{1}{2}\int_o^{t_f} \begin{bmatrix} \delta x^T & \delta u^T \end{bmatrix} \begin{bmatrix} H_{xx} & H_{xu} \\ H_{ux} & H_{uu} \end{bmatrix} \begin{bmatrix} \delta x \\ \delta u \end{bmatrix} dt \tag{8.1}$$

subject to

$$\delta\dot{x} = f_x\delta x + f_u\delta u , \quad \delta x(t_o) = 0 , \quad \psi_x\delta x(t_f) = 0 . \tag{8.2}$$

This problem is *identical to the time-varying continuous LQ terminal control problem with zero terminal error* treated in Section 5.4 if we make the following changes in nomenclature:

$$x(t) , \quad u(t) , \quad v \to \delta x(t) , \quad \delta u(t) , \quad dv ,$$
$$A , \quad B \to f_x(t) , \quad f_u(t) ,$$
$$Q , \quad N , \quad R \to H_{xx}(t) , \quad H_{xu}(t) , \quad H_{uu}(t) ,$$
$$x_0 , \quad \psi \to dx_0 , \quad d\psi ,$$
$$S_f , \quad M_f \to \Phi_{xx} , \quad \psi_x ,$$

where the derivatives f_x, f_u, H_{xx}, H_{xu}, H_{uu} are evaluated on the optimum path and are functions of t (only linear time-invariant systems were treated in Section 5.4). Thus the AMP EL equations are

$$\begin{bmatrix} \delta\dot{x} \\ \delta\dot{\lambda} \end{bmatrix} = \begin{bmatrix} \mathcal{A} & -\mathcal{B} \\ -\mathcal{C} & -\mathcal{A}^T \end{bmatrix} \begin{bmatrix} \delta x \\ \delta\lambda \end{bmatrix} , \tag{8.3}$$

where

$$\delta u = -H_{uu}^{-1}(f_u^T\delta\lambda + H_{ux}\delta x) , \tag{8.4}$$

and

$$\mathcal{A} \triangleq f_x - f_u H_{uu}^{-1} H_{ux} , \quad \mathcal{B} \triangleq f_u H_{uu}^{-1} f_u^T , \quad \mathcal{C} \triangleq H_{xx} - H_{xu} H_{uu}^{-1} H_{ux} . \tag{8.5}$$

The sweep (or DP) solution for the AMP is the solution by Riccati equation presented in Section 5.4

$$\dot{S} = -S\mathcal{A} - \mathcal{A}^T S + S\mathcal{B}S - \mathcal{C} , \quad \dot{M} = -M(\mathcal{A} - \mathcal{B}S) , \quad \dot{\mathcal{Q}} = -M\mathcal{B}M^T , \tag{8.6}$$

with boundary conditions

$$S(t_f) = \Phi_{xx} \triangleq \phi_{xx} + v^T\psi_{xx} , \quad M(t_f) = \psi_x , \quad \mathcal{Q}(t_f) = 0 . \tag{8.7}$$

The neighboring optimal perturbation control is

$$\delta u = u_x^o(t)\delta x \ ,\tag{8.8}$$

where

$$u_x^o(t) = -H_{uu}^{-1}(t)\left\{H_{ux}(t) + f_u^T(t)[S(t) + M^T(t)Q^{-1}(t)M(t)]\right\} \ .\tag{8.9}$$

Sufficient Conditions for a Local Minimum

If the feedback gain matrix $u_x^o(T)$ is finite over a stationary path, this *ensures that the nominal path is a local minimum.* Hence solving the AMP is to the calculus of variations as the second derivative test is to parameter optimization. This test originated with Jacobi in 1836 (Ref. Go), and inherently requires the Legendre-Clebsch condition that H_{uu} be positive definite over the whole path.

Example 8.2.1—Ship in a Linear Current; Nominal Optimal Path and NOC Feedback Gains

Consider the Zermelo problem of Example 7.2.1, namely finding min time paths for a ship to travel to $x = y = 0$ in the presence of a linear variation of current with y. A DP solution was obtained there by inverting the implicit equations below for (θ, θ_f) given (x, y):

$$y = \sec\theta_f - \sec\theta \ ,$$
$$2x = \sinh^{-1}(\tan\theta_f) - \sinh^{-1}(\tan\theta) - y\tan\theta + T\sec\theta_f \ ,$$

where the time-to-go is $T = \tan\theta_f - \tan\theta$. Thus we do not really need a NOC solution; however, it is instructive since it demonstrates the differences betwen DP and NOC. The gradients u_x^o and T_x^o can be found by perturbations of the solution above:

$$\begin{bmatrix}\delta x \\ \delta y\end{bmatrix} = \begin{bmatrix}x_\theta & x_{\theta_f} \\ y_\theta & y_{\theta_f}\end{bmatrix}\begin{bmatrix}\delta\theta \\ d\theta_f\end{bmatrix} \ .$$

where

$$x_\theta = -\sec\theta(\tan^2\theta - \sec\theta\sec\theta_f) \ , \qquad y_\theta = \sec\theta\tan\theta \ ,$$
$$x_{\theta_f} = -\sec\theta_f(\sec^2\theta_f - \tan\theta\tan\theta_f) \ , \quad y_{\theta_f} = -\sec\theta_f\tan\theta_f \ .$$

Inverting this relation, we obtain

$$\begin{bmatrix}\delta\theta \\ d\theta_f\end{bmatrix} = \frac{1}{D}\begin{bmatrix}y_{\theta_f} & -x_{\theta_f} \\ -y_\theta & x_\theta\end{bmatrix}\begin{bmatrix}\delta x \\ \delta y\end{bmatrix} \ .$$

where $D \overset{\Delta}{=} \sec^2\theta\sec^3\theta_f(\sin\theta_f - \sin\theta)$. This yields the desired perturbation feedback

$$\delta\theta = \theta_x\delta x + \theta_y\delta y \ ,$$

where $\theta_x \equiv y_{\theta_f}/D$ and $\theta_y \equiv -x_{\theta_f}/D$.

The point on the nominal optimum path to use as a reference for perturbations is the point that has the *same time-to-go as the point on the neighboring path*. Hence we consider differential changes of the time-to-go T:

$$dT = T_x \delta x + T_y \delta y ,$$

where

$$T_x = -\cos \theta_f , \quad T_y = -\tan \theta \cos \theta_f .$$

Given (x, y) we can find T by interpolation in a table of $T_x(T)$, $T_y(T)$, $x_N(T)$, $y_N(T)$, such that $dT = 0$, i.e., we find T so that

$$0 = T_x(T)[x - x_N(T)] + T_y(T)[y - y_N(T)] .$$

Then we use the neighboring optimum feedback law to find $\theta(t)$:

$$\theta = \theta_N(T) + \theta_x(T)[x - x_N(T)] + \theta_y(T)[y - y_N(T)] .$$

Equivalently, θ could be determined by interpolation on an (x, y) plot like the one shown in Figure 8.1, which was prepared for a nominal optimum path with $\theta_f = 240$ deg . Comparing this figure with the feedback charts in Figures 7.3 and 7.4, we see that the contours of constant θ and constant T are approximated locally on the nominal optimum path by straight-line segments tangent to the exact contours.

For implementation of NOC, a table like the one below could be stored in a computer.

NOC table for a Zermelo problem

x_N	y_N	θ_N	T_N	θ_x	θ_y	T_x	T_y
3.659	−1.864	105	5.464	−.0158	.0957	.5	−1.866
4.872	−.613	112.5	4.146	−.0373	.167	.5	−1.207
4.781	0	120	3.464	−.0625	.253	.5	−.866
4.124	.586	135	2.732	−.138	.455	.5	−.500
3.487	.845	150	2.309	−.238	.686	.5	−.289
2.920	.965	165	2.000	−.359	.926	.5	−.134
2.390	1.000	180	1.732	−.500	1.155	.5	0
1.861	.965	195	1.464	−.665	1.358	.5	.134
1.294	.845	210	1.155	−.887	1.536	.5	.289
.657	.586	225	.732	−1.362	1.784	.5	.500
0	0	240	0	−∞	∞	.5	.866

For example, suppose $x = 5.000$, $y = .500$. The point on the nominal optimum path that has the same time-to-go is found by interpolation to be at $T = 3.21$, which is 34% of the

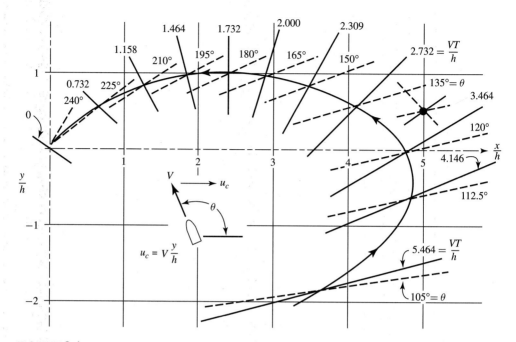

FIGURE 8.1   Perturbation Feedback Chart for Min Time Paths of a Ship through a Region with a Linear Variation in Current Velocity

way between the point at $\theta_N = 120$ deg and the point at $\theta_N = 135$ deg. The corresponding values of (x_N, y_N) are (4.56, .20) since

$$0 \cong .5(5.00 - 4.56) - .741(.50 - .20) \,,$$

and the corresponding value of θ_N is 125.1 deg.

Nominal feedforward plus neighboring optimum feedback with this time-to-go gives

$$\theta \cong 125.1 + \frac{180}{\pi}[-.0883(5.00 - 4.56) + .322(.50 - .20)] \,, \cong 128 \text{ deg} \,.$$

The exact values of T and θ obtained by solving the implicit equations are $T = 3.20$ and $\theta = 127$ deg. A MATLAB code for implementing NOC given an estimate of the current state x and a table containing the nominal optimum path $u_N(T_N)$, $x_N(T_N)$ and the feedback gains $K = -u_x(T_N)$, $K_t = -T_x(T_N)$ is listed in Table 8.1.

Example 8.2.2—Neighboring Min Time Paths for a Ship in a Linear Current; Simulation Using NOC

The code ZERM-NOC listed below contains a table of the nominal control, the nominal time-to-go, and the feedback gains; it calls the code listed in Table 8.1 to determine the NOC.

TABLE 8.1   A MATLAB Code for NOC

```
function [u,T]=noc(x,un,xn,Tn,Kt,K)
% Neighboring Optimum Control u and estimated time-to-go T given
    the current
% state x and a table containing un(i)=nominal control history (1 x N),
% xn(:,i)=nominal state history (ns x N),
    Kt(:,i)=gain for perturbation in T
% (ns x N), K(i)=gain vector for perturbation in control (ns x N), all as
% functions of Tn(i)=nominal time-to-go (1xN),
    where dT(i)=-Kt(i)'*[x-xn(i)],
% du(i)=-K(i)'*[x-xn(i)].
%
% Finds data points i=k and i=k+1 between which dT(i) changes sign and
    inter-
% polates a rough estimate of T:
%
N=length(Tn); for i=1:N, dT(i)=-Kt(:,i)'*(x-xn(:,i)); end
for i=1:N-1, if dT(i)*dT(i+1)<0, k=i; end; end
ep=dT(k)/(dT(k)-dT(k+1)); T=Tn(k)+ep*(Tn(k+1)-Tn(k));
%
% Refines T by also interpolating Kt and xn:
dTi=1; while abs(dTi)>1e-5,
 Kti=Kt(:,k)+ep*(Kt(:,k+1)-Kt(:,k)); xi=xn(:,k)+ep*(xn(:,k+1)-xn(:,k));
 dTi=-Kti'*(x-xi); T=T+dTi; ep=(T-Tn(k))/(Tn(k+1)-Tn(k));
end
%
% Finds neighboring optimum control:
ui=un(k)+ep*(un(k+1)-un(k)); Ki=K(:,k)+ep*(K(:,k+1)-K(:,k));
u=ui-Ki'*(x-xi);
```

```
            MATLAB Code for Simulation of Zermelo Problem Using NOC
function yp=zerm_noc(t,s)
% ZERMelo problem with NOC - ship in linear current; nominal optimal
    control
% un, state history sn, time-to-go Tn, and corresponding optimal feedback
% gains K and Kt.
%
un=(pi/180)*[100:5:235];
sn(1,:)=[-1.3786  3.6587  4.7366  4.8956  4.7811  4.5790  4.3524 ...
    4.1241   3.9025   3.6900   3.4866   3.2913   3.1030   2.9201 ...
    2.7414   2.5653   2.3905   2.2158   2.0397   1.8609   1.6781 ...
    1.4897   1.2945   1.0911   0.8786   0.6569   0.4287   0.2021];
sn(2,:)=[-3.7588 -1.8637 -0.9238 -0.3662  0.0000  0.2566  0.4443 ...
```

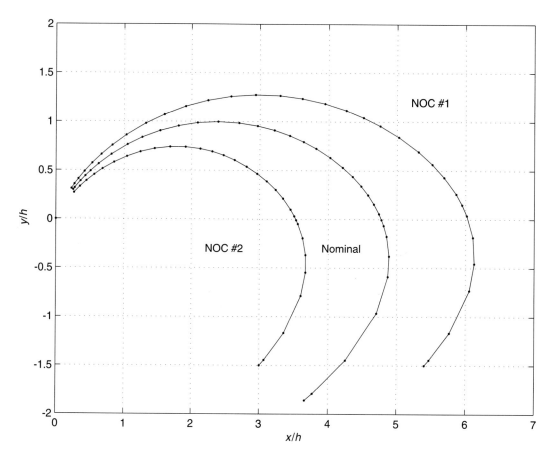

FIGURE 8.2   Simulated Paths Using NOC; Ship in a Linear Current

```
       0.5858    0.6946    0.7792    0.8453    0.8966    0.9358    0.9647 ...
       0.9846    0.9962    1.0000    0.9962    0.9846    0.9647    0.9358 ...
       0.8966    0.8453    0.7792    0.6946    0.5858    0.4443    0.2566];
Tn=[7.4033    5.4641    4.4795    3.8766    3.4641    3.1602    2.9238 ...
       2.7321    2.5712    2.4323    2.3094    2.1984    2.0960    2.0000 ...
       1.9084    1.8195    1.7321    1.6446    1.5557    1.4641    1.3681 ...
       1.2657    1.1547    1.0318    0.8930    0.7321    0.5403    0.3039];
Kt(1,:)=-.5*ones(1,28);
Kt(2,:)=[2.8356    1.8660    1.3737    1.0723    0.8660    0.7141    0.5959 ...
       0.5000    0.4195    0.3501    0.2887    0.2332    0.1820    0.1340 ...
       0.0882    0.0437    0.0000   -0.0437   -0.0882   -0.1340   -0.1820 ...
      -0.2332   -0.2887   -0.3501   -0.4195   -0.5000   -0.5959   -0.7141];
K(1,:)=[0.0071    0.0158  0.0281  0.0436  0.0625  0.0845  0.1096 ...
       0.1376    0.1684  0.2018  0.2377  0.2760  0.3165  0.3592 ...
```

```
     0.4039    0.4508    0.5000    0.5517    0.6065    0.6654    0.7297 ...
     0.8021    0.8873    0.9935    1.1383    1.3624    1.7894    3.0392];
K(2,:)=[-0.0563  -0.0957 -0.1419 -0.1944 -0.2526 -0.3160 -0.3838 ...
    -0.4555   -0.5302   -0.6074   -0.6863   -0.7661   -0.8462   -0.9257 ...
    -1.0041   -1.0806   -1.1547   -1.2259   -1.2938   -1.3583   -1.4196 ...
    -1.4784   -1.5368   -1.5988   -1.6736   -1.7839   -2.0000   -2.6783];
%
% Calls NOC to calculate optimal control be and time-to-go T:
[be,T]=noc(s,un,sn,Tn,Kt,K);
%
% State variable rates (sdot) for ODE23:
yp=[cos(be)-s(2); sin(be)];
```

The following command in MATLAB performs the simulation:

$$[t,s]=ode23('zerm-noc',[0 \; tf],s0)$$

where $s0 = [x(0) \; y(0)]^T$, t_f = final time. The outputs are a time vector t and a state variable history $s(t)$. The nominal and two neighboring optimal paths were generated in this manner and are shown in Figure 8.2. These paths are almost indistinguishable from the exact optimal paths, indicating the effectiveness of the NOC method.

8.3 The Discrete AMP (DAMP)

Consider the Mayer problem of Section 3.1, where final time t_f is specified. *First-order necessary conditions* are the EL equations (3.2 to 3.4) and (3.8 to 3.10). *Second-order necessary conditions* may be derived by considering perturbations of the first-order necessary conditions or, equivalently, by expanding the QPI to second-order and linearizing the plant equations about the nominal optimum path. This leads to the following DAMP:

$$\min_{du(i),dv} dJ^o \approx \frac{1}{2}dx^T(N)\Phi_{xx}dx(N) + \frac{1}{2}\sum_{i=0}^{N-1}\begin{bmatrix} dx^T(i) & du^T(i) \end{bmatrix}\begin{bmatrix} H_{xx} & H_{xu} \\ H_{ux} & H_{uu} \end{bmatrix}\begin{bmatrix} dx(i) \\ du(i) \end{bmatrix} \tag{8.10}$$

subject to

$$dx(i+1) = f_x dx(i) + f_u du(i), \quad dx(0) = dx_0, \quad \psi_x dx(N) = d\psi. \tag{8.11}$$

This problem is *identical to the discrete LQ terminal control problem* treated in Section 5.5 if we identify the following changes in nomenclature:

$$x(i), \; u(i), \; v \rightarrow dx(i), \; du(i), \; dv,$$
$$\Phi, \; \Gamma \rightarrow f_x(i), \; f_u(i),$$
$$Q_d, \; N_d, \; R_d \rightarrow H_{xx}(i), \; H_{xu}(i), \; H_{uu}(i),$$
$$x_0, \; \psi \rightarrow dx_0, \; d\psi,$$
$$S_f, \; M_f \rightarrow \Phi_{xx}, \; \psi_x,$$

where the derivatives $f_x, f_u, H_{xx}, H_{xu}, H_{uu}$ are evaluated on the optimum path and are functions of i.

Sufficient Conditions for a Local Minimum

If the feedback gain matrix $K(i)$ is finite for $i < N - n_t$ over a stationary path, this *ensures that the nominal path is a local minimum*. Hence solving the DAMP is to the discrete calculus of variations as the second derivative test is to parameter optimization. Note this requires the existence of $Z_{uu}^{-1}(i)$, *not* $H_{uu}^{-1}(i)$.

The MATLAB code DAMPC listed in Table 8.2 checks the sufficient conditions for a local minimum.

Example 8.3.1—Sufficient Conditions for DTDP for Max u_f with Specified (v_f, y_f)

The MATLAB script below tests Example 3.2.1; the result is that the path is at least a local maximum, since H_u is close to zero and the feedback gains are finite.

TABLE 8.2   A MATLAB Code for DAMPC

```
function [s,K,Hu]=dampc(name,u,nu,s0,tf)
% Disc. Accessory Min. Pb. w. term. Constr. using Riccati matrix S & aux.
% matrices M & Q; function file 'name' should be in the Matlab path and
% calculate f for flg=1, (Phi,Phis,Phiss) for flg=2, and
%    (fs,fu,fss,fus,fuu)
% for flg=3; inputs: u=opt. ctrl. sequence, nu=term. Lagrange mult. vector,
% s0=initial state, tf=final time; outputs: s=optimal state sequence, K=
% feedback gain sequence; iterate until Hu=0.
%
% Nominal optimum s(i) sequence:
N=length(u); ns=length(s0); s=zeros(ns,N+1); dt=tf/N; s(:,1)=s0;
for i=1:N, t=(i-1)*dt; s(:,i+1)=feval(name,u(i),s(:,i),dt,t,1); end;
%
[Phi,Phis,Phiss]=feval(name,u(N),s(:,N+1),dt,tf,2); nt1=length(Phi);
nt=nt1-1; phi=Phi(1); psi=Phi([2:nt1]); la=Phis'*[1; nu];
S=Phiss([1:ns],[1:ns]);
for j=1:nt, S=S+nu(j)*Phiss([j*ns+1:(j+1)*ns],:); end;
M1=Phis([2:nt1],:); Q=zeros(nt); K=zeros(N-nt,ns);
for i=N:-1:1, t=(i-1)*dt;
 [fs,fu,fss,fsu,fuu]=feval(name,u(:,i),s(:,i),dt,t,3);
 Hu(i)=la'*fu; Zss=zeros(ns); for j=1:ns,
  Zss=Zss+la(j)*fss([(j-1)*ns+1:j*ns],:); end;
 Zss=Zss+fs'*S*fs; Zus=la'*fsu+fu'*S*fs; Zsu=Zus';
 Zuu=la'*fuu+fu'*S*fu; M=M1*(fs-fu*(Zuu\Zsu'));
 if i<=N-nt, K(i,:)=Zuu\(Zus+fu'*M1'*(Q\M)); end;
 S=Zss-Zsu*(Zuu\Zus); Q=Q+M1*fu*(Zuu\fu')*M1'; M1=M; la=fs'*la;
end;
```

```
% Script e08_3_1.m; 2nd order sufficient conditions - DTDP for max uf with
% vf=0 and yf specified using DAMPC.
%
% Generate candidate path:
N=20; u0=(pi/3)*[1:-2/N:-1+2/N]; s0=[0 0 0 0]'; tf=1; k=-7;
tol=5e-5; [u,s,nu,la0]=dopc('dtdpc',u0,s0,tf,k,tol); t=[0:1/N:1];
%
% Test sufficient conditions:
[s,K,Hu]=dampc('dtdpc',u,nu,s0,tf);
```

8.4   An NR Algorithm for Discrete Systems

Here we develop an NR algorithm for discrete systems with soft terminal constraints, based on the DAMP, that finds the nominal optimum path starting with a non-optimal $[H_u(i) \neq 0]$ path. To do this we use a *backward sweep* identical to the ones described in Chapter 5 except we include an inhomogeneous term $h(i)$ to handle the forcing function $dH_u(i)$:

$$d\lambda(i) = S(i)dx(i) + h(i) . \tag{8.12}$$

Consider (8.12) with i replaced by $i + 1$ and substitute for $dx(i + 1)$ from (8.11):

$$d\lambda(i + 1) = S(i + 1)[f_x dx(i) + f_u du(i)] + h(i + 1) . \tag{8.13}$$

Substitute (8.13) into the third equation of (5.56) and solve for $du(i)$:

$$du(i) = -Z_{uu}^{-1}[Z_{ux}dx(i) + f_u^T h(i + 1) - dH_u^T(i)] , \tag{8.14}$$

where Z_{uu} and Z_{ux} were defined in (5.104).

Next, substitute (8.13) into the second equation of (5.56):

$$d\lambda(i) = Z_{xx}dx(i) + Z_{ux}^T du(i) + f_x^T h(i + 1) , \tag{8.15}$$

where Z_{xx} was defined in (5.104). Now substitute (8.12) and (8.14) into (8.15):

$$S(i)dx(i) + h(i) = Z_{xx}dx(i) - Z_{ux}^T Z_{uu}^{-1}[Z_{ux}dx(i) + f_u^T h(i + 1) - dH_u^T(i)] + f_x^T h(i + 1) . \tag{8.16}$$

For (8.16) to be valid for arbitrary $dx(i)$, the coefficient of $dx(i)$ must vanish. This gives a backward discrete Riccati equation for $S(i)$, namely, (5.105) and a linear equation for $h(i)$:

$$h(i) = f_x^T h(i + 1) - Z_{ux}^T Z_{uu}^{-1}[f_u^T h(i + 1) - dH_u^T(i)] , \quad h(N) = 0 . \tag{8.17}$$

If we put

$$dH_u^T(i) = -H_u^T(i) , \tag{8.18}$$

then we can simultaneously sequence the first-order necessary conditions and these second-order conditions and (8.15). This constitutes an NR procedure and the convergence is rapid near the nominal optimum path.

TABLE 8.3   MATLAB Code DOP0N2

```
function [u,s,K]=dop0n2(name,un,s0,tf,tol)
% Discrete OPtim. with 0 terminal constraints using a NR algorithm with 2nd
% deriv. propagated by bkwd Ricc. eqn; function file
    'name' should be in the
% MATLAB path and calculate f for flg=1, (phi,phis,phiss) for flg=2, and
% (fs,fu,fss,fus,fuu) for flg=3; un=init. guess of scalar control sequence;
% s0=initial state; tf=final time; tol=tolerance on duavg.
%
N=length(un); ns=length(s0); s=zeros(ns,N+1); sn=s; K=zeros(N,ns); dua=1;
it=0; dt=tf/N; s(:,1)=s0; disp('      Iter.      phi      duavg')
while max(dua) > tol,
  % Forward sequence, storing state histories:
    for i=1:N,
      u(i)=un(i)-K(i,:)*(s(:,i)-sn(:,i)); t=(i-1)*dt;
      s(:,i+1)=feval(name,u(i),s(:,i),dt,t,1);
    end;
  % Performance index and terminal boundary conditions:
    [phi,phis,phiss]=feval(name,0,s(:,N+1),0,0,2); la=phis'; S=phiss;
    h=zeros(ns,1);
  % Backward sequence, storing du(i) and K(i,:):
    for i=N:-1:1, t=(i-1)*dt;
    [fs,fu,fss,fus,fuu]=feval(name,u(i),s(:,i),dt,t,3); Hu=la'*fu;
    Zss=zeros(ns); for j=1:ns, Zss=Zss+la(j)*fss([(j-1)*ns+1:j*ns],:); end;
    Zss=Zss+fs'*S*fs; Zus=la'*fus+fu'*S*fs; Zsu=Zus';
    Zuu=la'*fuu+fu'*S*fu; K(i,:)=Zuu\Zus; du(i)=-Zuu\(fu'*h+Hu');
    la=fs'*la; S=Zss-Zsu*K(i,:); h=fs'*h+Zsu*du(i);
    end
  % New un and sn:
    dua=norm(du)/N; disp([it phi dua]); un=u+du; sn=s; it=it+1;
end
```

A by-product of this algorithm, from (8.14), is the *neighboring optimum feedback gain matrix* $K(i) \triangleq u_x^o$, where

$$u^o(i) = u_N^o(i) - K(i)[x(i) - x_N^o(i)] , \qquad (8.19)$$

and $K(i) = Z_{uu}^{-1} Z_{ux}$, $u_N(i)$ = nominal optimum control, $x_N(i)$ = nominal optimum state.

A MATLAB code for the NR algorithm described above is listed in Table 8.3. It requires a subroutine file for the EOM, the performance index with its first and second derivatives, and the first and second derivatives of f.

Since MATLAB 4 does not handle quantities with three subscripts like f_{ss}, (MATLAB 5 does), we enter it as an $n^2 \times n$ matrix, the first n rows being $f_{ss}^{(1)}$ and the last n rows being f_{ss}^n.

Example 8.4.1—DTDP for Max u_f with Specified v_f, y_f Using DOP0N2

A subroutine for finding the max final velocity u_f minus the quadratic penalty $[v(t_f)^2 + (y(t_f) - y_f)^2]$ using DOP0N2 is listed below (cf. Examples 3.1.1 and 3.2.1, where the terminal constraints were satisfied exactly). An advantage of the quadratic penalty approach is that the feedback gains do not approach infinity at the end of the path.

```
function [f1,f2,f3,f4,f5]=dtdp0n2(th,s,dt,t,flg)
% Subroutine for Ex. 8.4.1; DTDP for max uf with spec. (vf,yf); t in tf,
% (u,v) in a*tf, (x,y) in a*tf^2; s = [u,v,y,x]', th = control.
%
yf=.2; u=s(1); v=s(2); y=s(3); x=s(4); co=cos(th); si=sin(th); sf=2e5;
if flg==1,
  f1=s+dt*[co; si; v+dt*si/2; u+dt*co/2];                 % f1=f
elseif flg==2,
  f1=u-sf*(v^2+(y-yf)^2)/2;                               % f1=phi
  f2=[1 -sf*v -sf*(y-yf) 0];                              % f2=phis
  f3=[0 0 0 0; 0 -sf 0 0; 0 0 -sf 0; 0 0 0 0];            % f3=phiss
elseif flg==3,
  f1=[1 0 0 0; 0 1 0 0; 0 dt 1 0; dt 0 0 1];              % f1=fs
  f2=dt*[-si; co; dt*co/2; -dt*si/2];                     % f2=fu
  f3=zeros(16,4);                                         % f3=fss
  f4=zeros(4);                                            % f4=fsu
  f5=dt*[-co; -si; -dt*si/2; -dt*co/2];                   % f5=fuu
end
```

A code for the example is listed below along with an edited diary of a run. An approximate solution for the control history $\theta(i)$ is obtained first using DOPC; then only two iterations with DOP0N2 are required. The resulting nominal optimum path is identical to the one shown in Figure 3.2; Figure 8.3 shows the corresponding perturbation feedback gains that peak a few steps before the end of the path.

```
% Script e08_4_1.m; DTDP for max uf w. spec. (vf,yf) using DOP0N2;
%
N=10; th0=(pi/3)*[1:-2/N:-1+2/N]; s0=[0 0 0 0]'; tf=1; k=-7; tol=.001;
U=1; mxit=12; t=[0:1/N:1];
th=dopc('dtdpc',th0,s0,tf,k,U,tol,mxit); tol=1e-7;
[th,s,K]=dop0n2('dtdp0n2',th,s0,tf,tol);
u=s(1,:); v=s(2,:); y=s(3,:); x=s(4,:); thh=[th th(N)]; Kh=[K; K(N,:)];
```

```
e08_4_1
 Iter.     phi    norm(psi)    dua
     0    0.8262    0.0489    0.2229
1.0000    0.6961    0.0064    0.0531
     -         -         -         -
```

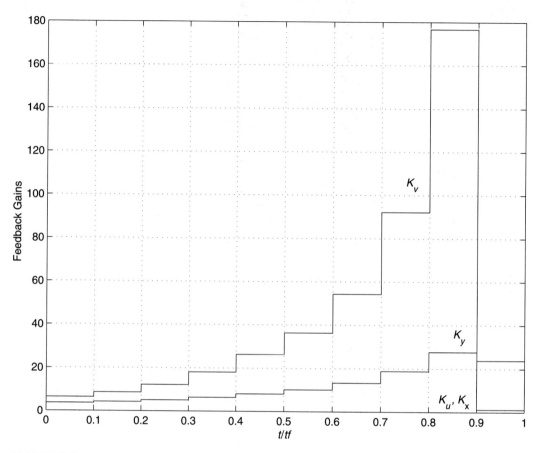

FIGURE 8.3   DTDP for Max Final Velocity with Specified v_f, y_f; Neighboring Optimum Feedback Gains

```
7.0000      0.6744      0.0000      0.0007
 Iter.       phi        duavg
    0        0.6744      0.0011
1.0000       0.6744      0.0001
2.0000       0.6744      0.0000
```

Problems

8.4.1 DVDP for Max Range with Gravity, Specified y_f: Use DOPON2 to solve a modification of Problems 2.2.1 and 3.2.1, namely, maximize

$$J = u(N) - s_y[y(N) - y_f]^2$$

for specified N. Let $x(0) = y(0) = 0$, $t_f = 4$ and $y_f = 0$. Use the exact ZOH EOM that are quadratic in ΔT (see Problem 2.1.1); use DOPC to obtain a good initial guess of the optimal control history.

8.4.2 to 8.4.10: Solve STC versions of Problems 3.2.2 to 3.2.10 using DOP0N2.

8.4.11 DVDP for Max Range with Gravity and Thrust, Specified y_f: Use DOP0N2 to solve a modification of Problem 3.2.11, namely, maximize

$$J = x(N) - s_y[y(N) - y_f]^2$$

for $a = .5$, $y_f = .10$, $t_f = 1$. The path is identical to the one in Problem 3.2.11 since the specified range is the max range for the specified time. Use the exact ZOH EOM which are quadratic in ΔT (see Problem 2.2.5); use DOPC to obtain a good initial guess of the optimal control history.

8.5 Sufficient Conditions and Convexity

Consider the Bolza formulation of the calculus of variations and let

$$H_{yy} \triangleq \begin{bmatrix} H_{xx} & H_{xu} \\ H_{ux} & H_{uu} \end{bmatrix} = L_{yy} + \lambda^T f_{yy} , \tag{8.20}$$

$$\Phi_{xx} \triangleq \phi_{xx} + \nu^T \psi_{xx} . \tag{8.21}$$

For minimization problems, if $H_{yy} > 0$ over the whole path, the space is said to be convex. If $\Phi_{xx} > 0$, the terminal manifold is said to be convex. If both the space and the terminal manifold are convex, then a stationary path is also a minimizing path, i.e., *stationarity and convexity are sufficient conditions to ensure a local minimum.*

It is necessary that $H_{uu} \geq 0$ over the whole path (Legendre-Clebsch condition). However, it is *not* necessary that $H_{yy} \geq 0$.

If H_{yy} is not ≥ 0, the *space is nonconvex.* For such systems, a family of stationary paths, going backward from the terminal manifold, will eventually converge to a point, known as a *conjugate point.* A stationary path, going backward beyond a conjugate point is not a minimizing path. Examples are given in Section 8.6.

If Φ_{xx} is not ≥ 0, the *terminal manifold is nonconvex.* For such systems, a family of stationary paths, going backward from the terminal manifold, will eventually converge to a point, known as a *focal point.* A stationary path, going backward beyond a focal point is not a minimizing path. Examples are given in Section 8.6.

It is somewhat easier to demonstrate nonconvexity for systems with discontinuities in the derivatives of the system equations or the terminal manifold. Such systems have sets of points where there are two or more minimizing paths to the terminal manifold.

Nonconvex Discontinuous Terminal Manifold

Discontinuities in a nonconvex terminal manifold produce points on a stationary path where a quite different path to that point has an equal value of the performance index. Such points are called *Darboux points.* This is illustrated by a simple example.

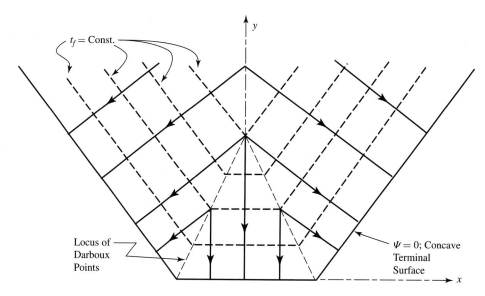

FIGURE 8.4   Min Distance Paths to a Nonconvex Terminal Surface with Discontinuities in Slope—An Example with Darboux Points

Example 8.5.1—Min Distance Paths to a Nonconvex Terminal Surface with Discontinuities in Slope

Consider min distance paths to the nonconvex terminal surface shown in Figure 8.4. A graphical solution is easily obtained; the min distance paths are straight lines perpendicular to the terminal manifold. Darboux points occur where there are two paths of equal length to the terminal manifold.

Nonconvex Discontinuous Space

Discontinuities in the system equations of a nonconvex system also produce points on a stationary path where a quite different path to that point has an equal value of the performance index. Such points are also called *Darboux points*. This is illustrated by a simple example.

Example 8.5.2—Min Distance Paths on a Box

Consider min distance paths on the surface of a box whose top and bottom are square with sides of length $2a$; the four sides of the box are rectangular with one side of length $2a$ and the other side of length $2b$ where $b < a$ (see Figure 8.5). We wish to find min distance paths to the point O located in the center of one of the sides.

This is easily done graphically by cutting and then unfolding the box onto a flat surface. This is done in Figure 8.5 for a box with $b/a = 3/5$. The Darboux points are on a line BC where there are two paths of equal length, one path going on the sides and the other going over the top of the box. The point C is the last point on the "equator" for which the equator is a min distance path; for points between C and A, the min distance path goes over the top.

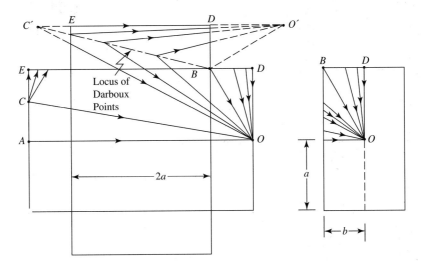

FIGURE 8.5 Example with Darboux Points; Min Distance Paths on a Box
(Unfolded View)

8.6 Focal Points; Nonconvex Terminal Manifolds

Continuous nonconvex terminal manifolds produce convergence of optimal paths going backward from the manifold. This shows up in the AMP as a point on a stationary path where the neighboring optimum feedback gains tend to infinity; such a point is called a *focal point* and the stationary path backward beyond such a point is not a minimizing path. This is illustrated by a simple example.

Example 8.6.1—Focal Point Example; Min Time Paths to a Parabola

Consider the Fermat problem of finding min time paths to the concave side of a parabola in a region where the velocity is constant as shown in Figure 8.6. A graphical solution is straightforward, namely, straight lines perpendicular to the parabola. The line $x = 0$ is a minimizing path only for $0 < y < 1$; there is a *focal point* at $y = 1$. For $y > 1$, $x = 0$ is not a minimizing path. Note the focal point is *not* at the focus of the parabola, which is at $y = .5$.

The contours of constant time-to-go have greater and greater curvature on $x = 0$ as one moves away from the parabola; in fact the curvature becomes infinite at $x = 0$, $y = 1$. On $x = 0$ for $y > 1$, there is a discontinuity in the slope of the constant time-to-go contours.

The problem may be posed as finding $\theta(t)$ to minimize t_f where

$$\dot{x} = -\sin\theta \ , \quad \dot{y} = -\cos\theta \ , \quad \psi = y_f - \tfrac{1}{2}x_f^2 = 0 \ .$$

It is straightforward to show that optimal $z \overset{\Delta}{=} \tan\theta$ satisfies the cubic equation

$$z^3 - 2(y-1)z + 2x = 0 \ .$$

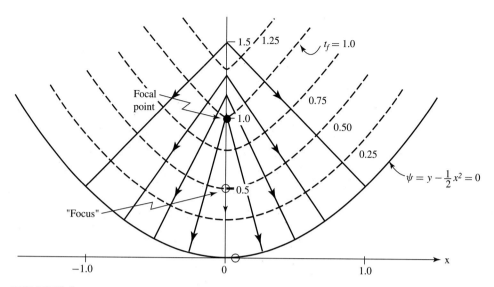

FIGURE 8.6 Focal Point Example; Min Time Paths to a Nonconvex Terminal Surface

For $x = 0$ there is only one real solution for $0 < y \leq 1$, but there are three real solutions for $y > 1$. The y-axis is not a minimizing path for $y > 1$; hence the solution is

$$\tan\theta = \pm[2(y-1)]^{1/2} \quad \text{for } y > 1, \quad \tan\theta = 0 \quad \text{for } 0 < y < 1,$$
$$t_f = (2y-1)^{1/2} \quad \text{for } y > 1, \quad t_f = y \quad \text{for } 0 < y < 1.$$

8.7 Conjugate Points; Nonconvex Spaces

Continuous nonconvex system equations produce convergence of optimal paths going backward from the terminal manifold. This also shows up in the AMP as a point on a stationary path where the neighboring optimum feedback gains tend to infinity; such a point is called a *conjugate point,* and the stationary path, backward beyond such a point is not a minimizing path. This is illustrated by a simple example.

Example 8.7.1—Min Distance between Points on the Equator of an Oblate Spheroid

Consider the problem of finding min distance paths on the equator of an oblate spheroid, a surface produced by rotating an ellipse about its minor axis (see Figure 8.7).

A point on the surface is located by its latitude θ and its longitude ϕ, where

$$x = a\cos\theta\cos\phi, \quad y = a\cos\theta\sin\phi, \quad z = b\sin\theta.$$

The differential element of distance on the surface is ds, where

$$ds^2 = dx^2 + dy^2 + dz^2,$$

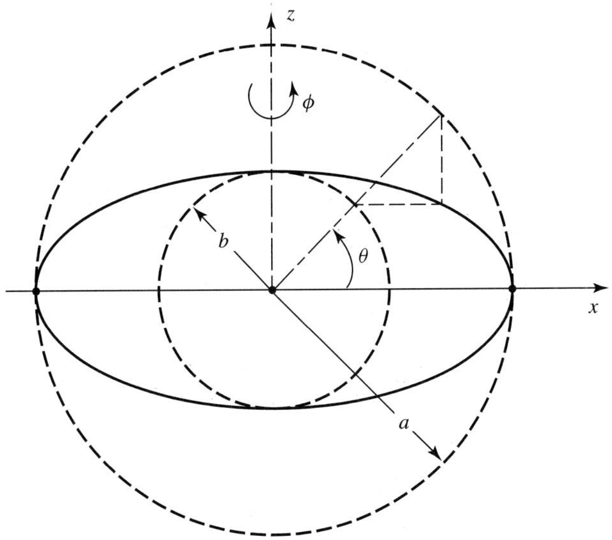

FIGURE 8.7 Nomenclature for Finding Min Distance between Points on the Equator of an Oblate Spheroid

or

$$ds^2 = h_\phi^2(d\phi)^2 + h_\theta^2(d\theta)^2 \ ,$$

where the metric coefficients are

$$h_\phi = a\cos\theta \ , \quad h_\theta = \sqrt{a^2\sin^2\theta + b^2\cos^2\theta} \ .$$

Let β be the heading angle north of east, so that

$$\tan\beta = \frac{h_\theta d\theta}{h_\phi d\phi} \ .$$

Then

$$ds^2 = h_\phi^2(d\phi)^2(1 + \tan^2\beta) \ ,$$

or

$$ds = a\cos\theta \sec\beta d\phi \ .$$

Thus the problem of finding min distance paths between two points on the surface having different longitudes may be stated as follows:

$$\min_{\theta(\phi)} J = a \int_{\phi_0}^{\phi_f} \cos\theta \sec\beta d\phi$$

subject to

$$\frac{d\theta}{d\phi} = \frac{\cos\theta}{\sqrt{1 - e^2 \cos^2\theta}} \tan\beta ,$$

with boundary conditions

$$\theta(\phi_0) = \theta_0 , \quad \theta(\phi_f) = \theta_f ,$$

where $e = \sqrt{1 - b^2/a^2}$ = the eccentricity of the ellipsoid.

For paths on and near the equator ($\theta = 0$) we may expand the integrand of J to second order in θ and β:

$$J \approx a\phi_f + \frac{a}{2} \int_{\phi_0}^{\phi_f} \left(-\theta^2 + \beta^2\right) d\phi .$$

The *AMP for a path on the equator* is to minimize this quadratic performance index with the linearized differential equation constraint

$$\frac{d\theta}{d\phi} = \frac{a}{b}\beta ,$$

and boundary conditions

$$\theta(\phi_0) = \theta_0 , \quad \theta(\phi_f) = 0 ,$$

The Hamiltonian for the AMP is

$$H = \frac{a}{2}(-\theta^2 + \beta^2) + \lambda\frac{a}{b}\beta .$$

Thus the *space is nonconvex* since $H_{\theta\theta} = -a < 0$.

The AMP is easily solved in this case. Let $(\ )' = d(\ )/d\phi$; then

$$\lambda' = -H_\theta = a\theta , \quad 0 = H_\beta = a\beta + \lambda a/b ,$$

Eliminating λ and β among these three equations gives

$$\theta'' + \frac{a^2}{b^2}\theta = 0 ,$$

which has the general solution

$$\theta = A\sin\left[\frac{a}{b}(\phi + \alpha)\right] ,$$

where A and α are constants. To satisfy the boundary condition $\theta(\phi_f) = 0$ requires that $\alpha = -\phi_f$. To satisfy the boundary condition $\theta(0) = \theta_0$ requires that $A = -\theta_0/\sin[a(\phi_f - \phi_0)/b]$. Thus the solution of the AMP is

$$\theta = \frac{\sin[a(\phi_f - \phi)/b]}{\sin[a(\phi_f - \phi_0)/b]}\theta_0 , \quad \beta = -\frac{\cos[a(\phi_f - \phi)/b]}{\sin[a(\phi_f - \phi_0)/b]}\theta_0 .$$

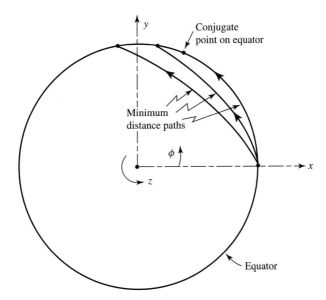

FIGURE 8.8 Min Distance Paths between Points on the Equator of an Oblate Spheroid—An Example of a Conjugate Point

Using the principle of optimality, θ_0 can be replaced by θ and ϕ_0 by ϕ, giving the neighboring optimum feedback law

$$\beta = -\frac{1}{\tan[a(\phi_f - \phi)/b]}\theta .$$

The feedback gain is negative for $0 \leq \phi_f - \phi \leq b\pi/2a$, positive for $b\pi/2a \leq \phi_f - \phi < b\pi/a$, and $\to \infty$ as $\phi_f - \phi \to b\pi/a$. Thus there is a *conjugate point at* $\phi_f - \phi = b\pi/a$, and the min distance path between two points on the equator with $\phi_f - \phi_0 > b\pi/a$ is *not* along the equator. Solution of the nonlinear problem shows that such paths go above or below the equator (see Figure 8.8). This is easily understood by considering a limiting case where $b/a \ll 1$; clearly, the min distance path between two points on the equator having a longitude difference of 180 deg is across the north or south pole, not along the equator.

Problems

8.7.1 Min Time Paths in a Region with Discontinuous Velocity: Consider the Fermat problem of finding min time paths to $x = y = 0$ in a region where the magnitude of velocity is equal to c for $0 < y < h$ and equal to $2c$ for $y > h$ as shown in Figure 8.9. A graphical solution is straightforward.

(a) Show that the locus of Darboux points (points where the direct path and the indirect path have equal times) is a parabola.

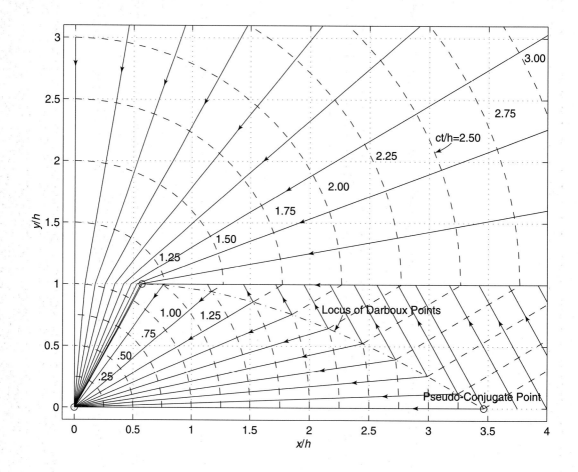

FIGURE 8.9   Fermat Problem with a Discontinuity in Velocity Magnitude

(b) If $T =$ time to go to $x = y = 0$, show that the corresponding point on the Darboux locus has the coordinates

$$y = \frac{3 - \sqrt{3}T}{2} + \left(\frac{\sqrt{3}T}{2} - \frac{3}{4} \right)^{1/2} , \quad x = 2T - 2\sqrt{3} + \sqrt{3}y ,$$

where (x, y) are in units of h and T is in units of h/c.

(c) Plot several min time paths and several contours of constant time t. In the region $y \geq 1$ use $\sin \theta_1 = \sin \theta_2 / 2$ (Snell's law), $t_1 = \sec \theta_1$, and the relations

$$x = \tan \theta_1 + 2(t - t_1) \sin \theta_2 , \quad y = 1 + 2(t - t_1) \cos \theta_2 .$$

Compare your results with Figure 8.9.

8.7.2 Min Distance to an Ellipse - Problem with a Focal Point: We wish to find min distance paths to an ellipse from points inside the ellipse (see Figure 8.10). This may be

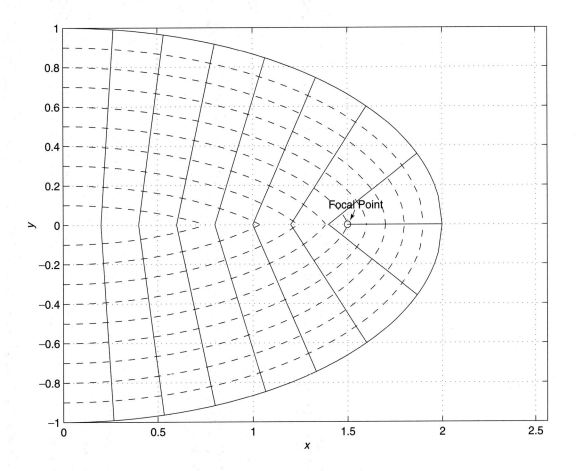

FIGURE 8.10   Min Distance to an Ellipse from Points on the Major Axis

formulated as

$$\min_{\theta(\mathcal{S})} \mathcal{S}_f \, ,$$

subject to

$$\frac{dx}{ds} = \cos\theta \, , \quad x(0) = x_o \, , \quad \frac{dy}{ds} = \sin\theta \, , \quad y(0) = y_o$$

with terminal constraint

$$\psi = 1 - \left(\frac{x}{a}\right)^2 - \left(\frac{y}{b}\right)^2 = 0 \, .$$

(a) Show that the min distance paths are straight lines ($\theta = $ constant) where

$$\tan\theta = \frac{a^2 y_f}{b^2 x_f} \, ,$$

and (x_f, y_f) is the final point on the ellipse.

(b) Show that

$$x_f - x = S \cos\theta , \quad y_f - y = S \sin\theta ,$$

where $S = $ distance to go to the ellipse.

(c) Show that $z \overset{\Delta}{=} \tan\theta$ satisfies the quartic equation

$$b^2 x^2 z^4 - 2b^2 xyz^3 + [a^2 x^2 + b^2 y^2 - (a^2 - b^2)^2]z^2 - 2a^2 xyz + a^2 y^2 = 0 ,$$

and that S is then given by

$$S = \frac{b^2 x \sin\theta - a^2 y \cos\theta}{(a^2 - b^2)\sin\theta \cos\theta} .$$

(d) For $y = 0$ show that the quartic equation in (c) can be solved explicitly:

$$z = 0 \quad \text{if} \quad |x| \geq a - b^2/a , \tag{8.22}$$

$$z = \pm\frac{a}{bx}[(a - b^2/a - x)(a - b^2/a + x)]^{1/2} \quad \text{if} \quad |x| < a - b^2/a . \tag{8.23}$$

Thus there are focal points at $x = \pm(a - b^2/a)$. Note that the focal points are NOT at the foci of the ellipse; the foci are at $x = \pm\sqrt{a^2 - b^2}$.

8.7.3 Zermelo Problem with a Conjugate Point: A boat moves with unit velocity with respect to the water through a region where the current is in the x-direction and its magnitude varies with y according to $u_c = u_o \sin^2 y$, as shown in Figure 8.11.

We wish to find paths starting from the origin that maximize $x(t_f)$ with $y(t_f) = 0$, where t_f is specified, i.e.,

$$\max_{\theta(t)} J = x(t_f) ,$$

subject to

$$\dot{x} = \cos\theta + u_o \sin^2 y , \quad x(0) = 0 , \quad \dot{y} = \sin\theta , \quad y(0) = 0 , \quad y(t_f) = 0 ,$$

where $u_o = $ max current velocity, $\theta = $ the ship's heading relative to the x-axis.

(a) Show that the x-axis is a maximizing path only for $t_f < t_c$, i.e., there is a conjugate point at $x = t_c$, $y = 0$, and find the value of t_c. Hint: Linearize the EOM about $y = 0$, $\theta = 0$, and expand the Hamiltonian to second order in (y, θ).

(b) For $t_f < t_c$ and $y(t_f) = 0$, find the NOC law for the case where the nominal optimum path is the x-axis.

(c) Plot a family of optimal paths starting from the origin. Compare your results with Figure 8.11.

8.7.4 Min Distance Between Two Points on the Outer Equator of a Torus: The metric coefficients on the surface of a torus are (see Figure 8.12)

$$h_\theta = b , \quad h_\phi = a + b \cos\theta .$$

Let β be the heading angle north of east so that

$$\tan\beta = \frac{h_\theta d\theta}{h_\phi d\phi} .$$

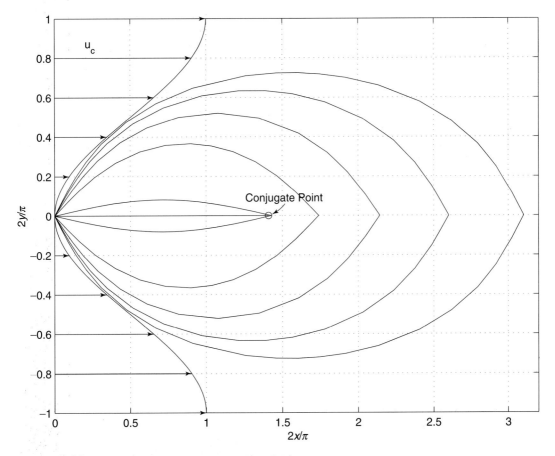

FIGURE 8.11   Current Variation with y in Problem 8.7.3

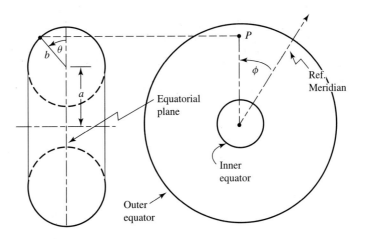

FIGURE 8.12   Nomenclature for Min Distance Paths on a Torus

Thus, to minimize the distance between two points on the torus with different longitudes, we must

$$\min_{\beta(\phi)} J = \int_{\phi_0}^{\phi_f} (a + b \cos \theta) \sec \beta d\phi$$

subject to

$$\frac{d\theta}{d\phi} = \left(\cos \theta + \frac{a}{b} \right) \tan \beta \, ,$$

with boundary conditions $\theta(\phi_0) = \theta_0$, $\theta(\phi_f) = \theta_f$.

(a) Show that the min distance path between two points on the outer equator of a torus is along the equator only if the longitude difference $\phi_f - \phi_0$ is less than $\pi/\sqrt{1 + a/b}$. *Hint:* Solve the AMP for a path on the equator. Start by showing that, for $|\theta| \ll 1$, $|\beta| \ll 1$,

$$J \approx (a + b)(\phi_f - \phi_0) + \frac{b}{2} \int_{\phi_0}^{\phi_f} (-\theta^2 + e\beta^2)d\phi \, ,$$

where $e = 1 + a/b$, and $d\theta/d\phi \approx e\beta$.

(b) For $|\theta| \ll 1$, show that min distance paths to the point $\theta = 0$, $\phi = \phi_f$ are obtained using the linear feedback law

$$\beta = -\frac{\theta}{\sqrt{e} \tan[\sqrt{e}(\phi_f - \phi)]} \, ,$$

which implies a conjugate point at $\theta = 0$, $\phi = \phi_f - \pi/\sqrt{e}$.

(c) Min distance paths starting from a point on the outer equator are shown in Figure 8.13 for the case $a/b = 1.5$. Verify these paths by forward integration of the EL equations.

8.7.5 A Fermat Problem with a Conjugate Point: Consider min time paths to $x = y = 0$ where $V(y) = V_o\sqrt{1 + y^2/h^2}$ and

$$\dot{x} = V(y)\cos \beta \, , \quad \dot{y} = V(y)\sin \beta \, .$$

(a) By solving an AMP along $y = 0$, show that there is a conjugate point at $x/h = \pi$, $y = 0$.

(b) Six min time paths are shown in Figure 8.14; verify them by integrating the system equations backward from $x = y = 0$ with the optimal $\beta(y)$ obtained from Snell's law (the first integral of the EL equations) combined with the optimality criterion $H_\beta = 0$. *Hint:* Differentiate $\beta(y)$ to get $\dot{\beta}$ and add β as a third state in the integration. The paths can also be calculated using FOPT or by using the Jacobi elliptic functions that are available in MATLAB.

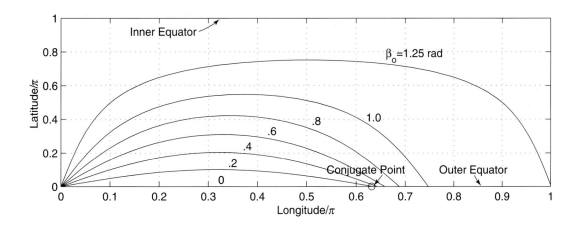

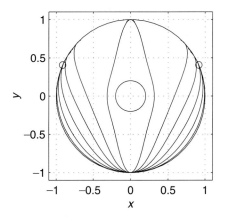

FIGURE 8.13   Min Distance Paths on a Torus from a Point on the Outer Equator

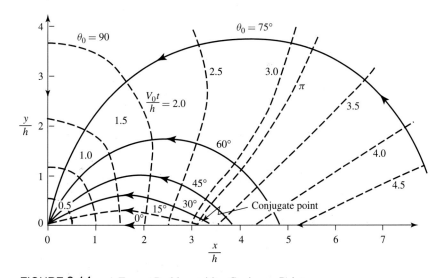

FIGURE 8.14   A Fermat Problem with a Conjugate Point

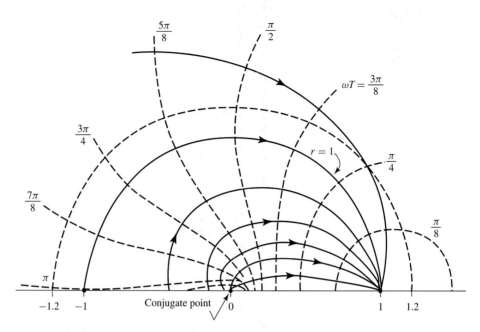

FIGURE 8.15   Min Time Paths in a Region with Velocity $= \omega r$

8.7.6 Min Time Paths with Velocity Proportional to Radius: Consider min time paths to $r = 1$, $\theta = 0$ where $V(r) = \omega r$ and

$$\dot{x} = V(r)\cos\beta , \quad \dot{y} = V(r)\sin\beta ,$$

and $r^2 = x^2 + y^2$, $\tan\theta = y/x$.

(a) By solving an AMP along $y = 0$, show that there is a conjugate point at $r = 0$.
(b) Verify Figure 8.15. Hint: the min time paths can be expressed in elementary functions using polar coordinates.

8.7.7 Sturm-Liouville Problems as Conjugate Point Problems: Sturm-Liouville problems occur in many branches of mathematical physics. They are eigenvalue problems where we wish to find the min value of t_f for which nontrivial solutions exist for the system

$$\frac{d}{dt}[r(t)\dot{x}] + q(t)x = 0 , \quad x(0) = 0 , \quad ax(t_f) + b\dot{x}(t_f) = 0 .$$

(a) Show that this problem may be interpreted as finding the conjugate point to $x(0) = 0$ of a single integrator system ($\dot{x} = u$), where we are minimizing the nonconvex QPI

$$J = \frac{1}{2}a[x(t_f)]^2 + \frac{1}{2}\int_0^{t_f}[-q(t)x^2 + r(t)u^2]dt ,$$

with $q(t)$ and $r(t)$ positive functions.
(b) For r and q positive constants and $a = 0$, show that the conjugate point is at $t_f = (\pi/2)\sqrt{r/q}$.

8.7.8 Family of Min Surface Area Curves with Conjugate Points: Consider Problem 3.3.14 (min area surface connecting two coaxial circular loops) with one loop of radius unity at $x = 0$ and the other loop at $x > 0$ (see Figure 8.16).

(a) Show that there is a one-parameter family of min area surfaces $r = r(x, a)$ from the unit radius loop at $x = 0$ given by

$$r = \frac{\cosh(x \cosh a - a)}{\cosh a} \, ,$$

where $H = 1/\cosh a$ = the Hamiltonian constant, and check the curves shown in Figure 8.16 for several values of a.

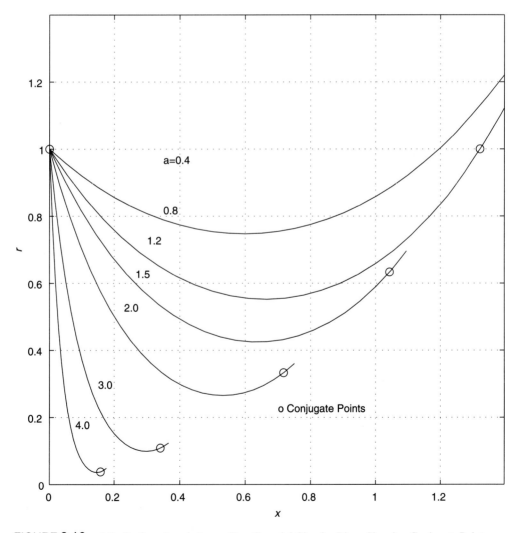

FIGURE 8.16   Min Surface Area between Two Co-axial Circular Rings Showing Conjugate Points

(b) The feedback law for finding neighboring min surface curves is

$$\delta\theta = -K(x,a)\delta r \ ,$$

where $dr/dx = \tan\theta$. Show that

$$K(x,a) = \frac{\cos^2\theta\,\cosh^2 a\,\cosh(x\cosh a - a)(x\sinh a - 1)}{\sinh a\,\cosh(x\cosh a - a) - \cosh a\,\sinh(x\cosh a - a)(x\sinh a - 1)} \ .$$

(c) Show that each "shape" curve has a *conjugate point* at the place where the curve is tangent to the envelope of the curves and check the locus of conjugate points shown in Figure 8.16. *Hint*: Find where the feedback gain blows up by putting the denominator of the expression for $K(x,a)$ to zero.

(d) Show that the min area surface of revolution has area

$$A_{\min} = \frac{\pi}{2\cosh^2 a}[\sinh(2x\cosh a - 2a) + \sinh(2a) + 2x\cosh a] \ .$$

(e) Show that $H = 0$ or $a \to \infty$ represents discontinuous minimizing surfaces, namely, two flat discs within the two circular loops connected by a hypothetical "thread" along $r = 0$. The locus where these discontinuous surfaces represent the global min area is shown in Figure 10.9 (see Problem 10.5.1). Note that it occurs *before* the smooth surfaces reach their conjugate points, so the *conjugate points are irrelevant in this problem* (Refs. Bo p. 398, GF p. 21).

8.8 Chapter Summary

Neighboring Optimal Paths and the Second Variation

- Perturbations about a nominal optimal path $\{H_u(t) = 0, \ \psi[x(t_f), t_f] = 0\}$ to second-order give

$$\delta^2 J = \frac{1}{2}\delta x^T(t_f)\Phi_{xx}\delta x(t_f) + \frac{1}{2}\int_{t_o}^{t_f}\begin{bmatrix}\delta x^T & \delta u^T\end{bmatrix}\begin{bmatrix}H_{xx} & H_{xu} \\ H_{ux} & H_{uu}\end{bmatrix}\begin{bmatrix}\delta x \\ \delta u\end{bmatrix}dt$$

$$\dot{\delta x} = f_x\delta x + f_u\delta u \ , \quad \delta x(t_o) = \delta x_o \ , \quad \psi_x\delta x(t_f) = 0 \ ,$$

where $\Phi = \phi + v^T\psi$.

- The solution is exactly the same as the LQ HTC Problem (Secion 5.4) if we replace (A, B) by $(f_x, f_u,)$, $(C^T QC, \ C^T QD, \ \bar{R})$ by $(H_{xx}, \ H_{xu}, \ H_{uu})$, M_f by ψ_x, and S_f by Φ_{xx}:

$$\delta u = -K(t)\delta x \ ,$$

where

$$K = H_{uu}^{-1}\left[H_{ux} + f_u^T(S + MQ^{-1}M^T)\right] \ ,$$

and the S, M, Q equations are the same as the LQ Terminal Controller Problem except that the coefficients are functions of time.

- Additional necessary conditions for minimum:
  1. $H_{uu}(t) \geq 0$, the *Legendre-Clebsch condition*.
  2. $J^o_{xx} \equiv S + MQ^{-1}M^T$ finite, $t_o \leq t \leq t_f$, the *Jacobi condition*.

- A point where $S + MQ^{-1}M^T \to \infty$ is a *focal point* or a *conjugate point* (produced by nonconvex terminal constraints or nonconvex ODEs, respectively).

- NOC (alias Differential DP) is obtained using the time-varying nominal control plus time-varying linear feedback on the state perturbations from nominal given above. If t_f is unspecified, index the nominal control and the feedback gains on estimated time-to-go $t_f - t$.

CHAPTER 9

Inequality Constraints

9.1 Introduction

Thus far we have considered only optimization problems with *equality constraints*. Obviously, many interesting optimization problems involve *inequality constraints*. Parameter optimization problems with inequality constraints (POIC) were formerly called "linear or nonlinear programming problems," but currently the term "programming" seems to be applied to all optimization problems. There is a large literature on POIC (e.g., Refs. KT, GMW, Lr, and Zo); we give a brief description of POIC in Section 9.2.

Dynamic optimization problems may be regarded as TPBVPs in which a parameter optimization problem must be solved at each time, namely, finding the control vector that minimizes the Hamiltonian while holding (x, λ, t) constant. Thus the concepts and methods of POIC are directly applicable to dynamic optimization problems with inequality constraints. Section 9.3 covers dynamic optimization problems with (a) control variable inequality constraints and (b) state variable inequality constraints. It also discusses an *inverse approach* to solving such problems, where the *output histories* rather than the input histories are determined so as to minimize a performance index. This method uses POIC codes, and is effective in solving problems with inequality constraints.

9.2 Static Optimization

Inequality Constraints

Parameter optimization problems involving inequality constraints require extension of the methods treated in Chapter 1. An important class of problems of this type involves finding a parameter vector y to minimize

$$L(y) \text{ subject to } f(y) \leq 0 , \tag{9.1}$$

where, in general, f and y are vectors of different dimensions. The dimension of f is often greater than the dimension of y, so that it is not possible to decompose y into state and decision variables.

Scalar Parameter y and Constraint f

Consider first the case in which y and f are scalars. There are two possibilities for the optimal value of y, which we denote as y^o:

$$f(y^o) < 0 \quad \text{or} \quad f(y^o) = 0 . \tag{9.2}$$

In the former case, the constraint is not active and can be ignored; this problem was treated in Section 1.1. In the latter case, consider small perturbations about y^o; if $L(y^o)$ is a minimum, then we have

$$dL = L_y dy \geq 0 , \tag{9.3}$$

for all feasible values of dy, which must satisfy

$$df = f_y dy \leq 0 . \tag{9.4}$$

In order that (9.3) and (9.4) be consistent, it is clearly necessary that either

$$\text{sgn} L_y = -\text{sgn} f_y , \quad \text{or} \quad L_y = 0 .$$

These two possibilities may be expressed in one relation as

$$L_y + \lambda f_y = 0 , \quad \lambda \geq 0 . \tag{9.5}$$

The two situations are shown geometrically in Figures 9.1(a) and (b). They may be treated analytically by adjoining (9.2) to (9.1):

$$H(y, \lambda) = L(y) + \lambda f(y) . \tag{9.6}$$

The necessary conditions become

$$H_y = 0 , \tag{9.7}$$

$$f(y) \leq 0 , \tag{9.8}$$

where

$$\lambda \begin{cases} \geq 0 , & f(y) = 0 , \\ = 0 , & f(y) < 0 . \end{cases} \tag{9.9}$$

Vector Parameter Vector *y* and Scalar Constraint *f*

When y is a vector but f is still a scalar, (9.3) to (9.5) remain applicable if we interpret the symbols in vector notation. We interpret (9.5) to mean

$$L_y \text{ parallel to } f_y \text{ and pointing in opposite directions .} \qquad (9.10)$$

The necessity of (9.10) can be established by contradiction. Assume (9.10) is not true as illustrated in 2D in Figure 9.2. Then the cross-hatched region represents a region of feasible y that will yield smaller L. This and the other situation (namely, when $f(y^o) < 0$) again can be summarized by the necessary conditions (9.7) to (9.9).

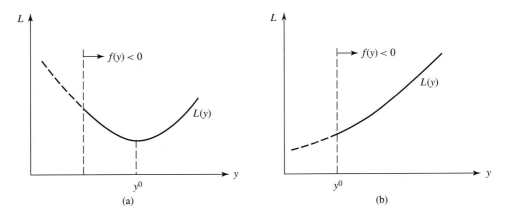

FIGURE 9.1   Two Possible Types of Minimum with an Inequality Constraint

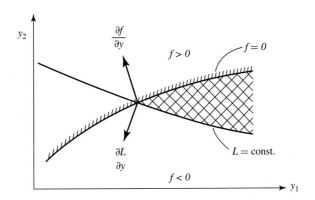

FIGURE 9.2   Example Showing Why f_y Must Be in Opposite Direction of L_y at a Minimum

Vector Parameter y and Constraint f

In the general case, when y and f are vectors, (9.4) and (9.5) still apply, noting that f_y is now a matrix, so that λ must be replaced by a vector λ^T. If only one component of f is active, the problem is the same as that just treated. If two components of f are active, the situation, in 2D, is as shown in Figure 9.3.

If y^o is an optimal point on $f_1 = f_2 = 0$, then L_y must lie between the negative gradients of f_1 and f_2 (recall the parallelogram construction of a resultant from two components). Analytically, this means that L_y can be expressed as a negative linear combination of f_y^1 and f_y^2. In general, when q components are active at a boundary optimal point, we must have

$$L_y + \lambda_1 f_y^1 + \cdots + \lambda_q f_y^q = 0 , \tag{9.11}$$

or

$$L_y + \lambda^T f_y = 0 , \tag{9.12}$$

where

$$\lambda \begin{cases} \geq 0 , \ f(y) = 0 , \\ = 0 , \ f(y) < 0 . \end{cases} \tag{9.13}$$

Equation (9.13) is to be understood in the component-by-component sense. Hence we may define a quantity $H \equiv L + \lambda^T f$ and write (9.12) as $H_y = 0$. Equations (9.12) and (9.13) are necessary conditions for minimality. For a maximum, the sign of λ must be changed in (9.13). In words, *the gradient of L with respect to y at a minimum must be pointed in such a way that decrease of L can occur only by violating the constraints.* A graphical way of saying this *for a minimum $-L_y$ must lie inside the cone of the constraint gradients f_y.*

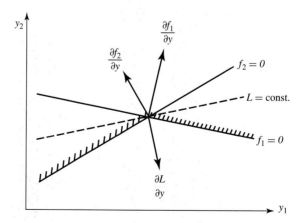

FIGURE 9.3   A Min Point of L Subject to Two Inequality Constraints; Both Active

Let us suppose that y has p components and that n components of the inequality constraint are active, i.e.,

$$f^i(y) = 0 , \quad i = 1, \dots, n . \tag{9.14}$$

The inactive constraints, $f^i(y) < 0$, $i = n+1...$, may be disregarded. It is clear that $p \geq n$. Next, take n of the components of y and call them x; let the remaining $p - n$ components be called u; i.e.,

$$y^T = (x_1, \dots x_n , \quad u_1, \dots, u_{p-n}) \triangleq (x^T, u^T) . \tag{9.15}$$

The choice must be such that

$$f_i(x, u) = 0 , \quad i = 1, \dots, n , \tag{9.16}$$

determines x when u is given. The *sufficient conditions* for a local minimum of $L(y)$, with $f(y) \leq 0$, are given in Section 1.4 to which we must add the condition that $\lambda_1, \dots, \lambda_n$ all be positive (for a precise statement, see e.g., Ref. Mc). The latter condition is easily interpreted from (9.8) since $-\lambda_i$ is equal to L_{f_i} holding u constant, which must be negative (i.e., $dL > 0$ for $df_i < 0$).

Example 9.2.1—A Min Point in 2D with Two Inequality Constraints

Consider $L(y_1, y_2)$ with $f_i(y_1, y_2) \leq 0$, $i = 1, 2$, and suppose the level curves are as shown in Figure 9.4.

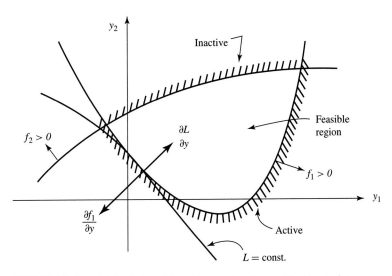

FIGURE 9.4 A Min Point of L Subject to Two Inequality Constraints

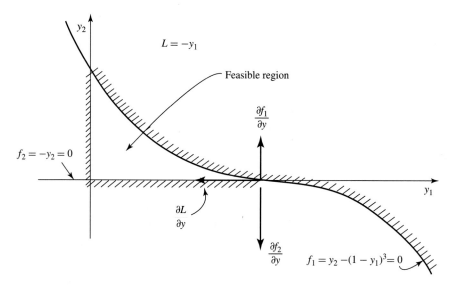

FIGURE 9.5   Example of Kuhn-Tucker Constraints Qualification

It is clear that the constraint $f_2 < 0$ is inactive while the constraint $f_1 \le 0$ is active, i.e., $f_1 = 0$. From Figure 9.4 we have

$$L_y + \lambda_1 f_y^1 = 0 , \quad \lambda_1 > 0 ,$$

i.e., grad L is parallel and in the opposite direction to grad f_1. Also, the "curvature" of L along $f_1 = 0$ is such that L increases on $f_1 = 0$ away from the minimum; to show this analytically, use the constraint to determine y_1 given y_2 and calculate $L_{y_1 y_1}$, which from Figure 9.4, is clearly positive. Equations (9.12) and (9.13) are the essence of the *Karush-Kuhn-Tucker theorem*. The precise statement of the condition requires the assumption of the so-called "constraints qualification" (Ref. KT, p. 483), which rules out geometric situations like the one shown in Figure 9.5. At the min point we see that $(y_1, y_2) = (1, 0)$ and L_y is not equal to any finite linear combination of f_y^1 and f_y^2.

Another approach to sufficiency is the *saddle-point theorem* of nonlinear programming. It is more elegant (but usually harder to apply) than the conditions given above since it does not require the arbitrary separation of y into x and u. Consider the function $H(y, \lambda) = L + \lambda^T f$. Suppose that it is possible to find y^o and λ^o such that they constitute a saddle point for H, i.e.,

$$H(y^o, \lambda) \le H(y^o, \lambda^o) \le H(y, \lambda^o) \tag{9.17}$$

for all $\lambda \ge 0$ and $f(y) \le 0$. Then we may conclude that y^o is a min point for $L(y)$ subject to $f(y) \le 0$, regardless of the nature of L and f.

Linear Programming (LP) Problems

If both the performance index and the inequality constraints are linear functions of y, the problem is called an *LP problem*. Clearly, in this case, the minimum, if it exists, must occur on the boundary since the curvature of L is zero everywhere. Let the problem be to choose y to minimize

$$L = c^T y \tag{9.18}$$

subject to

$$A^T y - b \leq 0 , \tag{9.19}$$

where y is an n-vector and b is an m-vector, $m > n$. If A is of rank n and c^T is not collinear with any of the rows of A^T or any negative linear combinations of $n - 1$ rows of A^T, the minimum, if it exists, must occur at a point determined by the simultaneous satisfaction of n of the constraints $A^T y - b = 0$. This result is not surprising to anyone with geometric intuition; it is the *fundamental theorem of LP*.

LP has found many practical uses. In the problem (9.18) and (9.19), we know that the minimum must occur at a point intersection of n hyperplanes whose normals are in the direction of rows of A^T (there are abnormal situations in which the minimum lies on an "edge" rather than at a "point"). We start the solution with a guess, selecting n equations out of (9.19) as the active constraints. If this point is feasible, we then examine the n edges leading away from this point (formed by the intersection of n sets of $n - 1$ hyperplanes we have chosen); these edges will have directions, away from the point, of e^1, e^2, ..., e^n, where e^i is a *unit n-vector* along the ith edge. The gradient of L is simply c^T, so we consider the projection of the edge directions on c^T, i.e., we consider the scalar products $c^T e^i$, $i = 1, ..., n$. If all the scalar products are positive, it is not possible to move along any edge to obtain an improvement (i.e., a smaller value of L); we have the optimal solution. On the other hand, if some of the scalar products are negative, let us choose the one with largest magnitude and move along the corresponding edge until we encounter another constraint. This new constraint and the $n - 1$ old constraints that form the edge determine a new point at which the value of L is necessarily smaller since we moved along a *projected gradient* $c^T e^i < 0$. The process is then repeated over and over until a point is found at which we have all $c^T e^i > 0$, i.e., no improvement is possible. This is the basis of the *simplex algorithm* proposed by Dantzig (Ref. Da), which uses, essentially, a method of feasible directions.

Example 9.2.2—LP Problem in 2D with Five Constraints

Minimize $L = -5y_1 - y_2$ subject to

$$f_1 = y_1 \geq 0 , \quad f_2 = y_2 \geq 0 , \quad f_3 = y_1 + y_2 - 6 \leq 0 ,$$
$$f_4 = 3y_1 + y_2 - 12 \leq 0 , \quad f_5 = y_1 - 2y_2 - 2 \leq 0 .$$

Figure 9.6 shows the feasible region, with contours of constant L.

Obviously, the minimum occurs at point A, where we have $3y_1 + y_2 - 12 = 0$ and $y_1 - 2y_2 - 2 = 0 \Rightarrow y_1 = 3\frac{5}{7}$, $y_2 = \frac{6}{7}$, $\Rightarrow L_{\min} = -19\frac{3}{7}$, and grad L can be expressed as a negative linear combination of n (but not $n - 1$) rows of A^T (namely, grad f_4 and grad f_5), as shown in Figure 9.6.

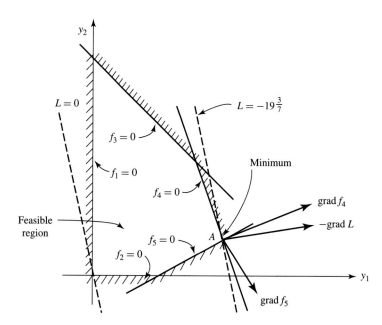

FIGURE 9.6   Solution of a Simple Linear Programming Problem

The implication of the fundamental theorem of LP for the numerical solution of LP problems is clear: Take n constraints at a time and treat them as equalities. Solving the equalities yields one solution (assuming admissibility), which is either optimal or nonoptimal. In the latter case, we can discard one of the constraints and, substituting another, repeat the process with the requirement that the new solution be feasible and better. Since there are only a finite number of such possibilities, this process must eventually arrive at the optimal combination. The method that accomplishes this is known as the *simplex method*. We shall say more about this in the next section.

This problem is easily solved using the command LP in the MATLAB Optimization Toolbox.

```
% Script e09_2_2.m; linear programming pb. in 2D with 5 constraints;
% min L=c*y with Ay <= b;
%
c=[-5 -1]; A=[-1 0; 0 -1; 1 1; 3 1; 1 -2]; b=[0 0 6 12 2]';
y=lp(c,A,b)
```

Example 9.2.3—A Feed Blending Problem

There are many *blending problems* that involve finding the cheapest mixture of several materials that contains at least a certain fraction of specified ingredients. A typical problem is to find the cheapest mixture of several feeds that contains at least certain specified amounts of nutrients (such as proteins, fats, and vitamins). Other blending problems occur in mixing fuel oils and fertilizers. Suppose that we are considering a mixture of three feeds and we

| Feed | Fraction of nutrient in each feed | | | Cost |
	1	2	3	
1	.06	.02	.09	15
2	.03	.04	.05	12
3	.04	.01	.03	8

FIGURE 9.7   Data for a Min Cost Feed
Blending Problem

have three inequality specifications on nutrients. Figure 9.7 shows the fraction of each of the three nutrients contained in each of the three feeds and the cost per unit amount of each of the three feeds. The problem is to find the cheapest mixture of the three feeds such that the fraction of nutrients one, two, three in the mixture is greater than or equal to .04, .02, and .07, respectively.

Let F_j = the fraction of feed j in the mixture, where $j = 1, 2,$ or 3; these are the quantities we are trying to find (our design parameters). Let N_i = fraction of nutrient i in the mixture, where $i = 1, 2,$ or 3; then we have

$$N_i = n_{i1} F_1 + n_{i2} F_2 + n_{i3} F_3 \; , \tag{9.20}$$

where n_{ij} = fraction of nutrient i in feed j (the data are given in Figure 9.7). For this problem we have $N_1 \geq .04$, $N_2 \geq .02$, $N_3 \geq .07$.

Let C = cost per unit amount of the mixture and c_j = cost per unit amount of feed j (also given in Figure 9.7). Then we have

$$C = c_1 F_1 + c_2 F_2 + c_3 F_3 \; . \tag{9.21}$$

The fractions of the three feeds in the mixture must add up to one:

$$F_1 + F_2 + F_3 = 1 \; . \tag{9.22}$$

The problem, then, is to find the two quantities F_1 and F_2 (using $F_3 = 1 - F_1 - F_2$) to minimize C and satisfy the inequalities. Several of these turn out to be redundant, i.e., other inequalities automatically cause them to be satisfied:

$$N_i \geq \bar{N}_i \; , \quad i = 1, 2, 3 \; , \quad 0 \leq F_j \leq 1 \; , \quad j = 1, 2 \; , \tag{9.23}$$

where $\bar{N}_i$ is the min allowable fraction of nutrient i in the mixture.

Figure 9.8 is a plot of F_1 vs. F_2 showing the inequalities as lines with arrows perpendicular to them pointing in the allowable directions.

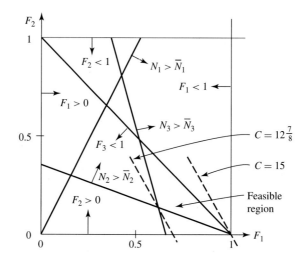

FIGURE 9.8   Solution of a Linear Programming Problem of Feed Mixing

In this problem, the inequalities are

(a) $N_1 = 0.6F_1 + .03F_2 + .04(1 - F_1 - F_2) \geq .04$ or $2F_1 - F_2 \geq 0$.
(b) $N_2 = .02F_1 + .04F_2 + .01(1 - F_1 - F_2) \geq .02$ or $F_1 + 3F_2 \geq 1$.
(c) $N_3 = .09F_1 + .05F_2 + .03(1 - F_1 - F_2) \geq .07$ or $3F_1 + F_2 \geq 2$.
(d) $0 \leq F_1 \leq 1$, $0 \leq F_2 \leq 1$, $0 \leq F_3 \leq 1$ or $F_1 + F_2 \leq 1$.

The *feasible region* is the region in the graph where all the inequalities are satisfied (see Figure 9.10). The lines of constant cost are given by setting $C =$ different constants in

$$C = 15F_1 + 12F_2 + 8(1 - F_1 - F_2) \quad \text{or} \quad C = 8 + 7F_1 + 4F_2 .$$

From Figure 9.8 it is clear that the cheapest feasible solution is one that occurs at the corner where $N_2 = \bar{N}_2$ and $N_3 = \bar{N}_3$. From above, this requires that

$$F_1 + 3F_2 = 1 , \quad 3F_1 + F_2 = 2 .$$

These two linear equations in two unknowns (F_1 and F_2) are readily solved to yield

$$F_1 = \tfrac{5}{8} \text{ and } F_2 = \tfrac{1}{8} .$$

The amount of feed number 3 is obtained by substituting the results above into $F_3 = 1 - F_1 - F_2$. This yields

$$F_3 = \tfrac{1}{4} .$$

The min cost is given by

$$C = 15 \left(\tfrac{5}{8} \right) + 12 \left(\tfrac{1}{8} \right) + 8 \left(\tfrac{1}{4} \right) = 12\tfrac{7}{8} \text{ per unit amount of mixture .}$$

The amount of nutrient one is above the minimum required fraction:

$$N_1 = .06\left(\tfrac{5}{8}\right) + .03\left(\tfrac{1}{8}\right) + .04\left(\tfrac{1}{4}\right) = .05125 > .04\ .$$

Notice that the *most expensive feasible solution* is to use feed number 1 all by itself.

Again, this problem is easily solved using the command LP in the MATLAB Optimization Toolbox.

```
% Script e09_2_3.m; linear prog. pb. in 2D with 8 constraints
% min L=c*y with Ay <= b.
%
c=[7 4]; A=[-2 1; -1 -3; -3 -1; 1 1; -1 0; 0 -1; 1 0; 0 1];
b=[2 -1 -2 1 0 0 1 1]'; y=lp(c,A,b);
```

Example 9.2.4—A Transportation Planning Problem

A grain dealer has purchased 50,000 bushels of wheat in Grand Forks, North Dakota, and 40,000 bushels in Chicago. He then sells 20,000 bushels to a customer in Denver, 36,000 bushels to a customer in Miami, and the remaining 34,000 bushels to a customer in New York. He wishes to determine the min-cost shipping schedule, given the freight rates (in cents per bushel) shown in Figure 9.9 in the upper right-hand corners of each square. Different modes of shipment cause the rates not to be proportional to the distance between the cities.

His problem is to find a *nonnegative amount* in each of the six squares so that (a) the amounts in the first row add up to 50,000 and the amounts in the second row add up to 40,000; (b) the amounts in the first, second, and third columns add up to 20,000, 36,000 and 34,000, respectively; (c) the total freight cost is a minimum; this cost is obtained by multiplying the amount in each square by the rate in the upper right-hand corner and adding these numbers together.

This problem is somewhat like a crossword puzzle, only harder, since it is not sufficient just to get the rows and columns to add up properly (a feasible solution); we must, in addition, minimize the total cost. By "cut and try" we might be able to find the solution. However,

42 x	55 y	60 $50 - x - y$	50
36 $20 - x$	47 $36 - y$	51 $x + y - 16$	40
20	36	34	

FIGURE 9.9 A Transportation Planning Problem

a systematic approach is apt to take less time, and for problems with more shipping points and more destinations, a systematic approach (an algorithm) and a computer are essential.

Suppose we designate the amount from Grand Forks to Denver as x in thousands of bushels. Then, clearly, the amount from Chicago to Denver must be $20 - x$. Similarly, let us designate the amount shipped from Grand Forks to Miami as y; then the amount from Chicago to Miami must be $36 - y$. Now the amount from Grand Forks to New York must be $50 - x - y$, and the amount from Chicago to New York must be $40 - (20 - x) - (36 - y) = x + y - 16$. We automatically satisfy the requirement that the total shipped to New York be 34,000 since the total amount sold equals the amount owned.

We have reduced the number of unknowns to two, x and y, which must satisfy six inequalities:

$$x \geq 0, \quad y \geq 0, \quad 50 - x - y \geq 0, \quad 20 - x \geq 0,$$
$$36 - y \geq 0, \quad x + y - 16 \geq 0.$$

We can conveniently plot all these inequalities on a (x vs. y) graph (see Figure 9.10). As in the previous example, there is a *feasible region* where all the inequalities are satisfied.

Next, we calculate the cost in terms of x and y:

$$C = 1000/100 \times [42x + 55y + 60(50 - x - y) + 36(20 - x) + 47(36 - y + 51(x + y - 16)] ,$$

or

$$C = 45{,}960 - 30x - 10y \quad \text{(in dollars)} .$$

Lines of constant cost are shown in Figure 9.10 (left side) as dashed lines. Clearly, the min cost solution is $x = 20$, $y = 30$, and is shown in table form in Figure 9.10 on the right.

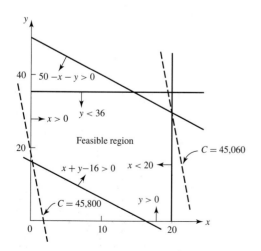

FIGURE 9.10   Solution of a Linear Programming Problem of Transportation Planning

Note that no wheat should be shipped from Chicago to Denver, even though such a shipment would involve the lowest rate per bushel. The difference between the best and the worst feasible solutions is only $740 out of about $45,000. However, this 1.6% difference could be a substantial percent of the profit involved in the sales.

As shown below, this problem is easily solved using the command LP in the MATLAB Optimization Toolbox:

```
% Script e09_2_4.m; linear programming pb. in 2D with 6 constraints;
%
c=[-30 -10]; A=[1 1; 1 0; 0 1; -1 -1; -1 0; 0 -1];
b=[50 20 36 -16 0 0]'; y=lp(c,A,b)
```

Nonlinear Programming Problems (NLP)

Numerous texts exist on this subject (e.g., Refs. GMW, Lr, and Zo), so we shall describe only the main features of the *method of feasible directions* or *gradient projection*. This method is divided into two steps:

1. *Finding a feasible solution.* With reference to (9.2), locating a value of y such that $f(y) \leq 0$ is often not a trivial task. In problems with equality constraints, finding a feasible solution is usually straightforward since there are more variables (x and u) than there are constraint equations [$f(x, u) = 0$]. In problems with inequality constraints, there are often more constraint equations (components of f) than variables (components of y). Finding a feasible solution may be approached by guessing a value for y, then considering a small perturbation, dy, that will change f according to

$$df = f_y dy . \tag{9.24}$$

If certain components of $f(y)$ are greater than zero, i.e., infeasible, we require a dy such that the corresponding components of df are less than zero. In other words, $f(y + dy)$ should be an improvement toward a feasible solution, i.e.,

$$\bar{f}_y dy \leq 0 , \tag{9.25}$$

where $\bar{f}_y$ contains only those rows of f_y corresponding to infeasible values of f. The problem is thus reduced to finding feasible solutions to successive linear inequalities (instead of nonlinear inequalities).

2. *Finding a feasible improvement.* If a feasible y can be found, the next step is to find a dy that is not only feasible but that also improves the performance index, i.e., we must have $f(y + dy) \leq 0$ and $L(y + dy) < L(y)$. This gives rise to another set of linear inequalities like those above:

$$\begin{bmatrix} L_y \\ f_y \end{bmatrix} dy \leq 0 . \tag{9.26}$$

Example 9.2.5—A Quadratic Programming Problem

Find $(y_1,\ y_2)$ to minimize

$$L = \frac{(y_1 - 2)^2}{4} + (y_2 - 1)^2 \equiv \frac{y_1^2}{4} + y_2^2 - y_1 - 2y_2 + 2$$

subject to

$$3y_1 + 2y_2 - 6 \le 0\ ,\ \ y_1 > 0\ ,\ \ y_2 > 0\ .$$

A sketch of the feasible region with contours of constant L is shown in Figure 9.11. If we guess $y_1 = y_2 = \frac{1}{2}$ (point A) to start with, we find that

$$L_{y_1} = -\frac{3}{4}\ ,\ \ L_{y_2} = -1\ .$$

Since we are minimizing, the greatest improvement will be in the direction of the *negative gradient* that is shown at A.

We proceed in that direction until we reach a minimum (which may be on a boundary). In this case we reach a minimum at point B in the *interior* of the feasible region. Following the direction of the negative gradient at B, we reach point C on the constraint $3y_1 + 2y_2 - 6 = 0$. Here the negative gradient points *out* of the feasible region, so we take the component of the negative gradient along the *constraint boundary*, which is *up* in this case. Moving along the constraint, we finally arrive at the minimum at D, $y_1 = 1.4$, $y_2 = .9$, where the negative gradient points out of the feasible region and is perpendicular to the boundary.

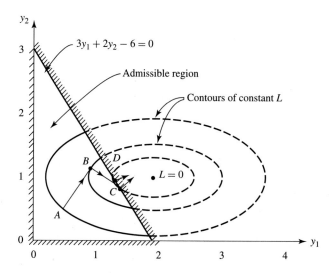

FIGURE 9.11   Solution of a Quadratic Programming Example

As shown below, this problem is easily solved using CONSTR in the MATLAB Optimization Toolbox.

```
% Script e09_2_5.m; 2D quadratic programming pb. with 1 constraint;
% J=.5*x'*H*x+f'*x, A*x<=b.
%
y0=[.5 .5]'; optn(1)=1;lb=[0 0]'; ub=[2 3]';
   y=constr('qpp',y0,optn,lb,ub);
%
function [f,g]=qpp(y)
f=(y(1)-2)^2/4+(y(2)-1)^2; g(1)=3*y(1)+2*y(2)-6; g(2)=-y(1); g(3)=-y(2);
```

Problems

9.2.1—QPI and Linear Inequality Constraint: Find $(x, \ y)$ to minimize

$$L = x^2 - xy + y^2 - 3x \ \text{ subject to } \ f = 3x + y - 3 \le 0 .$$

If the constraint is active, verify that the Karush-Kuhn-Tucker condition is satisfied.

Problem 9.2.2—QPI and Quadratic Inequality Constraints: Find $(x, \ y)$ to minimize

$$L = x^2 - y^2 ,$$

subject to

$$f_1 = (x - 3)^2 + (y - 1)^2 - 5 \le 0 ,$$
$$f_2 = (x - 3)^2 + (y + 1)^2 - 5 \le 0 .$$

If the constraints are active, verify that the Karush-Kuhn-Tucker conditions are satisfied.

9.3 Dynamic Optimization

The shooting method of solving dynamic optimization problems (Section 4.5), combined with the Minimum Principle of Chapter 7, is capable of handling inequality constraints on the control variables and inequality constraints that involve both the control variables and the state variables. However constraints that involve only the state variables require an additional concept, namely *tangential entry*. A state-constrained arc must be anticipated, i.e., the control histories must be modified so that the plant arrives at the entry point to the state-constrained arc with zero state-constraint rate; otherwise the system cannot stay on the constraint. This often requires that higher time derivatives of the state constraint also be zero at entry. Exact solution methods exist but they require preknowledge of the number and sequence of the constrained arcs, and they require solution of multipoint boundary value problems due to the interior boundary conditions that occur at the entry points to state-constrained arcs (Ref. BH, Sections 3.11 to 3.13 and 7.9 to 7.11).

A sometimes effective (but rather inefficient) method of solving dynamic optimization problems with state variable inequality constraints (SVICs) is the *slack variable method* (Refs. VT, Zh). It converts SVIC problems into unconstrained problems by introducing slack variables that are measures of how far the system is from the constraint boundaries; thus the slack variables are nonzero off the constraint and zero on the constraint. This may introduce singularities into the calculations (see Chapter 10).

Inverse Dynamic Optimization

A simple method of solving dynamic optimization problems with SVICs was suggested by Seywald (Ref. Se). It may be regarded as a form of *inverse control* where points on one or more state variable histories are used as unknown parameters to be determined so that the path is optimal. The *control variable histories are found by numerical differentiation* of the state variable histories. It makes use of professional nonlinear programming (NLP) codes (e.g., Refs. GMSW and Gr) so that it is relatively simple to use. Some of the advantages over the exact and slack variable methods are

1. The number and sequence of constrained arcs do not have to be guessed as in the exact method.
2. State variable histories are usually easier to guess than control variable histories.
3. NLP codes handle inequality constraints more efficiently than the slack variable method.
4. NLP codes use quasi-Newton methods that produce second-order convergence.
5. NLP codes find gradients numerically so that analytic expressions for them are not required.
6. Singular arcs (see Chapter 10) are handled without any special coding. The number and location of singular arcs does not have to be guessed.

Some disadvantages are

1. One must be able to solve explicitly for the controls in terms of the states and the state rates.
2. For more than about 25 time steps the solution without analytical gradients takes a long time on current personal computers (1998) due to the inversion of large matrices.
3. For less than about 25 time steps, singular arcs and the "corners" of constrained arcs are not determined very accurately.

Items 2 and 3 can be overcome by supplying analytical gradients or by using the NLP solutions as initial guesses for exact methods, since the number and sequence of constrained arcs and singular arcs are determined by the NLP algorithm.

The *first main idea* of the method is to discretize the differential equations $\dot{x} = f(x, u)$ as follows:

$$\dot{x}(k) \approx \frac{x(k+1) - x(k)}{\Delta t} , \tag{9.35}$$

$$\approx f[\bar{x}(k), u(k)] , \tag{9.36}$$

where

$$\bar{x}(k) \triangleq \frac{x(k+1) + x(k)}{2} . \tag{9.37}$$

Equation (9.37) is an essential feature since (9.35) approximates the time derivative in the middle of the time step and (9.37) also approximates x in the middle of the time step. Use of (9.35) to (9.37) makes it difficult to solve the difference equations sequentially given the control history.

The *second main idea* is to *determine the control histories from time derivatives of one or more key state histories*. The other states can then be found by the usual forward sequencing with the key state histories and the derived control histories as inputs. The *key state histories form the unknown parameter vector in the NLP code*.

Example 9.3.1—Max Range for a Double Integrator Plant with Bounded Acceleration (Ref. Se)

We wish to maximize the range in a given time t_f with bounded acceleration, beginning and ending with zero velocity. Measuring acceleration a in units of maximum acceleration $a_{\max}$, velocity v in units of $a_{\max}t_f$, distance x in units of $a_{\max}t_f^2$, the EOM are

$$\dot{y} = v , \quad \dot{v} = a ,$$

with $|a(t)| \leq 1$, $y(0) = v(0) = 0$, $v(1) = 0$, and we wish to maximize $y(1)$.

As the parameter vector p we take points on the velocity history $v(t)$, excluding the initial and final velocities that are prescribed to be zero. The displacement $y(t)$ can be obtained by integrating $v(t)$ and the control $a(t)$ can be obtained by differentiating $v(t)$. A MATLAB function file that solves the problem using the MATLAB code CONSTR is listed below; it calculates the performance index 'f' and the constraint vector 'g'. This problem can also be solved using LP since it is linear with linear inequality constraints; this takes substantially less computer time.

```
function [f,g]=dblin_fg(p,N)
% Range & acceleration constraints for DouBLe INTegrator plant.
%
v=[0 p 0]; y(1)=0; dt=1/N; a=(v(2:N+1)-v(1:N))/dt; vb=(v(2:N+1)+v(1:N))/2;
for i=1:N,
  y(i+1)=y(i)+dt*vb(i); g(i)=abs(a(i))-1; end    % Disp. & inequal. constr.
f=-y(N+1);                                        % Performance index
```

A script for solving the problem using CONSTR is listed below.

```
% Script e09_3_1.m; finds max range for double integrator plant with bounded
% acceleration using MATLAB code CONSTR.
%
N=20; un=ones(1,N-1); p=.25*un; lb=-2*un; ub=2*un; optn(1)=1; optn(4)=1e-5;
optn(14)=200; p=constr('dblin_fg',p,optn,lb,ub,[],N);
```

Figure 9.12 is a plot of the converged solution to this problem, which confirms the intuitive and well-known result that one should use $a_{\max}$ for $0 < t < t_f/2$ and $-a_{\max}$ for $t_f/2 < t < t_f$; this type of control is called *bang-bang control*.

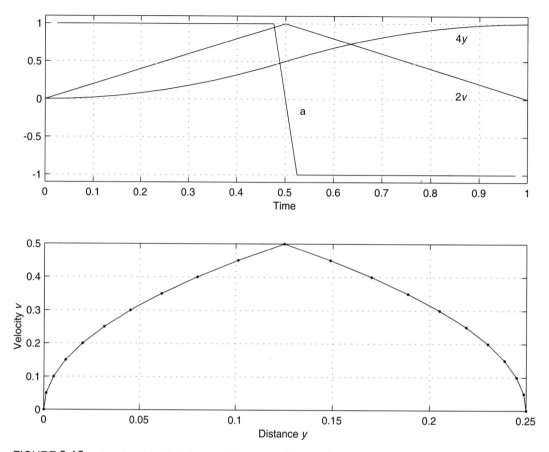

FIGURE 9.12   Acceleration, Velocity, and Distance vs. Time and Velocity vs. Distance for Maximum Range Double Integrator Plant

Example 9.3.2—Maximum Range of a Glider with Altitude ≥ 0

Given an initial velocity $V(0)$, an initial altitude $h(0)$, and an initial flight-path-angle $\gamma(0)$, we wish to determine the angle-of-attack history $\alpha(t)$ to maximize the horizontal range $x(t_f)$ where t_f is open.

This problem is straightforward without constraints (see Problem 3.4.22), *but the glider goes below* $h = 0$ *and uses very large angles-of-attack near the end.* Thus we must put in the state variable constraint $h \geq 0$ and the control variable constraint $\alpha \leq \alpha_{max}$, which makes the problem considerably harder to solve. The problem can be stated as

$$\max_{\alpha(t)} J = x(t_f) ,$$

subject to

$$mV\dot{\gamma} = L - mg\cos\gamma , \quad m\dot{V} = -D - mg\sin\gamma , \quad \dot{h} = V\sin\gamma , \quad \dot{x} = V\cos\gamma ,$$

with $h \geq 0$ and $\alpha \leq \alpha_{max}$, where

$$D = \tfrac{1}{2}(C_{D_o} + \eta C_{L_\alpha}\alpha^2)\rho V^2 S \ , \quad L = \tfrac{1}{2}C_{L_\alpha}\alpha\rho V^2 S \ ,$$

and $V = $ velocity, $\gamma = $ flight-path-angle, $h = $ altitude, $x = $ horizontal range, $\alpha = $ angle-of-attack, $t = $ time, $m = $ mass, $g = $ gravitational force per unit mass, $L = $ lift, $D = $ drag, $\rho = $ air density, $S = $ reference area for lift and drag coefficients, $C_{L_\alpha} = dC_{L_\alpha}/d\alpha = $ slope of lift-coefficient curve, $C_{D_o} = $ zero-lift drag coefficient, $\eta = $ aerodynamic efficiency factor. Note that $\alpha = \alpha_{max}$ corresponds to stalling the aircraft.

We shall measure (x, h) in units of ℓ, V in units of $(g\ell)^{1/2}$, t in units of $(\ell/g)^{1/2}$, where $\ell = 2m/\rho S C_{L_\alpha}$. The equations of motion then become

$$\dot{V} = -\eta(\alpha^2 + \alpha_m^2)V^2 - \sin\gamma \ , \quad \dot{\gamma} = \alpha V - \frac{\cos\gamma}{V} \ , \quad \dot{h} = V\sin\gamma \ , \quad \dot{x} = V\cos\gamma \ ,$$

where $\alpha_m = (C_{D_o}/\eta C_{L_\alpha})^{1/2}$ is the angle-of-attack for min drag-lift ratio. In these normalized variables the *max range steady-state glide* occurs with $\alpha = \alpha_m$ and

$$\gamma = -\tan^{-1}(2\eta\alpha_m) \ , \quad V^2 = \frac{\cos\gamma}{\alpha_m} \ .$$

This is a reasonable initial guess, but the glider obviously hits the ground without flaring. Flaring, followed by flying just above the ground until the aircraft stalls, adds considerable range.

We choose as the key state histories $V(t)$ and $\gamma(t)$, since if they are known, the control $\alpha(t)$ can be determined from the $\dot{\gamma}$ equation, while $h(t)$ and $x(t)$ can be obtained by simple integrations; the $\dot{V}$ equation is used as an equality constraint. $V(0)$, $\gamma(0)$, and $\gamma(t_f)$ are specified, so they must be excluded from the unknown parameter vector p. The final portion of the decelerating flight along $h = 0$ can be calculated analytically backward from the final point, yielding the total horizontal distance traveled (the range) as

$$x_f = x_1 + \frac{1}{4\eta\alpha_m^2}\log\frac{1 + \alpha_m^2 V_1^4}{1 + \alpha_m^2 V_f^4} \ ,$$

where x_1, V_1 are the values of x, V at the end of the integration. Along $h = 0$ we have $\gamma = 0$, $\dot{\gamma} = 0$, so that $\alpha = 1/V^2$; thus $V_f^2 = 1/\alpha_{max}$. We have taken $\alpha_{max} = 3\alpha_m$ in the example calculated below so that $\alpha_m^2 V_f^4 = 1/9$. A MATLAB code to calculate the performance index 'f' and the constraint vector 'g' is listed below.

```
function [f,g,al,x,h,ga,V,tf]=glid_fg(p,N,h0,V0)
% Subroutine for Ex. 9.3.2; max range glide with h>=0 and al<=3*alm using
% inverse dynamic optim; s = [V ga h x]'; control=al=angle-of-attack.
%
alm=1/12; eta=.5; V=[V0 p(1:N)]; ga=[0 p(N+1:2*N-1) 0]; tf=p(2*N); dt=tf/N;
gad=(ga(2:N+1)-ga(1:N))/dt; gab=(ga(2:N+1)+ga(1:N))/2;
```

```
Vd=(V(2:N+1)-V(1:N))/dt;  Vb=(V(2:N+1)+V(1:N))/2;
al=gad./Vb+cos(gab)./Vb.^2;  un=ones(1,N);  h(1)=h0;  x(1)=0;
for i=1:N,
 h(i+1)=h(i)+dt*Vb(i)*sin(gab(i));  x(i+1)=x(i)+dt*Vb(i)*cos(gab(i));
end;
g=[Vd+eta*(al.^2+alm^2*un).*Vb.^2+sin(gab) -h al-3*alm*ones(1,N)];
f=-x(N+1);
```

A script for solving the problem using CONSTR is listed below.

```
% Script e09_3_2.m; max range glide with h >=0 and al < almax using NLP.
%
N=25; alm=1/12; eta=.5; gas=atan(2*eta*alm); Vs=sqrt(cos(gas)/alm);
un1=ones(1,N-1); un=ones(1,N); h0=1.0; V0=1.1*Vs;
p0=[3.7272  3.6566  3.5962  3.5469  3.5071  3.4736  3.4421  3.4081 ...
    3.3673  3.3164  3.2539  3.1813  3.1046  3.0267  2.9477  2.8674 ...
    2.7855  2.7015  2.6150  2.5255  2.4323  2.3345  2.2310  2.1203 ...
    2.0000  0.0000 -0.0078 -0.0195 -0.0315 -0.0414 -0.0475 -0.0488 ...
   -0.0452 -0.0370 -0.0256 -0.0129 -0.0027       0  0.0000  0.0000 ...
    0.0000  0.0000  0.0000  0.0000  0.0000  0.0000  0.0000  0.0000 ...
    0.0000 22.6863];                           % Converged solution
% p0=[3*un -.04*un1 10];                  % Rough initial guess
vlb=[2*un -.05*un1 22.4]; vub=[V0*un 0*un1 23]; optn(1)=1;
optn(4)=1e-5; optn(13)=N; optn(14)=12;
p=constr('glid_fg',p0,optn,vlb,vub,[],N,h0,V0);
[f,g,al,x,h]=glid_fg(p,N,h0,V0); V=[V0 p(1:N)]; ga=[0 p(N+1:2*N-1) 0];
tf=p(2*N);
```

Figures 9.13 through 9.15 are plots of the max range flight path from the converged results. The first part of the path oscillates (at the phugoid frequency) about the optimal steady-state glide path and ends tangent to the ground. The second part is a glide just above the ground, losing velocity and increasing angle-of-attack α until it reaches α_{max}.

Problems

9.3.1 VDP for Max y_f with Gravity and Thrust: Consider a variation on Problem 2.3.5 where we wish to maximize y_f with x_f open. This problem is *singular* (see Chapter 10) since the control γ enters the problem only as $\sin\gamma$ so that, effectively, the control is $\sin\gamma$ and it enters the problem *linearly*.

(a) Show that the optimal control is *bang-bang* and consists of a segment with $\sin\gamma = -1$ followed by a segment with $\sin\gamma = +1$. The explanation of this curious path (going down first to maximize height going up) is that the specific force a does work on the system going down, storing up kinetic energy, and then trades it for potential energy in going up (while still doing work on the system).

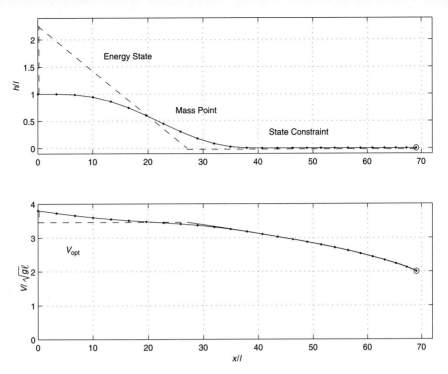

FIGURE 9.13   Altitude and Velocity vs. Range for Max Range Glide

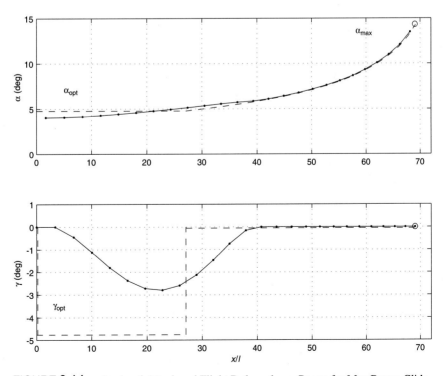

FIGURE 9.14   Angle-of-Attack and Flight Path angle vs. Range for Max Range Glide

367

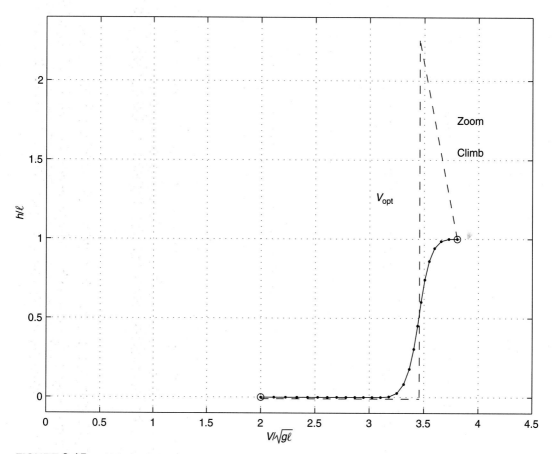

FIGURE 9.15   Altitude vs. Velocity for Max Range Glide

(b) Measuring time in units of t_f, V in gt_f, y in gt_f^2, a in g, show that the switching time $t_s = 1/(2 + a)$ and that the maximum $y_f = a(a + 1)/[2(2 + a)]$. Show also that $y(t_s) = -(a + 1)/[2(2 + a)^2]$.

9.3.2 Min Time Path to $\dot{x} = x = 0$ with Bounded Control: Consider the double integrator plant

$$\ddot{x} = u , \quad \text{where} \ -1 \le u \le 1 .$$

(a) Show that the min time path from an arbitrary point in the $(\dot{x}, x)$ space to $\dot{x} = x = 0$ involves at most only one switch.

(b) Plot the switching curve in the $(\dot{x}, x)$ space and show a few minimum time paths.

(c) Plot the contours of constant time-to-go in the $(\dot{x}, x)$ space and show a few min time paths.

9.3.3 Min Time Control of Position with Damping: The EOM for a mass with a viscous damping force and a control force is

$$\ddot{x} + \dot{x} = u ,$$

where x is the position of the mass and $-1 \le u \le 1$.

(a) Show that the force history to take the mass from an arbitrary $(\dot{x}, x)$ to $\dot{x} = 0$, $x = 0$ in minimum time is bang-bang with at most one switch.
(b) Plot the switching curve in the $\dot{x}$, x space and a few min time paths.

9.3.4 Min Fuel Path to $\dot{x} = x = 0$ with Bounded Control: For the double integrator plant of Problem 9.3.2 with $|u| \le u_o$, find the path from a given point in the $(\dot{x}, x)$ space to $\dot{x} = x = 0$ which minimizes

$$J = \int_0^{t_f} |u| dt ,$$

where t_f is specified. Measure $\dot{x}$ in units of $u_o t_f$, x in $u_o t_f^2/2$, u in u_o. Plot the family of paths in the normalized $(\dot{x}, x)$ space.

9.3.5 Min Time Path to a Line Segment with Bounded Control: For the double integrator plant of Problem 9.3.2, find the min time path from a given point in the $(\dot{x}, x)$ space to the terminal line

$$\dot{x} = 0 , \quad -b \le x \le b ,$$

where b is a specified distance. Plot the paths in the $(\dot{x}, x)$ space.

9.3.6 Min Time Path for a NMP Second-Order Plant with Bounded Control: The EOM for a plant are

$$\dot{q} = u , \quad \dot{\theta} = q - \tau u ,$$

where $|u| \le u_{max}$.

(a) Show that the min time path from an arbitrary point in the (q, θ) space to $q = \theta = 0$ involves at most only one switch.
(b) Plot the switching curve in the (q, θ) space and show a few min time paths.
(c) Plot the contours of constant time-to-go in the (q, θ) space and show a few min time paths.

9.3.7 Min Fuel Horizontal Translation of a S/C with Bounded Control: Find the tilt angle history $\theta(t)$ to translate a S/C horizontally over the surface of the Moon through a specified distance ℓ using min fuel (see Fig. 9.16). The thrust T is varied so that the vertical component of thrust, $T \cos \theta$, equals gravitational force mg at all times, while $|\theta|$ is constrained to be less than θ_o. The problem may be stated as finding $\theta(t)$ and t_f to minimize

$$J = \int_o^{t_f} \sec \theta \, dt$$

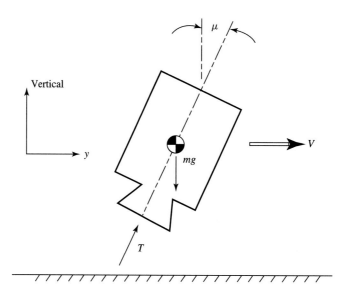

FIGURE 9.16   Min Fuel Horizontal Translation of a S/C above the Surface of the Moon

subject to

$$\dot{y} = v , \quad y(0) = 0 , \quad y(t_f) = 1 , \quad \dot{v} = \tan\theta , \quad v(0) = v(t_f) = 0 , \quad |\theta| \le \theta_o ,$$

where $\theta_o < \pi/2$, y is in units of ℓ, v in $\sqrt{g\ell}$, and time in $\sqrt{\ell/g}$.

(a) Clearly $\theta(0) > 0$ and $\theta(t_f) < 0$; assuming this, show that

$$-\lambda_v(t) = \frac{1 - 2t/t_f}{\sin\theta_o}$$

and

$$\theta = \begin{cases} -\theta_o , & \lambda_v > \sin\theta_o \\ \sin^{-1}(-\lambda_v) , & -\sin\theta_o < \lambda_v < \sin\theta_o , \\ \theta_o , & \lambda_v < -\sin\theta_o \end{cases}$$

From this show that the optimal path starts with a period where $\theta = \theta_o$, followed by a period where $-\theta_o < \theta < \theta_o$, and ends with a period where $\theta = -\theta_o$.

(b) In the unconstrained period, show that

$$v = \frac{t_f}{2} \sin\theta_o \cos\theta , \quad \sin\theta = \left(1 - \frac{2(t - t_1)}{t_2 - t_1},\right) \sin\theta_o .$$

where (t_1, t_2) are the switchtimes.

(c) Show that t_f is determined by

$$1 = \frac{t_f^2}{4} [\sin\theta_o \cos^3\theta_o + \sin^2\theta_o(\theta_o + \sin\theta_o \cos\theta_o)] ,$$

and $t_1 = t_f - t_2 = \frac{t_f}{2} \cos^2\theta_o$.

9.3.8 Min Control Effort to Reverse Velocity in a Bounded Space: A particle leaves $y = 0$ with velocity v_o. We wish to apply specific force a to stop the particle and bring it back to $y = 0$ in time t_f with velocity $-v_o$. We should like to do this without having the particle travel beyond $y = \ell$, and we should like to use as little effort as possible.

If we measure v in units of v_o, time in units of t_f, a in units of v_o/t_f, and (y, ℓ) in units of $v_o t_f$, then the problem may be stated as

$$\min_{u(t)} J = \int_0^1 a^2 dt$$

subject to

$$\dot{y} = v , \quad y(0) = y(1) = 0 , \quad \dot{v} = a , \quad v(0) = -v(1) = 1 ,$$

where $y \le \ell$.

(a) Solve this problem using inverse optimization and the MATLAB code CONSTR for the case where $\ell < 1/6$.
(b) Let $\epsilon = t_1 - t$, where t_1 is the time when the path goes on to the constraint boundary $y = \ell$; the analytical solution gives $t_1 = 3\ell$ and the path from $t = 0$ to $t = t_1$ is given by

$$y = \ell - \frac{(\epsilon/3)^3}{\ell^2} , \quad v = -\frac{dy}{d\epsilon} = \frac{(\epsilon/3)^2}{\ell^2} , \quad a = -\frac{dv}{d\epsilon} = -\frac{2\epsilon}{(3\ell)^2} ,$$

Compare your solution from (a) with this analytical solution (Ref. BH, pp. 120–123).

9.3.9 DVDP for Min Time to $x = x_f$ with a State Constraint: Consider DVDP for min time to $x = x_f$ (the dual of Problem 2.1.1) with the state constraint $y \ge h + x \tan\theta$. Measure (x, y) in units of x_f, V in $\sqrt{gx_f}$, t in $\sqrt{x_f/g}$.

Use CONSTR in the the MATLAB Optimization Toolbox with $\tan\theta = .5$ and plot the minimum time paths from $x = y = 0$ for $h/x_f = [0\ .1\ .2\ .284]$. Compare your results with those in Fig. 9.17. Note the constraint is not active for $h/x_f \ge .284$.

9.3.10 Min Time Control of an Inverted Pendulum with Bounded Torque: The EOM for an inverted pendulum with torque Q at its base is

$$\ddot{\theta} - \theta = Q ,$$

where θ = angle of the pendulum from vertical and $-1 \le Q \le 1$.

(a) Show that the torque history to go from an arbitrary $(\dot{\theta}, \theta)$ to $\dot{\theta} = 0$, $\theta = 0$ in minimum time is bang-bang with, at most, one switch.
(b) Plot the switching curve in the the $\dot{\theta}$, θ space and a few min time paths. Identify the controllable and uncontrollable regions in this space.

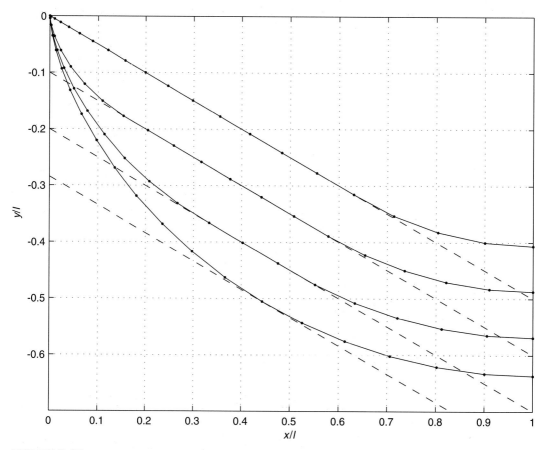

FIGURE 9.17　Min Time with Gravity to $x = x_f$ and $y \geq h + x/2$

9.3.11 Min Time Control of a Pendulum with Bounded Torque:　The EOM for a pendulum with torque Q at its base is

$$\ddot{\theta} + \theta = Q \,,$$

where θ = angle of the pendulum from vertical and $-1 \leq Q \leq 1$.

(a) Show that the torque history to go from an arbitrary $(\dot{\theta}, \theta)$ to $(\dot{\theta} = \theta = 0)$ in min time is bang-bang with a finite number of switches (Bushaw 1958, Ref. Bus).

(b) Show that the the switching curve in the $\dot{x}, x$ space is made up of semi circular arcs of radius 1. Plot these arcs and show a few min time paths (see Ref. AF, pp. 572–579 for a complete solution).

9.3.12 Max Range of a Glider Using the Energy State Approximation:　In Example 9.3.2 we considered the problem of max range for a glider with the constraint that altitude should always be ≥ 0. A good approximation to the solution of this problem can be obtained using the *energy state approximation*, which was described in Section 4.4. This much simpler solution gives physical insight into the nature of the constrained solution.

(a) Using the energy state approximation ($|\gamma| \ll 1, |\dot{\gamma}| \ll 1$), with energy E as the independent variable, range x as the state variable, and V as the control variable, show that the problem is simplified to

$$\max_{V(E)} J = x(E_f)$$

subject to

$$\frac{dx}{dE} = -\eta \left(\frac{1}{V^2} + \alpha_m^2 V^2 \right) ,$$

where E *decreases* monotonically from $E_o \equiv h_0 + V_0^2/2$ to $E_f \equiv V_f^2/2$, with $x(E_0) = 0$, $V^2 \leq 2E$ and $\alpha \approx 1/V^2$.

(b) Show that the optimal velocity V_{ss} for (a) is

$$V_{ss}^2 = \frac{1}{\alpha_m} \quad \text{if } E > \frac{\alpha_m}{2} ,$$

$$= 2E \quad \text{if } E \leq \frac{\alpha_m}{2} .$$

(c) In general, (b) implies a rapid initial change in velocity (a constant energy zoom climb or dive), then a steady glide at $V = V_{ss} \overset{\Delta}{=} 1/\sqrt{\alpha_m}$ (corresponding to maximum L/D) to $h = 0$, and then a gradual decrease in velocity to α_{max} on $h = 0$. Sketch the solution in the $(h, V^2/2)$ space where contours of constant E are straight lines with a slope of -1.

(d) Using (b), show that optimal constrained path $x(E)$ is

$$x = \frac{E_0 - E}{\alpha_m/2} \quad \text{for } E > \frac{\alpha_m}{2} ,$$

$$= x_1 + \frac{1}{4} \log \frac{\alpha_m^2/2}{E^2 + \alpha_m^2/4} \quad \text{for } E < \frac{\alpha_m}{2} ,$$

where $x_1 = 2E_0/\alpha_m - 1/4$.

(e) From (d) the max range is

$$x_{max} \equiv x(E_f) = \frac{2E_0}{\alpha_m} - \frac{1}{4} + \frac{1}{4} \log \frac{\alpha_m^2/2}{E_f^2 + \alpha_m^2/4} .$$

Find x_{max} using the data in Example 9.3.2 ($\alpha_m = 1/12$, $\eta = .5$, $h_0 = 1.0$, $V_0 = 1.1 V_{ss}$, $\alpha_{max} = 3\alpha_m$). Compare these approximate results with the more precise results in Figs. 9.11 to 9.13.

9.3.13 Erection of an Inverted Pendulum by Motion of a Cart (Ref. Vg):   A pendulum of length ℓ and mass m is mounted on a cart of mass M. It is mounted so that it can swing through to an inverted position (straight up). The maximum magnitude of the horizontal force on the cart is u_{max}. Let θ = angle of pendulum from the vertical ($= 0$ when it is straight down), and let y = distance of the cart from a reference point.

(a) Show that the EOM are

$$\begin{bmatrix} 1 & \epsilon \cos\theta \\ \cos\theta & 1 \end{bmatrix} \begin{bmatrix} \ddot{y} \\ \ddot{\theta} \end{bmatrix} = \begin{bmatrix} \epsilon\dot{\theta}^2 \sin\theta + u \\ -\sin\theta \end{bmatrix},$$

where y is in units of ℓ, time in $\sqrt{\ell/g}$, u in $(M+m)g$, and $\epsilon \overset{\Delta}{=} m/(M+m)$.

(b) Find $u(t)$ to take the system from $y = \dot{y} = \dot{\theta} = \theta = 0$ to $y = \dot{y} = \dot{\theta} = 0,\ \theta = \pi$ *in min time* for the case where $\epsilon = .5,\ u_{max} = 1$. Compare your solution with the one

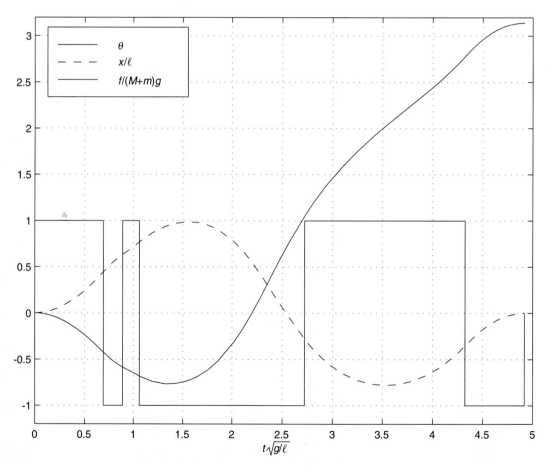

FIGURE 9.18   Min Time Erection of an Inverted Pendulum using Bounded Horizontal Force on a Cart; Force $f(t)$, Cart Position $x(t)$ and Pendulum Angle $\theta(t)$

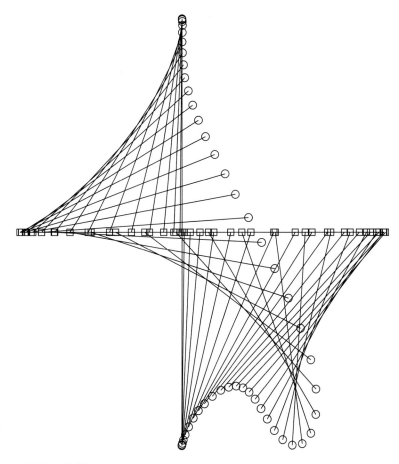

FIGURE 9.19 Min Time Erection of an Inverted Pendulum using Bounded Horizontal Force on a Cart; Stroboscopic Movie at Equal Time Intervals

shown in Figs. 9.18 and 9.19. An initial guess can be obtained using inverse dynamic optimization with CONSTR; a precise solution can then be found using CONSTR with the switching times as the unknown parameters, integrating between switching times with ODE23.

9.3.14 Min Time Pick-and-Place Path for a Robot Arm (Ref. Me): Consider a robot arm modeled as two rigid links of mass m and length L and a tip load of mass μm. There are torquers at the shoulder and at the elbow. There are four state variables: θ_s = angle of the inner link with respect to a reference line, θ_e = angle between the outer link and an extension of the inner link, ω_s = angular velocity of the inner link, and ω_e = angular velocity of the outer link. The two controls are the shoulder and elbow torques that can vary between $-Q_{max}$ and Q_{max}.

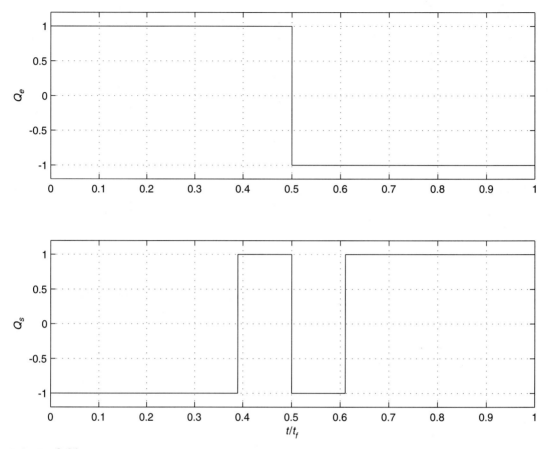

FIGURE 9.20   Min Time Path for Two-Link Robot Arm; Elbow and Shoulder and Torques (Q_e, Q_s) vs. Time

(a) Show that the EOM are

$$Q_e = (\mu + \tfrac{1}{2})\dot{\omega}_s \cos\theta_e + (\mu + \tfrac{1}{3})\dot{\omega}_e + (\mu + \tfrac{1}{2})\omega_s^2 \sin\theta_e ,$$

$$Q_s = Q_e + (\mu + \tfrac{4}{3})\dot{\omega}_s + (\mu + \tfrac{1}{2})\dot{\omega}_e \cos\theta_e - (\mu + \tfrac{1}{2})\omega_e^2 \sin\theta_e,$$

$$\dot{\theta}_s = \omega_s , \quad \dot{\theta}_e = \omega_e - \omega_s .$$

where (Q_e, Q_s) are in units of $Q_{\max}$, t is in units of $t_c = \sqrt{mL^2/Q_{\max}}$, ($\omega_e$, ω_s) are in units of $1/t_c$.

(b) We wish to find the min time "pick-and-place" motion between two points A and B on a horizontal surface where D = distance between A and B (measured in units of L), and the optimum orientation of the robot relative to the line AB. In other words, find $Q_e(t)$, $Q_s(t)$ to take the system from $\omega_e(0) = \omega_s(0) = \theta_s(0) = 0$, $\theta_e(0) = \theta_{eo}$ to $\omega_e(t_f) = \omega_s(t_f) = 0$ with the tip traveling a distance D, and $|Q_e| \le 1$, $|Q_s| \le 1$.

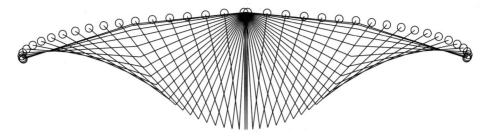

FIGURE 9.21   Min Time Path for Two-Link Robot Arm; Stroboscopic Movie at Equal Time
Intervals

The initial elbow angle θ_{eo} determines the most efficient orientation of the robot
with respect to the workspace, so it is an unknown parameter to be determined optimally. An initial guess can be obtained using inverse dynamic optimization with
CONSTR; the two angular velocities can be used as the "key" state variables, since their
derivatives determine the control histories, and their integrals determine the two angle
histories.

(c) Obtain a more precise solution using CONSTR using the switching times, θ_{eo}, and
t_f as the unknown parameters, integrating between switching times with ODE23.
Compare your results with Figs. 9.20 and 9.21. The elbow torque is at $Q_{\max}$ for
the first half of the path and at $-Q_{\max}$ for the second half. The shoulder torque has
three switches and is also antisymmetric about $t_f/2$; it has to delay to let the elbow
"catch up."

9.3.15 F4 Min Time Climb Path with $h > 0$ Constraints: Consider Example 4.4.3
with the SVIC $h \geq 0$, $T(t) = T_{\max}(h, V)$, and initial conditions $V(0) = 440$ ft/sec,
$h(0) = 0$, $\gamma(0) \geq 0$, $W(0) = 41{,}998$ lb. Use inverse optimization with key state variable
histories $V(t)$, $\gamma(t)$, obtaining $x(t)$, $h(t)$, $m(t)$ by integration and $\alpha(t)$ by differentiation.
Verify that the optimal path stays on $h = 0$ until about $V = .9$ kft/sec, then begins an almost
constant Mach number climb as shown in Figures 4.5 and 9.22. Note that this checks the
energy-state solution of Example 4.4.4.

9.3.16 Max Time Glide Path in the Horizontal Plane with a SVIC: Consider Problem
4.4.16 (and 4.6.16) with the SVIC $|y(t)| \leq y_{\max}$ (e.g., the glider is in a narrow canyon
with steep walls) with $y_{\max} = .4$. Use inverse optimization with key state variable histories
$V(t)$, $\psi(t)$, obtaining the bank angle $\sigma(t)$ by differentiation and $x(t)$, $y(t)$ by integration.
Compare your results with Fig. 9.23; the max time is reduced by only about 1 percent, but
the path is quite different, requiring a very steep bank to turn around and head back to the
desired final point.

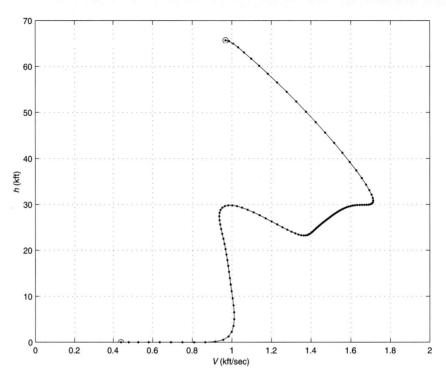

FIGURE 9.22   F4 Min Time to Climb Path with SVIC $h \geq 0$ and $V(0) = 440$ ft/sec

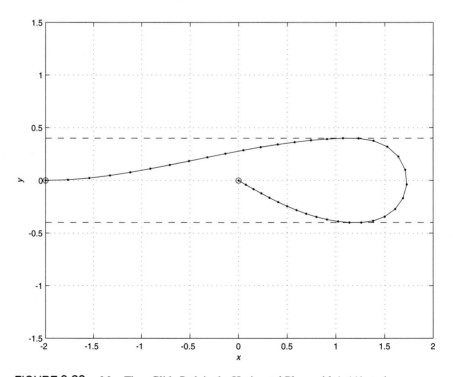

FIGURE 9.23   Max Time Glide Path in the Horizontal Plane with $|y(t)| \leq .4$

9.4 Chapter Summary

Parameter Optimization Problems with Inequality Constraints

These are sometimes called *nonlinear programming problems* (NLP). Minimize $L(y)$ with $f(y) \leq 0$, where y is $(p \times 1)$, f is $(n \times 1)$ and p may NOT be $< n$.

- Necessary condition for a minimum *(Karush-Kuhn-Tucker condition)*: $-L_y$ must lie in the cone of active constraints, i.e.,

$$-L_y = \lambda^T f_y^A , \quad \lambda_i \geq 0 ,$$

where f^A is the subset of active constraints.

Function Optimization Problems with Control Variable Inequality Constraints

The solution involves finding

$$\min_u [H(x, u, t, \lambda)] \text{ with } C(u) \leq 0 ,$$

which is an NLP at every time $t_o \leq t < t_f$.

- A necessary condition is

$$-H_u = \mu^T C_u^A , \quad \mu_i \geq 0 ,$$

where C^A is the subset of active constraints (Kuhn-Tucker and McShane-Pontryagin).
- *Bang-bang control* if H and C are linear in x and $u \Rightarrow$ an LP problem at every time $t_o \leq t < t_f$.
- *Bang-off-bang control* if H and C are linear in x, u, and $|u|$. If no bounds on $u \Rightarrow$ impulse-off-impulse solution.

Function Optimization Problems with State Variable Inequality Constraints

These problems are more difficult to solve than problems with control constraints. Exact methods involve differentiating the state constraint and substituting for the state derivatives from the equations of motion until the control appears; this reduces it to a problem with a control constraint. However, the state constraint and the derivatives that do not involve the control are now *interior boundary conditions at the entry point* onto the state constraint, since otherwise the path could not stay on the arc ("tangential entry"). This method also requires preknowledge of the number and sequence of the constrained arcs.

Seywald's method of *Differential Inclusion* solves dynamic optimization problems with control or state variable inequality constraints by converting them into parameter optimization problems and using available NLP codes like NPSOL or CONSTR in the MATLAB Optimization Toolbox. The parameters p are points on the "key" state variable histories. This method does NOT require preknowlege of the number and sequence of the constrained arcs; however, it becomes slow on current PCs if there are more than about 25 points.

CHAPTER 10

Singular Optimal
Control Problems

10.1 Introduction

In some optimal control problems, extremal arcs ($H_u = 0$) occur on which the matrix H_{uu} is only semidefinite, i.e., it has one or more zero eigenvalues; such arcs are called *singular arcs*. In the classical calculus of variations an optimal path that contains one or more singular arcs is called a *broken extremal* (Ref. Bo) since the control variables have discontinuities ("corners") at one or both ends of the singular arcs. Singular arcs may appear in optimal control problems involving *cascaded dynamic systems* if the direct effect of the control and the effect of a strongly controlled state have opposite signs. Cascaded linear systems of this type are called *nonminimum phase (NMP) systems* and are characterized by having RHP transmission zeros in the transfer function matrix.

In Section 10.2, we treat LQ controllers for NMP systems with no penalty on the control. Optimal paths consist of impulses and singular arcs. If hard bounds (inequality constraints) are placed on the control, the impulses are replaced by finite-time pulses where the control is on the bound; such a solution is said to be "bang-singular-bang."

In Section 10.3 we treat nonlinear dynamic optimization problems that contain singular arcs. As in linear problems, the solutions involve impulses if there are no bounds on the control, and finite pulses otherwise. As an example we discuss the famous Goddard Problem of finding the thrust history to maximize the altitude of a sounding rocket (a rocket launched vertically upward). With atmospheric drag, the optimal path is ideally "impulse-singular-coast." The more realistic solution, with a bound on thrust magnitude, is "bang-singular-coast."

Section 10.4 summarizes the methods used to solve optimal control problems that contain singular arcs.

10.2 LQ Controllers for NMP Systems

NMP systems have transfer functions with RHP TZs and are fundamentally deficient for fast control (see Section 6.2). Even with a very small penalty using integral-square-control, the fastest possible response is a mode that corresponds to a pole near the smallest reflected RHP TZ (Ref. KS). Furthermore, output initially moves in the wrong direction. With no penalty on integral-square control, the optimal response involves control impulses and a singular arc (or arcs). With hard bounds on the control, the impulses are replaced by periods where the control is on the boundary. The example below involves a simple NMP system having one RHP zero (Ref. JG).

Example 10.2.1—A Simple Second-Order NMP System

We consider a second-order linear system with transfer function

$$\frac{y(s)}{u(s)} = \frac{s-1}{s^2} \tag{10.1}$$

that has a double pole at $s = 0$ and *RHP zero at* $s = 1$. A state variable realization of (10.1) is

$$\dot{x}_1 = x_2 + u , \quad \dot{x}_2 = -u , \quad y = x_1 . \tag{10.2}$$

We wish to find $u(t)$ to minimize

$$J = \frac{1}{2} \int_0^{t_f} x_1^2 dt , \tag{10.3}$$

starting with arbitrary $x_1(0)$ and $x_2(0)$ and ending with $x_1(t_f) = x_2(t_f) = 0$.

The Hamiltonian is *linear in* u *and* x_2 but quadratic in x_1:

$$H = \tfrac{1}{2}x_1^2 + \lambda_1(x_2 + u) - \lambda_2 u . \tag{10.4}$$

The EL equations are

$$\dot{\lambda}_1 = -H_{x_1} \equiv -x_1 , \quad \dot{\lambda}_2 = -H_{x_2} \equiv -\lambda_1 , \tag{10.5}$$

and

$$H_u \equiv \lambda_1 - \lambda_2 = 0 . \tag{10.6}$$

The problem is *singular* since

$$H_{uu} \equiv 0 , \tag{10.7}$$

and (10.6) does *not determine* u. However, if (10.6) is to hold for a finite period of time, then its *time derivative must be zero*:

$$0 = \dot{H}_u = \dot{\lambda}_1 - \dot{\lambda}_2 \equiv -x_1 + \lambda_1 . \tag{10.8}$$

Equation (10.8) still does not determine u, so we *take another time derivative*:

$$0 = \ddot{H}_u = -\dot{x}_1 + \dot{\lambda}_1 \equiv -x_2 - x_1 - u , \tag{10.9}$$

which does determine u as a *linear state feedback*:

$$u = -x_1 - x_2 \ . \tag{10.10}$$

Using (10.6), (10.8), and (10.9) in (10.4) gives

$$H = x_1 \left(x_2 + \tfrac{1}{2} x_1 \right) \ , \tag{10.11}$$

and $H = $ constant since it does not explicitly involve the time. Equation (10.11) represents a one-parameter *family of hyperbolas* in the state space (see Figure 10.1).

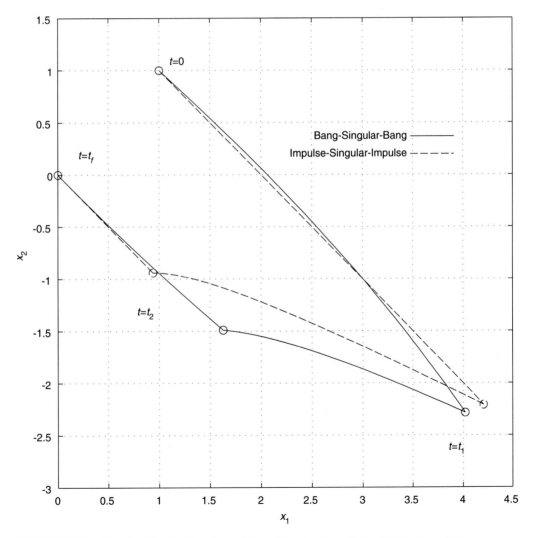

FIGURE 10.1   Impulse-Singular-Impulse and Bang-Singular-Bang Optimal Paths for a NMP System—x_1 vs. x_2

Adding the two equations in (10.2) eliminates the control u:

$$\frac{d}{dt}(x_1 + x_2) = x_2 . \tag{10.12}$$

Thus, while an impulse in u changes both x_1 and x_2 it *does not change* $x_1 + x_2$. In Figure 10.1 this means the system can be moved instantaneously along the lines $x_1 + x_2 = $ constant by impulses in u with *no effect on the performance index* since the performance index does not involve u. A positive impulse moves the state down and right while a negative impulse moves the state up and left. A graphical solution is now evident: if $x_1(0) + x_2(0) > 0$ then apply a positive impulse to move the system to one of the hyperbolas in the fourth quadrant; then move along the hyperbola (a singular arc) until it reaches the line $x_1 + x_2 = 0$; then apply a negative impulse to move the system to the origin. This is an "impulse-singular-impulse" solution. Despite the fact that x_1 is the controlled quantity, the optimal solution *first increases x_1 before decreasing it;* this is typical behaviour for NMP systems.

The particular hyperbola (i.e., the value of H) is determined by the fact that the time moving on it must equal t_f. It is straightforward to show that

$$H = -2c^2 \frac{e^{-2t_f}}{(1 - e^{-2t_f})^2} , \tag{10.13}$$

where $c = x_1(0) + x_2(0)$. Also

$$x_1(t) = x_1(0+)e^{-t} , \quad x_2(t) = x_1(0+) \sinh t + x_2(0+)e^t , \tag{10.14}$$

where $x_1(0+), \; x_2(0+)$ are the state variables after the initial impulse in u:

$$x_1(0+) = \frac{2c}{1 - e^{-2t_f}} , \quad x_2(0+) = -c \coth t_f . \tag{10.15}$$

Note $x_1(t_f) + x_2(t_f) = 0$, so that a final impulse in u can bring the states to zero. Figures 10.1 to 10.3 show the solution for $x_1(0) = x_2(0) = 1, \; t_f = 1.5$.

If u *is bounded* then the impulses are replaced with finite pulses (finite time arcs with the control on the bounds), a bang-singular-bang solution. Figures 10.1 to 10.3 show the solution of the same LQ problem as above except that $|u| \leq 8$. This solution was found using the FSOLVE command in MATLAB to determine the switch times t_1 and t_2 to bring the two states to zero in the given time t_f in a bang-singular-bang sequence.

If the problem is discretized, it is no longer singular (see Problem 10.2.1).

Example 10.2.2—A/C Turn Using Rudder Only

This is a well-known physical example of a NMP linear system. We wish to bring an A/C from a steady banked turn back to level flight using only the rudder, keeping the bank angle as small as possible, and observing realistic bounds on rudder deflection (many unmanned A/C do not have ailerons since that requires two additional actuators, which adds cost and weight). We first use an integral-square penalty on both bank angle and rudder deflection.

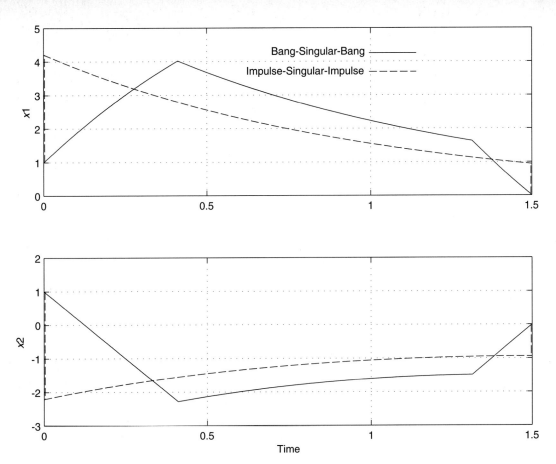

FIGURE 10.2   Impulse-Singular-Impulse and Bang-Singular-Bang Optimal Paths for a NMP
System—x_1 and x_2 vs. Time

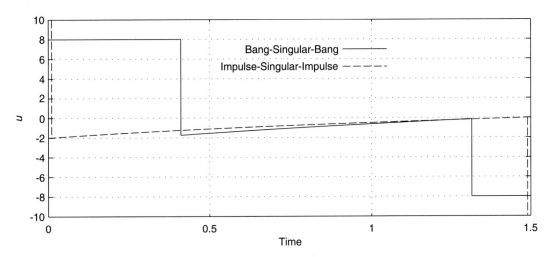

FIGURE 10.3   Impulse-Singular-Impulse and Bang-Singular-Bang Optimal Paths for a NMP
System—Control vs. Time

Then we change to a hard bound on rudder deflection and show that the solution is bang-bang-bang-singular.

We use data for the Navion A/C (Ref. HJ) with state vector $x = [v\ r\ p\ \phi]^T$, where $v =$ sideslip velocity in ft/sec, $r =$ yaw angular velocity in crad/sec, $p =$ roll angular velocity in crad/sec, $\phi =$ bank angle in crad, and control $\delta r =$ rudder angle in crad. The EOM are

$$\dot{x} = Ax + B\delta r \ , \tag{10.16}$$

where

$$A = \begin{bmatrix} -0.254 & -1.760 & 0 & .322 \\ 2.55 & -.76 & -.35 & 0 \\ -9.08 & 2.19 & -8.40 & 0 \\ 0 & 0 & 1 & 0 \end{bmatrix}, \quad B = \begin{bmatrix} .1246 \\ -4.60 \\ 2.55 \\ 0 \end{bmatrix}. \tag{10.17}$$

The transfer function from δr to ϕ is

$$\frac{\phi(s)}{\delta r(s)} = \frac{2.55(s + 3.608)(s - 6.988)}{(s + 8.433)(s + .009)[(s + .486)^2 + (2.333)^2]} \ , \tag{10.18}$$

which has a *RHP zero at s* = 6.988.

We first consider the problem with an integral-square penalty on rudder deflection, and then we will treat it below with a hard bound on rudder deflection. We wish to find $\delta r(t)$ to minimize

$$J = \frac{1}{2} \int_0^\infty (Q\phi^2 + \delta r^2)dt \ , \tag{10.19}$$

subject to (10.16), bringing the system to zero from an initial steady turn.

The SRL vs. Q (see Figure 10.4) shows that as $Q \to \infty$, two of the closed-loop poles tend to $-\sqrt{Q} \pm j\sqrt{Q}$, one pole goes to the zero at $s = -3.608$ where it is canceled, and the fourth pole goes to the *reflected zero* at $s = -6.988$ where it is *not* canceled. Thus, for large Q, ϕ would be reduced rapidly with the fast complex mode; then the other states would attenuate with the two real modes (ϕ would attenuate only with the mode corresponding to the reflected zero). Thus there is a bandwidth limitation with state feedback; even with large rudder deflections, the turn could not be attenuated any faster than the pole at the reflected zero (Ref. KS).

However, large values of Q give unattainable rudder deflections. By interpolation on Q we found that $Q = 5$ gives attainable rudder deflections; $Q = 5$ corresponds to closed-loop poles that are still a long way from the zeros (see Figure 10.4).

Figures 10.5 and 10.6 show responses for $Q = 5$, chosen so that the maximum rudder deflection was about 13 deg. Figure 10.5 shows roll angle ϕ and rudder angle δr vs. time while Figure 10.6 shows yaw rate r and sideslip velocity v vs. time. This is a typical NMP response to an output command; the output *first becomes slightly larger and then decays*. The reason for this is that *the rudder can produce large rolling moments only indirectly* by first yawing the A/C; this yaw rate r produces a sideslip v, which then produces a large rolling moment due to the dihedral of the wings. However, the rudder deflection for the

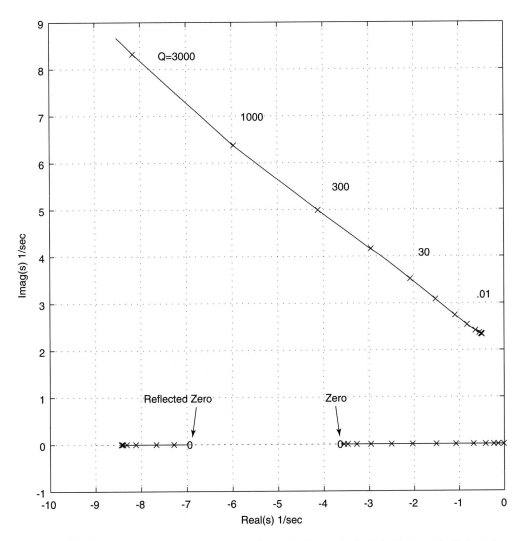

FIGURE 10.4   Symmetric Root Locus of Closed Loop Poles vs. Q; the Pole Going to the Reflected Zero is *not* Canceled and Dominates the Response for Large Q

desired yaw produces a small rolling moment of the wrong sign (since the rudder force is above the A/C centerline), which starts the bank angle in the wrong direction. A short time later the sideslip produces a large rolling moment of the correct sign to bring the bank angle to zero (thus the term "cascaded system").

Now we consider the same problem with hard bounds on rudder deflection, which makes it nonlinear (no simple feedback solution):

$$|\delta r| \leq \delta_{ro} \,. \tag{10.20}$$

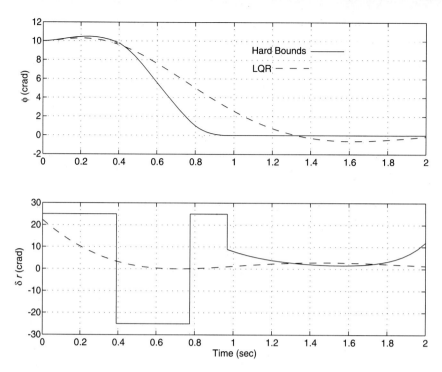

FIGURE 10.5 Roll Angle ϕ and Rudder Angle δr vs. Time; LQR with $Q = 5$ (Dashed) and with Hard Bounds

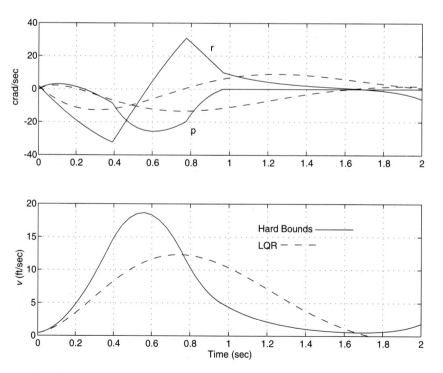

FIGURE 10.6 Yaw Angular Velocity r and Sideslip Velocity v vs. Time; LQR with $Q = 5$ (Dashed) and with Hard Bounds

However, we keep the integral-square performance index on bank angle:

$$J = \frac{1}{2} \int_0^\infty \phi^2 dt \ .$$

(10.21)

This is a problem with a final *singular arc* and three initial arcs where the rudder deflection is at the hard bounds (i.e., a bang-bang-bang-singular solution). A bang-bang solution was anticipated from the discussion above for $Q \to \infty$, which involves impulses and doublets (corresponding to the fast complex poles). A singular arc was anticipated since the Hamiltonian is now linear in the control δr and the transfer function from control to output is NMP. The sequence of arcs was found using the NLP inverse-control method of Section 9.3; the accuracy of $\delta r(t)$ on the singular arc was poor, so the solution was refined by the following method: three switching times $(t_1, \ t_2, \ t_3)$ and a scalar constant c were determined so that the four-state vector $x(t_3)$ was equal to cV, where V is the eigenvector corresponding to the closed-loop pole that canceled the reflected zero above (see below). λ and δr on the singular arc were determined as linear functions of the state x by the *optimality condition and its first four time derivatives*, also shown below.

The Hamiltonian is

$$H = \tfrac{1}{2}\phi^2 + \lambda^T (Ax + B\delta r) \ .$$

(10.22)

The EL equations are:

$$\dot{\lambda}^T = -x^T Q - \lambda^T A \ ,$$

(10.23)

$$0 = B^T \lambda \ ,$$

(10.24)

where $Q \triangleq$ diagonal [0 0 0 1]. The optimality condition (10.24) does not determine the control δr, so we take three time derivatives of (10.24) and substitute for $\dot{x}$ and $\dot{\lambda}$ from (10.16) and (10.23), respectively. These time derivatives still do not contain δr but they do determine λ:

$$\begin{bmatrix} B^T \\ (AB)^T \\ (A^2 B)^T \\ (A^3 B)^T \end{bmatrix} \lambda = \begin{bmatrix} 0 \\ 0 \\ -B^T A^T Q \\ T \end{bmatrix} x \ ,$$

(10.25)

or

$$\lambda = Gx \ ,$$

(10.26)

where $T \triangleq B^T A^T (QA - A^T Q)$. The coefficient matrix of λ in (10.25) is the *controllability matrix*. Taking a fourth time derivative of (10.26) finally determines δr as a linear feedback on the state x:

$$\delta r = -Kx \ ,$$

(10.27)

where

$$K = (TB)^{-1}[U(Q + A^T G) + TA] , \tag{10.28}$$

and $U \triangleq (A^3 B)^T$.

The closed-loop system matrix $A - BK$ has poles at the *zeros and the reflected zeros* of (10.18). The eigenvector corresponding to the pole at the real zero $s = -3.608$ has zero components for p and ϕ, so this is the desired V mentioned above.

The solution was obtained as described above using FSOLVE in MATLAB to find the four parameters t_1, t_2, t_3, c to satisfy the relation $x(t_3) = cV$. The resulting optimal path is shown in Figures 10.5 and 10.6 for $\delta_{ro} = 25$ crad, alongside the results from LQR with $Q = 5$ (dashed). The bang-bang-bang-singular solution shows the basics of turning an A/C using rudder only: (1) Use max rudder for a short time Δ to move the A/C nose in the desired turn direction; turn rate r increases nearly linearly with time and sideslip v increases nearly quadratically. (2) Use max rudder in the other direction for about the same length of time Δ; r decreases and overshoots to the opposite direction while v reaches a maximum and starts to decrease. (3) Switch the rudder again and hold it for about $\Delta/2$ until the roll rate and bank angle are both zero; both r and v tend toward zero. (4) Decrease rudder to about .4 maximum and then decrease it slowly to hold zero bank angle while r and v tend to zero (the singular arc). In other words, *sideslip is approximately the pseudo-control for roll rate, and sideslip is approximately the double integral of rudder*. The approximate system is

$$\dot{r} \approx -4.60\delta r , \quad \dot{v} \approx -1.760r , \quad \dot{p} \approx -8.40p - 9.08v , \quad \dot{\phi} = p . \tag{10.29}$$

This is approximately a doubly-cascaded system, while the example of the last section was a singly-cascaded system.

The LQR solution is qualitatively similar to the bang-bang-bang-singular solution, but is slower since it does not use the max available control.

Problems

10.2.1 Discretization of Example 10.2.1:   Consider the discretized version of the Example 10.2.1.

(a) With sample time T_s, show that the discretized plant is

$$\begin{bmatrix} x_1(k+1) \\ x_2(k+1) \end{bmatrix} = \begin{bmatrix} 1 & T_s \\ 0 & 1 \end{bmatrix} \begin{bmatrix} x_1(k) \\ x_2(k) \end{bmatrix} + \begin{bmatrix} T_s - T_s^2/2 \\ -T_s \end{bmatrix} u(k) ,$$

and the discretized performance index is

$$J = \frac{1}{2} \sum_{k=1}^{N} \begin{bmatrix} x_1(k) & x_2(k) & u(k) \end{bmatrix} \begin{bmatrix} Q_d & N_d \\ N_d^T & R_d \end{bmatrix} \begin{bmatrix} x_1(k) \\ x_2(k) \\ u(k) \end{bmatrix} ,$$

where

$$Q_d = \begin{bmatrix} T_s & T_s^2/2 \\ T_s^2/2 & T_s^3/3 \end{bmatrix} , \quad N_d = \begin{bmatrix} T_s^2/2 - T_s^3/6 \\ T_s^3/3 - T_s^4/8 \end{bmatrix} , \quad R_d = T_s^3/3 - T_s^4/4 + T_s^5/20 .$$

(b) Use TDLQS to find the optimal path from $x_1 = x_2 = 1$ to $x_1 = x_2 = 0$ for $t_f = 1.5$ using six steps. Compare your results with Figures 10.1 to 10.3.

10.2.2 An Oscillatory NMP System: Consider Example 10.2.1 with the transfer function (10.1) replaced by

$$\frac{y(s)}{u(s)} = \frac{s-1}{s^2+1} ,$$

with the state variable realization

$$\dot{x}_1 = x_2 + u , \quad \dot{x}_2 = -x_1 - u , \quad y = x_1 .$$

We wish to find $u(t)$ to minimize

$$J = \frac{1}{2} \int_0^{t_f} x_1^2 dt ,$$

starting with arbitrary $x_1(0)$ and $x_2(0)$ and ending with $x_1(t_f) = x_2(t_f) = 0$.

(a) Show that the solution is impulse-singular-impulse.
(b) Add the constraint $|u| \le u_0$ so that the solution becomes bang-singular-bang, i.e., $u = u_0$ until $t = t_1$, then a singular arc until $t = t_2$, then $u = -u_0$ until $t = t_f$. Find the switch times t_1 and t_2 for the case $u_0 = 15$, $x_1(0) = x_2(0) = 2$, $t_f = 1.5$ and plot x_1, x_2, , u vs. time and x_1 vs. x_2.
(c) Find and plot the solution for the case where u is unbounded and $x_1(0) = x_2(0) = 2$, $t_f = 1.5$.

10.3 Nonlinear Problems with Singular Arcs

Nonlinear optimization problems with singular arcs may involve impulses in the control if no bounds are placed on the control. If bounds are imposed the solution will involve periods with the control on the constraint boundary and singular arcs. The control is often discontinuous when switching from a constrained arc to a singular arc or vice-versa; these are called "corners" in the classical calculus of variations and such paths are called "broken extremals" (Refs. Bo1 and Bo2). The numerical solution of such problems is often quite complicated.

Example 10.3.1—Max Altitude of a Sounding Rocket

This problem was first posed by the American rocket pioneer R. H. Goddard in 1919 when he was building a rocket to be fired vertically to reach high altitudes (a "sounding rocket"). The problem was formulated as a calculus of variations problem by G. Hamel in 1927 (Ref. Hm) and a solution was given by Tsien and Evans (Ref. TE) in 1951. A complete discussion of the problem was later given by Garfinkel (Ref. Ga). It can be stated as follows: Find the thrust history to maximize the final altitude of a sounding rocket, given the initial mass, the fuel mass, and the drag characteristics of the rocket.

The EOM are

$$m\dot{v} = F(t) - D(v, r) - mg(r) , \quad \dot{r} = v , \quad c\dot{m} = -F(t) , \qquad (10.30)$$

where v = vertical velocity, r = radial distance from the center of the Earth, m = mass of the rocket, F = rocket thrust, D = aerodynamic drag, g = gravitational force per unit mass, c = specific impulse of the rocket fuel (a constant). The fuel mass is $m_0 - m_f$. We wish to find $F(t)$ to maximize $r(t_f)$ with $v(0) = 0$, $r(0) = r_0$, $m(0) = m_0$, and $m(t_f) = m_f$ and bounds on the rocket thrust:

$$0 \leq F \leq F_{\max} . \qquad (10.31)$$

The Hamiltonian of the problem is

$$H = \lambda_v \left(\frac{F - D}{m} - g \right) + \lambda_r v - \lambda_m \frac{F}{c} , \qquad (10.32)$$

which is linear in the control variable F. Since H is not an explicit function of time, a first integral of the problem is H = constant; since the final time is not specified and ϕ does not depend explicitly on t, this constant is zero:

$$H(t) = 0 . \qquad (10.33)$$

The equations for the adjoint functions are

$$\dot{\lambda}_v = \lambda_v \frac{D_v}{m} - \lambda_r , \quad \dot{\lambda}_r = \lambda_v \left(\frac{D_r}{m} - g_r \right) , \quad \dot{\lambda}_m = \lambda_v \frac{F - D}{m^2} . \qquad (10.34)$$

F is given by maximizing H with respect to F, i.e., by maximizing

$$\left(\frac{\lambda_v}{m} - \frac{\lambda_m}{c} \right) F \qquad (10.35)$$

with respect to F. There are three possible maxima:

$$F = F_{\max} \text{ if } sf > 0 , \quad 0 \leq F \leq F_{\max} \text{ if } sf = 0 , \quad F = 0 \text{ if } sf < 0 , \qquad (10.36)$$

where $sf \overset{\Delta}{=} \lambda_v/m - \lambda_m/c =$ the switching function. The case $sf = 0$ corresponds to a singular arc. If $sf = 0$ for a finite time, then $d(sf)/dt = 0$; differentiating sf and substituting from (10.30) and (10.34) gives

$$\lambda_v \left(D_v + \frac{D}{c} \right) - \lambda_r m = 0 , \tag{10.37}$$

which still does not determine the thrust $F(t)$. Hence we differentiate (10.37) and substitute from (10.30) and (10.34), which gives another relation homogeneous in λ_v and λ_r:

$$
\begin{aligned}
0 = \lambda_v & \left[\frac{D_v}{m} \left(D_v + \frac{D}{c} \right) + \left(D_{vv} + \frac{D_v}{c} \right) \left(\frac{F - D}{m} - g \right) \right. \\
& \left. + v \left(D_{vr} + \frac{D_r}{c} \right) - m g_r - D_r \right] + \lambda_r \left(\frac{F - D}{c} - D_v \right) .
\end{aligned}
\tag{10.38}
$$

Consistency between (10.37) and (10.38) requires that the determinant of the coefficients of λ_v and λ_r be zero, which gives the following *nonlinear feedback law*:

$$F = D + mg - \frac{mg}{D + 2cD_v + c^2 D_{vv}} \left\{ D + cD_v + \frac{c^2}{g} \left[vD_{vr} - \left(1 - \frac{v}{c} \right) D_r - mg_r \right] \right\} . \tag{10.39}$$

The locus of possible singular arcs is defined in the state space by requiring consistency between $H = 0$ (using $sf = 0$) and (10.37), which yields the locus of possible points in the state space where singular arcs can occur:

$$D + mg - \frac{v}{c} D - v D_v = 0 , \tag{10.40}$$

which is a surface in the 3D (v, r, m) space.

The case considered in Ref. TE assumed that the drag was quadratic in v and that the atmospheric density varied exponentially with altitude, both quite good approximations:

$$D \sim v^2 \exp \beta(r - r_0) . \tag{10.41}$$

If we use (10.41) and $g \sim 1/r^2$, then (10.39) becomes

$$F = D + mg + \frac{mg}{D(1 + 4c/v + 2c^2/v^2)} \left\{ D \left[\frac{\beta c^2}{g} \left(1 + \frac{v}{c} \right) - 1 - 2\frac{c}{v} \right] - 2\frac{c^2}{gr} \right\} , \tag{10.42}$$

while (10.40) becomes

$$mg = D \left(1 + \frac{v}{c} \right) . \tag{10.43}$$

The solution, in most instances, can be found by forward integration, assuming three arcs: (1) An arc where $F = F_{max}$. (2) A singular arc where $0 \leq F \leq F_{max}$. (3) A coasting arc where $F = 0$. The integration of the first arc is continued until the singular surface (10.43) is reached; then the nonlinear feedback law (10.42) is used until the fuel is used up, i.e., when $m(t) = m_f$; finally the coasting arc with $F = 0$ is integrated until $v = 0$. Thus the original TPBVP has been converted into an initial value problem with a nonlinear feedback (10.42).

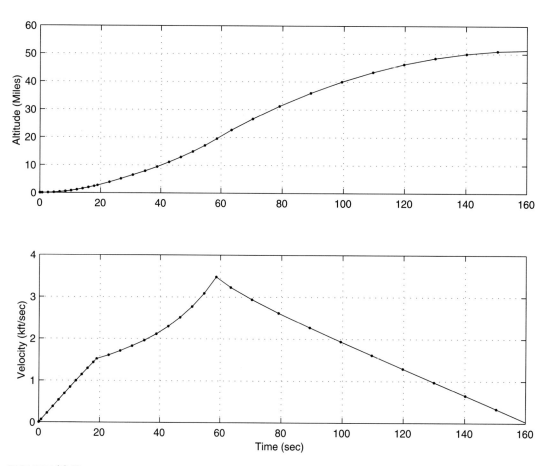

FIGURE 10.7   Max Altitude of a Sounding Rocket—Altitude and Velocity vs. Time

The singular arc corresponds to thrust slightly larger than drag plus weight so that the rocket is accelerating upward but not wasting fuel to overcome the very large drag it would have encountered if maximum thrust had been used.

Figures 10.7 and 10.8 show a solution for the case where $g = g_0 r_0^2 / r^2$, $F_{max} = 3.5 m_0 g_0$, $\beta = 500/r_0$, $m_f = .6 m_0$, $c = .5\sqrt{g_0 r_0}$, $C_D \rho_0 S r_0 / m_0 = 620$, and $g_0 = $ gravitational force per unit mass at the Earth's surface, $r_0 = $ radius of the Earth.

Problems

10.3.1 Min Surface Area between Two Co-axial Rings: Problem 3.3.12 is a classical problem in the calculus of variations, namely, the min surface area between two coaxial circular rings (Bolza's Example 1, which is treated 12 times throughout Ref. Bo). For small distances between the rings, the minimizing surface is a smooth catenary, which was found in Problem 3.3.12. However, when the distance between the rings becomes sufficiently large the min surface consists of two flat circular areas, one in each ring (easily demonstrated with soap bubbles); this is a *broken extremal* or problem with a singular arc. Verify the plot

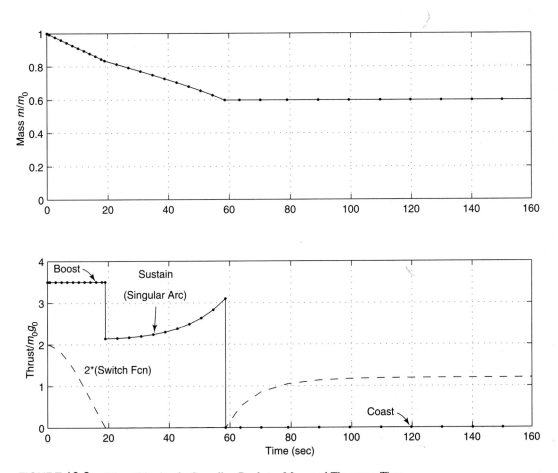

FIGURE 10.8   Max Altitude of a Sounding Rocket—Mass and Thrust vs. Time

shown in Figure 10.9. Note that the conjugate points are irrelevant since the discontinuous solutions are minimizing before the conjugate points are reached.

10.3.2 Max Range for a Ground-to-Air Rocket: Some of the first *numerical solutions* for optimal rocket trajectories were given in Refs. BR, Bk, and OE. In Ref. BR the thrust history was calculated for max range paths to different final altitudes for a short-range rocket with zero lift, i.e., a "ballistic" or "free-fall" path. The parameters used in these calculations were approximately those of the Raytheon Hawk missile of that period, namely, initial weight 1000 lb, fuel weight 500 lb, $C_D S = .47$ ft^2, $c = 6698$ ft/sec, where $C_D = $ drag coefficient, S is a reference area, and $c = $ fuel specific impulse. The EOM are

$$m\dot{V} = T - D(h, V) - mg \sin \gamma \ , \quad V\dot{\gamma} = -g \cos \gamma \ ,$$

$$\dot{m} = -\frac{T}{c} \ , \quad \dot{h} = V \sin \gamma \ , \quad \dot{x} = V \cos \gamma \ ,$$

where $V = $ velocity, $\gamma = $ flight-path-angle, $m = $ mass, $h = $ altitude, $x = $ horizontal range,

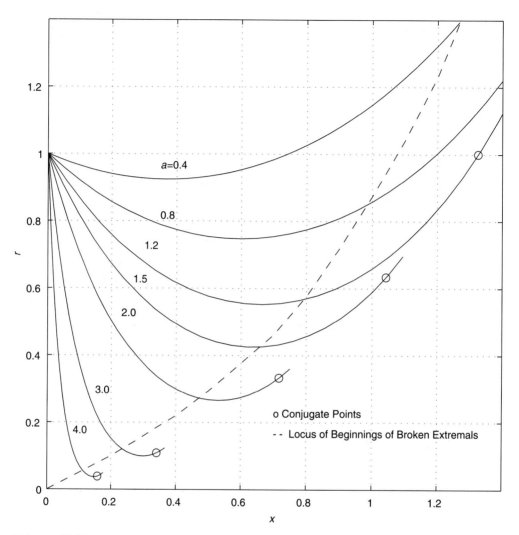

FIGURE 10.9   Min Surface Area between Two Co-axial Circular Rings, Showing Beginning of Broken Extremals (Singular or Discontinuous Solutions)

g = gravitational force per unit mass, $D = C_D S \rho V^2 / 2$ = drag, $\rho = \rho_0 \, \exp(-\beta h)$ = air density where $1/\beta = 30,000$ ft and $\rho_0 = .0021$ slug/ft^3.

(a) Write the EOM in normalized variables with $(V, \; c)$ in units of $\sqrt{g\ell}$, m in m_f = final mass, $(h, \; x, \; 1/\beta)$ in ℓ, t in $\sqrt{\ell/g}$ and $(T, \; D)$ in $m_f g$, where $\ell \overset{\Delta}{=} 2m_f/(C_D S \rho_0)$ = a characteristic length.

(b) Determine the max range paths to final altitudes of 0, 50, 100, 120, 138 kft, approximating the initial high thrust period (the boost phase) as an instantaneous impulse so that

$$m(0+) = m_0 \exp \left(-\frac{V(0+)}{c} \right) \, ,$$

where m_0 = initial mass, $m(0+)$ = mass after the boost, $V(0+)$ = velocity after the boost. Use inverse optimization with $V(t)$ as the key state variable and $[\gamma(0), t_f]$ as parameters. Note that $[V(t), \gamma(0)]$ determine $\gamma(t)$; then $[V(t), \gamma(t)]$ determine $[h(t), x(t)]$ by quadratures; then $\mu(t) \triangleq m(t)[\exp(V(t)/c)]$ is determined by another quadrature, using $\exp[(1/c)\int_0^t \sin\gamma(t)dt]$ as an integrating factor. Finally $m(t) = \mu(t)\exp[-V(t)/c]$ and $T(t) = -c\dot{m}(t)$, which is not determined very accurately on the singular arc by inverse optimization. Compare your results with Figures 10.10 and 10.11.

(c) Determine the max altitude using the vehicle as a sounding rocket. This is the Goddard problem discussed above $[\gamma = \pi/2, \ x(t) = 0]$. Compare your results with Figure 10.10.

10.3.3 Backing Up a Truck with a Trailer (D. H. Nguyen):

A truck has a single axle trailer that is pivoted about the center point of the rear axle of the tractor (see Figure 10.12). Let ψ = the angle between the centerline of the tractor and a reference x-axis, α = the angle between the trailer and tractor centerlines, and (x, y) = the coordinates of the center point of the tractor rear axle.

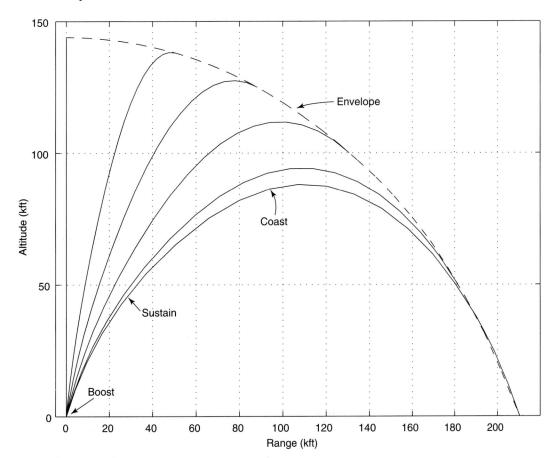

FIGURE 10.10   Max Range Paths for a Short Range Rocket (Zero Lift)

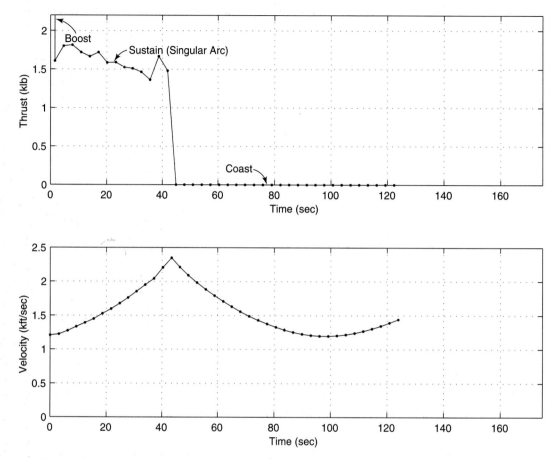

FIGURE 10.11   Max Range Paths for a Rocket; Velocity and Thrust vs. Time for Final Altitude = 100 kft

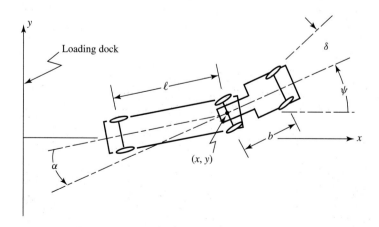

FIGURE 10.12   Nomenclature for Truck-Trailer Problem

(a) Verify that the equations of *forward* motion (purely kinematic) are

$$\psi' = u; \quad \alpha' = u - \sin\alpha; \quad x' = \cos\psi, \quad y' = \sin\psi,$$

where the independent variable is d = the distance traveled by the center point of the tractor rear axle; (d, x, y) are measured in units of ℓ = distance from the pivot point to the rear axle of the trailer; $u = (b/\ell)\tan\delta$, where δ is the steering angle of the front wheels of the tractor, and b = distance between axles of the tractor.

(b) We wish to minimize d_f = the distance traveled in going from a point where all state variables are zero (e.g., a loading dock), to an arbitrary final state, while keeping $|u| \leq u_{\max}$. The latter is a control variable inequality constraint. Show that the back-up problem is simply the forward problem with the sign of d reversed, and set up the forward optimization problem.

(c) Find the min distance forward path to $x = 1.3$, $y = -.3$, $\psi = -3\pi/4$, $\alpha = 0$ with $u_{\max} = 5/3$, and compare your results with Figures 10.13 and 10.14. Note that this path appears to be bang-singular-bang-bang with chattering junctions.

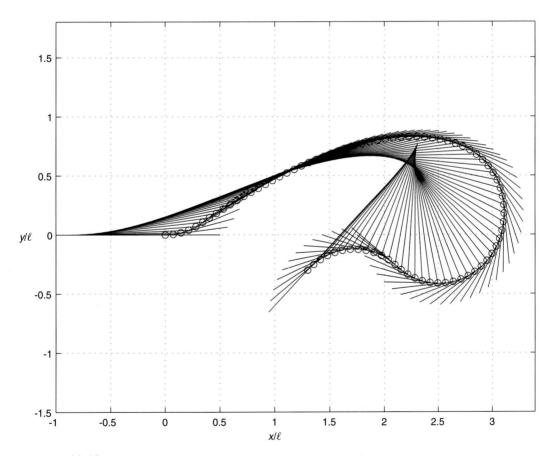

FIGURE 10.13   Min Distance Path for Tractor-Trailer Back-Up Problem

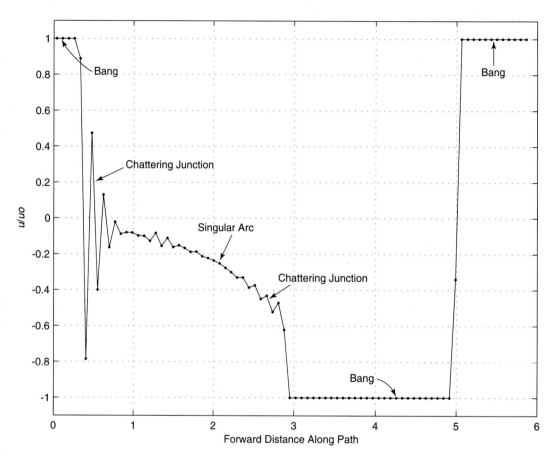

FIGURE 10.14   Control vs. Distance for Tractor-Trailer Problem (Bang-Singular-Bang-Bang)

10.4 Chapter Summary

If $H_{uu} \equiv 0$ for a finite interval of time, this is called a *singular arc*. Such problems are similar to problems with state constraints. They can be viewed as cascaded systems where some state variable is the real control and the original control is obtained by differentiating that state variable.

- Differentiate $H_u = 0$ (which does not involve u) with respect to time and replace derivatives from the EL equations until the control appears; $H_u = 0$ and its derivatives that do not involve the control are now *interior boundary conditions at the entry point* onto a singular arc since otherwise the path could not stay on the arc (tangential entry). This method also requires preknowledge of the number and sequence of the singular arcs.
- Example—Max altitude of a sounding rocket with given fuel mass (Goddard-Hamel-Tsien) $\Rightarrow$ impulse-singular arc-coast solution.

- Example—LQ problems with $R \to 0$ for systems with RHP zeros in $y(s)/u(s)$. Poles that tend to reflected zeros are not canceled $\Rightarrow$ impulse, singular arc, impulse solutions.

The inverse optimization/NLP method described in Chapter 9 determines the controls on singular arcs with only moderate precision. To find the controls more precisely, use the knowledge about where the singular arcs occur and piece together the bounded arcs and singular arcs using the time derivatives of the optimality condition ($H_u = 0$) as described above.

Appendix—History of Dynamic Optimization

Introduction

This is a brief history of dynamic optimization from 1950 to 1985.[1] The subject deserves a book, not just a chapter. However, the author hopes it may convey the admiration he has for the many people who helped to create the theory and also for those who showed how to apply it to engineering problems.

Roots in the Calculus of Variations

Dynamic optimization (DO) is one of several applications and extensions of the calculus of variations (CV). It deals with finding control time functions (histories) or control feedback gains that minimize a performance index with differential equation constraints. CV also deals with functions of more than one variable and is used to postulate variational principles in physics.

Herman H. Goldstine, a former assistant to Gilbert A. Bliss at the University of Chicago, has written an excellent, scholarly history (Ref. Go) of CV from its beginnings to the Chicago school in the early twentieth century.

Goldstine suggests that CV started with Pierre de Fermat (1601–1665) when he postulated his principle that light travels through a sequence of optical media in minimum time

---

[1] A slightly expanded version entitled 'History of Optimal Control' appeared in the June 1996 issue of the *IEEE Control Systems Magazine,* a special issue on the history of control.

(1662). Galileo Galilei (1564–1642) posed two problems in 1638 that were later solved by CV: (1) The brachistochrone problem of finding the shape of a wire such that a bead sliding along it traverses the distance between the two end points in minimum time. (2) The shape assumed by a chain hanging between two points. However, Galileo's conjectures on the solutions were incorrect. Johannes Bernoulli (1667–1748) used Fermat's ideas to solve a discrete-step version of the brachistochrone problem in 1697. Isaac Newton (1642–1727) invented CV in 1685 to find the minimum drag nose shape of a projectile, but did not publish his method until 1694. Bernoulli challenged his colleagues to solve the continuous brachistochrone problem in 1699; not only did he solve it himself, but so did Leibniz (coinventor of the calculus with Newton), his brother Jacob, l'Hospital, and Newton (anonymously, because he disliked controversies).

Leonard Euler (1707–1783), inspired by Johannes Bernoulli, published a treatise in 1744 called *The Method of Finding Curves that Show Some Property of Maximum or Minimum*. He treated many special problems and gave the beginnings of a real theory of CV. Jean Louis Lagrange (1736–1813) corresponded with Euler, and invented the method of "variations" that Euler generously praised and that gave the subject its name. Lagrange also invented the method of multipliers (not published until 1762); in modern nomenclature these multipliers are sensitivities of the performance index to changes in the states. Euler adopted this idea too and gave the first-order necessary conditions for a stationary solution, which today we call the Euler-Lagrange (EL) equations.

Adrien-Marie Legendre (1752–1833) was the first to treat the second variation (1786). However, it was not until 1836 that Karl Gustav Jacob Jacobi (1804–1851) gave a more insightful treatment and discovered conjugate points in fields of extremals. Jacobi showed that the partial derivatives of the performance index with respect to each parameter of a family of extremals (which today we call states) obeyed a certain differential equation. At almost the same time William Rowan Hamilton (1805–1865) published his work on least action in mechanical systems that involved two partial differential equations. Jacobi criticized Hamilton's work in 1838, showing that only one partial differential equation was required. The result is the Hamilton-Jacobi equation, which is the basis of dynamic programming, developed by Bellman over 100 years later (see below).

Karl Wilhelm Theodor Weierstrass (1815–1897) put CV on a more rigorous basis and discovered his famous "condition" involving an "excess-function" that is the predecessor of the maximum principle of Bellman and Pontryagin in this century. In this period, Alfred Clebsch (1833–1872) gave a sharper interpretation of Legendre's condition (the Legendre-Clebsch condition), which, in modern language, states that the second derivative matrix of the Hamiltonian with respect to the controls must be positive definite (assuming no active control or state constraints).

Oskar Bolza (1857–1942) and Gilbert A. Bliss (1876–1951) built on the work of Weierstrass at the University of Chicago and gave CV its present rigorous mathematical structure (Refs. Bo1, Bo2, Bl). Both were elected to the National Academy of Sciences but Bolza lost membership when he became a German citizen in 1911. Hestenes, another former assistant to Bliss, states (Ref. He) that "The maximum principle in control theory is equivalent to the conditions of Euler-Lagrange and Weierstrass in the classical theory. The development given here is an outgrowth of a method introduced by McShane in 1939

(Ref. McS) and later modified and extended to optimal control theory by Pontryagin and his school." McShane (1904–1989), still another former assistant to Bliss, became one of the prominent American mathematicians of this century.

Placido Cicala (Ref. Ci) was one of the first to write a clear, straightforward monograph on the possible uses of CV for engineering design. Derek Lawden (Ref. La) was among the first to see the uses of CV for optimal S/C trajectories.

Roots in Classical Control

DO also has roots in *classical control theory*. Classical control is based largely on cut-and-try methods of synthesis. A type of feedback control compensation was postulated such as proportional-integral-derivative, lead, or lag, and the gains were adjusted until the performance of the closed-loop system was "satisfactory." An excellent history of classical control is given in Ref. MAG.

During and after WWII, analytical methods based on Laplace/Fourier transforms and complex variables were developed for predicting stability and performance of closed-loop control systems. Gradually, performance criteria became more quantitative and, since the available theory was in the frequency domain (Black, Nyquist, Bode, and Nichols), it was natural for these criteria to be expressed as frequency response criteria, such as gain and phase margins. Evans developed his root locus method of synthesis about 1950 and root locus plots in the complex s-plane became as common as Nyquist and Bode plots. Analog computers also became available in the 1950s so that time-response criteria were easier to check, such as overshoot and settling time for a step command.

Integral-square error as a control design performance index appeared in the book by Newton, Gould, and Kaiser (Ref. NGK) in 1957. A constraint on integral-square control is mentioned but no clear algorithm was given. They took a position somewhere between classical control and optimal control by postulating the form of the compensation and using a constraint on bandwidth to determine the optimal gains. Chang (Ref. Ch) in 1961 clearly states the need for a constraint on integral-square control and adjoins it to the integral-square error with a positive weighting factor k^2. He also proposed a "root-square locus" vs. k^2 in the complex s-plane, an important connection to classical control theory.

In 1960 Kalman (Ref. Ka1) introduced an integral performance index that had a quadratic penalty on output errors and control magnitudes, and used CV to show that the optimal controls were *linear feedbacks of the state variables*. His theory applied to time-varying linear systems and to multiple-input/multiple-output (MIMO) systems. This was a very significant contribution since MIMO problems had previously been designed by "successive loop-closure," which can easily give results that are far from optimum, e.g., poorly coordinated controls that fight each other, thus wasting control authority. Athans (Ref. At) later named this the *Linear Quadratic Regulator, or LQR*. Kalman also showed that the optimal state-feedback gain matrix could be obtained by solving a *backward Riccati equation* to steady-state (see below). In his papers he introduced the concept of state and control variables and proposed a compact vector-matrix notation that became standard in OC. State variables were inherent in the use of analog computers (early 1950s), since one state variable was associated with each integrator.

Roots in Linear and Nonlinear Programming

DO also has roots in linear and nonlinear programming (NLP), i.e., parameter optimization with inequality and/or equality constraints, which were developed shortly after WWII (Ref. Da, KT). In particular, Kuhn and Tucker (Ref. KT) gave a simple necessary condition for the system to be on a constraint boundary, namely that the gradient of the performance index must lie inside the "cone" of the constraint gradients. Professional codes have since been developed that solve NLP problems with thousands of parameters (Ref. GMSW). The steady-state LQ control problem can be solved using NLP by optimizing the parameters of an assumed compensator form (Ref. Ly).

For numerical solutions of optimal trajectory problems the control history must be approximated by values at a finite number of time points, so collocation methods using NLP can be used to solve such problems (Ref. HP). While these methods do not take advantage of the sequential dynamics, they allow the use of professional NLP codes that reliably handle inequality constraints on the controls and states (see below). Optimal trajectory problems can be solved using NLP codes by parametrizing the control histories (Ref. HP) or the output histories using the concept of "inverse" DO (Ref. Se).

Algorithms and the Digital Computer

There is little question that the truly enabling technology for DO is the *digital computer* that appeared in the middle 1950s. Before that, only rather simple problems could be solved, so CV, DO, and NLP were little used by engineers.

To use digital computers for solving DO problems, one needs *algorithms* and reliable codes for these algorithms. This is perhaps the main difference between DO and CV. Knuth (Ref. Kn), among others, pointed out that development of efficient algorithms is a challenging intellectual activity. However few mathematicians other than Knuth have been interested in numerical methods and algorithms, leaving this field to applied mathematicians, computer scientists, and engineers.

Dynamic Programming and the Maximum Principle

Dynamic Programming (DP), a new view and an extension of Hamilton-Jacobi theory, was developed by Bellman and his colleagues starting in the 1950s (Ref. Be). It deals with *families of extremal paths* that meet specified terminal conditions. An "optimal return function" $V(x, t)$ was defined as the value of the performance index starting at state x and time t, and proceeding optimally to the specified terminal conditions. Associated with V is an optimal control function $u(x, t)$ that, in control terminology, is a feedback on the current state x and the time t. Hence another name for DP might be *nonlinear optimal feedback control*. Bellman extended Hamilton-Jacobi theory to discrete-step dynamic systems and combinatorial systems (discrete-step dynamic systems with quantized states and controls). The partial derivatives of $V(x, t)$ with respect to x are identical to Lagrange's multipliers and a very simple derivation of the EL equations can be made using DP (Ref. Dr).

However, the Bellman school underestimated the difficulty of solving realistic problems with DP algorithms. The "curse of dimensionality" (Bellman's own phrase) causes DP algorithms to exceed the memory capacity of current computers when the system has more than two or three state variables. However, if the state space is limited to a region close to a nominal optimum path, the DP problem can often be well approximated by an LQ problem, i.e., a problem with linear (time-varying) dynamics and a QPI whose (time-varying) weighting matrices are the second derivatives of the Hamiltonian with respect to the states and the controls (Refs. BH, DM, JM). This is the classical *Accessory Minimum Problem (AMP),* the basic problem for examining the second variation in CV, and it was well understood by 1900. However, the AMP was not easily accessible to engineers since the CV treatises were written in rigorous mathematical language and contained few examples of the "controls" type; indeed, few interesting examples can be calculated without computers. The AMP can be formulated as a *time-varying linear* TPBVP, but it is often not a trivial task to solve such problems since the obvious "shooting" method often fails due to the inherent instability of the EL equations for both forward and backward integration.

The *Maximum Principle* is an extension of Weierstrass' necessary condition to cases where the control functions are bounded (Ref. GF, p. 225). It was developed by Pontryagin and his school in the USSR (Ref. PBGM). In DO terminology, it states that a minimizing path must satisfy the EL equations where the optimal controls maximize the Hamiltonian within their bounded region at each point along the path (Pontryagin used the classical definition of the Hamiltonian that is opposite in sign from the one commonly used today). This transforms the CV problem to a NLP problem at each point along the path. Letov (Ref. Lt) and his students were among the first to attempt some engineering applications of CV in the USSR.

The Maximum Principle deals with one extremal at a time while Dynamic Programming deals with families of extremals. The Maximum Principle is inherent in DP since the Hamilton-Jacobi-Bellman equation includes finding the controls (possibly bounded) that minimize the Hamiltonian at each point in the state space.

Calculating Nonlinear Optimal Trajectories

An important use of DO is for finding optimal trajectories for nonlinear dynamical systems, particularly for A/C, S/C, and robots. The American rocket pioneer Robert H. Goddard (1882–1945) posed one of the first aerospace DO problems in 1919; given a certain mass of rocket fuel, what should the thrust history be for the rocket to reach maximum altitude? The problem was formulated as a CV problem by Hamel in 1927, and an analytical solution was given by Tsien and Evans in 1951 (Ref. BH, p. 253). W. Hohmann determined the optimal impulsive transfer between circular orbits in 1925 (Ref. Ho). George Leitmann edited the first authoritative book on DO (Ref. Le) in 1962, which contained chapters by himself, Richard Bellman, John Breakwell, Ted Edelbaum, Henry Kelley, Richard Kopp, Derek Lawden, Angelo Miele, and other pioneers of DO. Athans and Falb (Ref. AF) authored the first textbook on DO in 1966.

Some of the first *numerical solutions* for optimal rocket trajectory problems were given by Bryson and Ross (Ref. BR), Breakwell (Ref. Bk), and Okhotsimskii and Eneev (Ref. OE). Figure 10.10 shows max range paths for a short-range rocket with drag (Ref. BR).

The parameters used in these calculations were approximately those of the Hawk missile of that period.

These papers used the *shooting method* of guessing the initial values of the Lagrange multipliers $\lambda(t_o)$, integrating the EL equations forward, and then interpolating on the elements of $\lambda(t_o)$ until the final conditions are satisfied. This method is feasible for conservative systems (e.g., trajectories in space) but it is usually not feasible for nonconservative systems (e.g., aircraft trajectories). The reason for this is that the EL equations are unstable for non-conservative systems for both forward and backward integration, causing loss of numerical accuracy for computer solutions. To avoid the instability problem, the initial values of the Lagrange multipliers from a gradient code (see below) may be used as initial guesses. This is of interest only if a very precise solution is desired; gradient solutions are often sufficiently accurate for engineering purposes. Another way to get around the instability problem is to use a "multishooting" algorithm (Ref. Kr) that divides the path into shorter segments. A multishooting FORTRAN code (BNDSCO) was developed by Bulirsch (Ref. Bu) and his students at the University of Munich (Ref. OG).

Gradient algorithms were proposed by Kelley and his colleagues at Grumman (Refs. Ke1, Ke2, Ba) and by Bryson and Denham at Raytheon (Ref. BD). These algorithms eliminate the instability problem of the shooting method but they require reasonable initial guesses of the control histories. The EOM are integrated forward and the trajectory is stored; then the adjoint equations (Lagrange multipliers) are integrated backward over this nominal trajectory, which is a stable integration. This determines the impulse response functions (the "gradients") of the performance index and the terminal constraints with respect to perturbations in the control histories. The control histories are then changed in the direction of the negative gradients (for a minimum problem) and the procedure is repeated until the terminal conditions are satisfied to a satisfactory accuracy and the performance index is no longer decreasing significantly. Reference Ke1 used penalty functions for handling the terminal constraints whereas Reference BD used a projected gradient method. A general-purpose MATLAB gradient code (FOPT) was developed at Stanford University by Hur and Bryson (Ref. Br2).

One of the first S/C applications of the gradient method was made by Moyer and Pinkham (Ref. MP). They found the thrust direction program for a low-thrust spacecraft to go from Earth to Mars in min time. A recomputation of that path is shown in Fig. 3.16.

One of the first A/C applications of the gradient method was made in Reference BD. Raytheon was interested in determining how rapidly the supersonic F4 (Phantom) fighter could reach a high altitude and get into a position to launch their Sparrow missile.

Using aerodynamic data from McDonnell and thrust data from General Electric, Denham calculated (Ref. BD) the min time-to-climb path to an altitude of 20 km, Mach 1, and level flight, using angle-of-attack as the control variable (see Figure 4.5). The path was tested in January of 1962 at the Patuxent River Naval Air Station. The copilot had a card with the optimal Mach number tabulated for every 1000 ft of altitude, which he read off to the pilot as they went through that altitude. The pilot then moved the stick forward or backward to get as close to this Mach number as he could. They got to the desired flight condition in 338 sec where the predicted value was 332 sec. This was a substantially shorter time to that flight condition than had been achieved by cut-and-try.

A few years later a simpler approximate method using only two A/C states (energy per unit mass and mass) was used to calculate the same optimal flight path using velocity as the control variable (Ref. BDH). Figure 4.5 shows these computations are very close to the more precise five-state (mass point approximation) computations. The energy-state method shows the reason for the unusual flight path: the excess power (thrust minus drag) as a function of altitude and velocity has the usual high "ridge" just below Mach 1 from sea level to about 30 kft; this is the place to rapidly add *potential energy* (altitude); however, because the thrust increases so much with speed for these engines, another high "ridge" appears between 20 and 30 kft for Mach number between 1 and 2; this is the place to rapidly add *kinetic energy* (velocity), which can then be traded for potential energy in a "zoom climb" to 20 km and Mach 1.

During WW II, Kaiser (Ref. Ks) in Germany suggested ways to take advantage of the new jet engines for better climb performance. In the 1950s, Lush (Ref. Lu) and Rutkowski (Ref. Ru) introduced the concept of "energy climb" that inspired the work in Ref. BDH.

There are also NR algorithms that are related to the AMP and neighboring optimum perturbation feedback control mentioned above (Refs. BSB, KKM1, JM, DM). However, these involve substantially more programming than the first-order (gradient) methods and require analytic expressions for the second (as well as the first) derivatives of the system equations and the terminal boundary conditions.

Other important applications of optimal control methods are determining optimal aerodynamic shapes (Ref. Mi) and optimal structural shapes (Ref. HA).

Inequality Constraints

Control variable inequality constraints can be handled by the maximum principle using a shooting method. They can also be handled using penalty functions or "slack variables" with shooting or gradient methods. The latter idea was suggested by Valentine (Ref. Vt), another member of the Bliss school.

State variable inequality constraints are more difficult to handle, since the optimal path must enter "tangentially" onto a constrained arc, i.e., one or more time derivatives of the constraint must be zero at entry points. Also, the number of constrained arcs is not known ahead of time. The Maximum Principle does not apply in the form given by Pontryagin. Gamkrelidze gave necessary conditions for such problems (Ref. PBGM) but did not give a method of solution. Later Dreyfus (Ref. Dr) and Speyer (Ref. SB1) gave gradient methods for solving DO problems with state variable inequality constraints, where the number and sequence of constrained arcs are assumed known. Collocation methods using generalized gradients and NLP codes are the most reliable methods for solving problems with state variable inequality constraints, since they do not assume the number and sequence of constrained arcs (Refs. MD, HP, Se).

Singular Problems

Some DO problem solutions contain *singular arcs,* i.e., arcs where the second derivative matrix of the Hamiltonian with respect to the controls is only positive semidefinite, e.g., Goddard's problem and Ross's problem (see above) where the controls enter the EOM linearly. Extensions of the necessary conditions for optimality and some methods of numerical solution for such problems have been found but precise solutions are still difficult to find (Ref. BJ). Approximate solutions can be found using collocation methods, generalized gradients, and NLP codes (Refs. MD, Se).

Inverse DO

Inverse control methods were developed in the 1970s (Ref. MC) for finding control histories to produce desired ouput histories of linear and nonlinear dynamic systems. However, with simple choices of output histories, the resulting control histories are often infeasible. This gave rise to the idea of *inverse optimal control* or "differential inclusion" (Ref. Se) where the output histories, instead of the control histories, are iterated using collocation and NLP codes to minimize a performance index until the controls are feasible. The control histories are obtained by numerical differentiation of the output histories. This method is attractive for several reasons: (1) Approximate output histories are usually easier to guess than the control histories (to start the iterative computation). (2) Many NLP codes find gradients numerically so that analytical gradients do not need to be entered. (3) For problems with state/control constraints and singular arcs, the number and sequence of constrained and singular arcs does not need to be known ahead of time. Currently this method is limited to relatively short histories if the gradients are found numerically.

Riccati Equations

Kalman (Refs. Ka1, Ka2) showed that the MIMO LQ DO problem (essentially the AMP except that the weighting matrices are chosen by the designer) can be solved numerically in an elegant, efficient manner with a "backward sweep" of a *matrix Riccati equation*. Jacopo Francesco Riccati (1676–1754) gave the scalar form of his equation for solving linear second-order TPBVPs. Kalman was influenced by the work of Carathéodory (Ref. Ca). Gelfand and Fomin (Ref. GF) gave a clear description of the sweep method, which was translated into English in 1963.

MacFarlane (Ref. Mf) proposed an algorithm for solving the *steady-state Riccati equation* for time-invariant dynamic systems that used eigenvector decomposition of the EL equations. This method has many similarities to the Weiner-Hopf technique used earlier in Reference NGK. However eigensystem codes available at that time were slow and not very accurate. Kalman and Englar (Ref. KE) proposed integrating the Riccati equation backward to steady-state; this is often quite slow since the time-step required for an accurate solution is very small. Wilkinson et al. (Ref. WMP) developed an efficient code for the QR algorithm of Francis (Ref. Fr), that finds the eigensystem of linear dynamic systems with complex eigenvalues. Hall (Ref. Ha) used this code with MacFarlane's algorithm to develop a code he called OPTSYS. This allowed routine solution of steady-state Riccati equations, and is the basis for many of the professional codes now available (e.g., MATLAB, Matrix-X, and Control-C). However, the QR algorithm does not handle repeated eigenvalues so OPTSYS was not quite as reliable as desired. This restriction was later removed in some professional codes by using Schur decomposition instead of eigenvector decomposition (Ref. Lb). Figures 6.9 and 6.10 (Ref. HB) show the closed-loop path followed by a 747 airliner to make a last minute "S-turn" to line up with the center of the runway; the two controls, aileron and rudder, are well coordinated during the maneuver; the feedback gains were computed using OPTSYS and are now easily reproduced using the LQR command in MATLAB.

Summary

Dynamic optimization had its origins in the *calculus of variations* in the seventeenth century (Fermat, Newton, Liebnitz, and the Bernoullis). The calculus of variations was developed further in the eighteenth century by Euler and Lagrange and in the nineteenth century by Legendre, Jacobi, Hamilton, and Weierstrass. In the early twentieth century, Bolza and Bliss put the final touches of rigor on the subject. In 1957, Bellman gave a new view of Hamilton-Jacobi theory that he called *dynamic programming,* essentially a nonlinear feedback control scheme. McShane (1939) and Pontryagin (1962) extended the calculus of variations to handle control variable inequality constraints, the latter enunciating his elegant *maximum principle.* The truly enabling element for use of optimal control theory was the *digital computer* that became available commercially in the 1950s. In the late 1950s and early 1960s, Lawden, Leitmann, Miele, and Breakwell demonstrated possible uses of the calculus of variations in *optimizing aerospace flight paths* using shooting algorithms, while Kelley and Bryson developed gradient algorithms that eliminated the inherent instability of shooting methods. Also in the early 1960s, Simon, Chang, Kalman, Bucy, Battin, Athans, and many others showed how to apply the calculus of variations to design optimal state *feedback logic for linear dynamic systems* using digital control.

References

[AF]   M. Athans and P. L. Falb, *Optimal Control,* McGraw-Hill, New York, 1966.

[At]   M. Athans (Ed.), Special Issue on Linear-Quadratic Problems, *IEEE Trans. Auto. Ctrl.,* Vol. 16, No. 6, 1971.

[BCMD]   A. E. Bryson, F. J. Carroll, K. Mikami, and W. F. Denham, "Determination of Lift or Drag Programs to Minimize Reentry Heating," *Jour. Aero. Sci.,* Vol. 29, No. 4, April 1962.

[BCS]   G. Brauer, D. Cornick, and R. Stevenson, "Capabilities and Applications of the Program to Optimize Simulated Trajectories (POST)," NASA CR-2770, Feb. 1977.

[BD]   A. E. Bryson and W. F. Denham, "A Steepest Ascent Method for Solving Optimum Programming Problems," *ASME Jour. Appl. Mech.,* Vol. 29, Series E, June 1962.

[BDH]   A. E. Bryson, M. N. Desai, and W. C. Hoffman, "The Energy-State Approximation in Performance Estimation of Supersonic Aircraft," *Jour. Aircraft,* Vol. 6, No. 6, pp. 481–487, 1969.

[Be]   R. Bellman, *Dynamic Programming,* Princeton University Press, Princeton, NJ, 1957.

[Be1]   S. Bennett, "A History of Control Engineering 1800–1930," IEE Control Engineering Series 8, Peter Pereginus, London, 1979.

[Be2]   S. Bennett, "A History of Control Engineering 1930–1955," IEE Control Engineering Series 47, Peter Pereginus, London, 1993.

[BH]   A. E. Bryson and Y. C. Ho, *Applied Optimal Control,* Blaisdell, Waltham, MA, 1969; revised printing Hemisphere, Washington, D.C., 1975.

[BHa]   A. E. Bryson Jr. and W. E. Hall Jr., "Modal Methods in Optimal Control Synthesis," *Control and Dynamic Systems,* Vol. 16 (Ed. C. T. Leondes), Academic Press, New York, 1980, pp. 53–80.

[BJ]   D. J. Bell and D. H. Jacobson, *Singular Optimal Control Problems,* Academic Press, New York, 1975.

[Bk]   J. V. Breakwell, "The Optimization of Trajectories," *SIAM Jour.,* Vol. 7, 1959.

[Bl]   G. H. Bliss *Lectures on the Calculus of Variations,* University of Chicago Press, Chicago, IL 1946.

[BN]   A. V. Balakrishnan and L. Neustadt (Eds.), *Conference on Computing Methods on Optimization Problems,* Academic Press, New York, 1964.

[Bo1]   O. Bolza, *Lectures on the Calculus of Variations,* Second Edition, Chelsea Publishing, 1904.

[Bo2]   O. Bolza, *Vorlesungen Uber Variationsrechnung,* Chelsea Publishing, New York, 1909.

[Br1]   A. E. Bryson, "Nonlinear Feedback Solution for Minimum Time Rendezvous with Constant Thrust Acceleration," 16th Int. Astro. Congress, Athens, Greece, Sept. 1965.

[Br2]   A. E. Bryson, *Lecture Notes on Dynamic Optimization,* Dept. of Aeronautics and Astronautics, Stanford University, 1992–1998.

[BR]   A. E. Bryson and S. E. Ross, "Optimum Rocket Trajectories with Aerodynamic Drag," *Jet Propulsion,* July 1958.

[BSB]   J. V. Breakwell, J. Speyer, and A. E. Bryson, "Optimization and Control of Nonlinear Systems Using the Second Variation," *SIAM Jour. Control,* Ser. A, Vol. 1, No. 2, pp. 193–223, 1963.

[Bu]   R. Bulirsch, "Die Mehrzielmethode," Carl Cranz Gesellschaft, Oberpfaffenhoven, Germany, 1971; reprint 1993, Dept. Mathematics, Munich University of Technology.

[Bw]   J. A. Breakwell, "Optimal Feedback Slewing of Flexible Spacecraft," *Jour. Guidance and Control,* Vol. 4, No. 5, 1981, pp. 472–479.

[BWC]   T. P. Bauer, L. J. Wood, and T. K. Caughey, "Low-Thrust Perturbation Guidance," Proc. of AIAA/AAS Astrodynamics Conf., San Diego, CA, 1982.

[Ca]   C. Carathéodory, *Calculus of Variations and Partial Differential Equations of First Order,* translated by R. Dean, Holden-Day, Boca Raton, FL, 1967 (originally in German, 1935).

[CH]   R. Courant and D. Hilbert, *Methoden der Mathematischen Physik,* Springer, Berlin, 1937, Vol. II, Chapter 2.

[Ch]   S. S. L. Chang, *Synthesis of Optimal Control Systems,* McGraw-Hill, New York, 1961.

[Ci]   P. Cicala, *An Engineering Approach to the Calculus of Variations,* Levrotto and Bella, Torino, Italy, 1957.

[Da]   G. Dantzig, *Linear Programming and Extensions,* Princeton University Press, Princeton, NJ, 1963.

[DM]   P. Dyer, and S. R. McReynolds, *The Computation and Theory of Optimal Control,* Academic Press, New York, 1970.

[Dr]   S. E. Dreyfus, *Dynamic Programming and the Calculus of Variations,* Academic Press, New York, 1966.

[DS]   J. C. Doyle and G. Stein, "Multivariable Feedback Design; Concepts for a Classical/Modern Synthesis," *IEEE Trans.,* Vol. AC-26, No. 1, 1981.

[FPE]   G. F. Franklin, J. D. Powell, and A. Emami-Naeini, *Feedback Control of Dynamic Systems,* 2nd ed., Addison-Wesley, Reading, MA, 1994, Chapter 7—State Space Design.

[Fr]   J. G. F. Francis, "The QR Transformation, Parts I and II," *Comput. J.,* Vol. 4, pp. 265–271, 1961 and Vol. 5, pp. 332–345, 1962.

[Ga]   B. Garfinkel, "A Solution of the Goddard Problem," *SIAM Jour. of Control,* Vol. 1, No. 3, pp. 349–368, 1963.

[GF]   I. M. Gelfand and S. V. Fomin, *Calculus of Variations,* translated by R. A. Silverman, Prentice-Hall, Englewood Cliffs, NJ, 1963.

[GMW]   P. E. Gill, W. Murray, and M. H. Wright, *Practical Optimization,* Academic Press, New York, 1981.

[GMSW]   P. E. Gill, W. Murray, M. A. Saunders, and M. H. Wright, "User's Guide for NPSOL (Version 4.0): A Fortran Package for Nonlinear Programming," Dept. of Operations Research, Stanford University, Stanford, CA, January 1986.

[Go]   H. H. Goldstine, *A History of the Calculus of Variations from the 17th through the 19th Century,* Springer-Verlag, New York, 1980.

[Gr]   A. Grace, "MATLAB Optimization Toolbox," The MathWorks, Inc., Cochituate Place, 24 Prime Park Way, Natick, MA, December 1992.

[HA]   E. J. Haug and J. S. Arora, *Applied Optimal Design,* Wiley, New York, 1979.

[Ha]   W. E. Hall Jr., "Optimal Control and Filter Synthesis by Eigenvector Decomposition," Ph.D. Thesis, Stanford University Dept. of Aero./Astro. Rpt. 436, 1971.

[HB]   W. E. Holley and A. E. Bryson, "Wind Modelling and Lateral Control for Automatic Landing," *Jour. Spacecraft and Aircraft,* Vol. 14, No. 2, pp. 65–72, 1977.

[He]   M. R. Hestenes, *Calculus of Variations and Optimal Control Theory,* Wiley, New York, 1966.

[HJ]   R. K. Heffley and W. F. Jewell, "Aircraft Handling Qualities," Systems Technology Inc. Technical Report 1004–1, Hawthorne, CA, 1972.

[Hm]   G. Hamel, "Uber eine mit dem Problem der Rakete Zusammenhangende Aufgabe der Variationsrechnung," *ZAMM,* Vol. 7, No. 6, pp. 451–452, 1927.

[Ho]   W. Hohmann, *Die Erreichbarkeit der Himmelskorper,* Oldenbourg, Munich, 1925.

[HP]   C. R. Hargraves and S. W. Paris, "Direct Trajectory Optimization Using Nonlinear Programming and Collocation," *Jour. Guidance, Control and Dynamics,* Vol. 10, No. 4, 1987.

[JG]   C. D. Johnson and J. E. Gibson, "Singular Solutions in Problems of Optimal Control," *IEEE Trans. Auto. Control,* Vol. 8, No. 1, 1963.

[JL]   D. H. Jacobson and M. M. Lele, "A Transformation Technique for Optimal Control Problems with a State Variable Inequality Constraint," *IEEE Trans. Auto. Control,* Vol. 14, No. 5, pp. 457–464, 1969.

[JM]   D. H. Jacobson and D. Q. Mayne, *Differential Dynamic Programming,* American Elsevier, New York, 1970.

[Ka1]   R. E. Kalman, "Contributions to the Theory of Optimal Control," *Bol. de Soc. Math. Mexicana,* p. 102, 1960.

[Ka2]   R. E. Kalman, "The Calculus of Variations and Optimal Control Theory," Chapter 16 in *Mathematical Optimization Techniques,* R. Bellman (Ed.), University of California Press, Berkeley, 1963.

[Ke1]   H. J. Kelley, "Gradient Theory of Optimal Flight Paths," *Amer. Rocket Soc. Jour.,* Vol. 30, pp. 947–954, 1960.

[Ke2]   H. J. Kelley, "Method of Gradients," Chapter 6 in *Optimization Techniques,* G. Leitmann (Ed.), Academic Press, New York, 1962.

[Ke3]   H. J. Kelley, "Guidance Theory and Extremal Fields," *Proc. IRE,* p. 75, 1962.

[KE]   R. E. Kalman and T. Englar, "The Automatic Synthesis Program," NASA Contractor Report CR-475, June 1966.

[KKM1]   H. J. Kelley, R. E. Kopp, and G. Moyer, "A Trajectory Optimization Technique Based upon the Theory of the Second Variation," *Progress in Astronautics,* Vol. 14, Academic Press, New York, 1964.

[KKM2] H. J. Kelley, R. E. Kopp, and A. G. Moyer, "Singular Extremals," *Topics in Optimization,* Vol. II, G. Leitmann (Ed.), Chapter 3, Academic Press, New York, 1966.

[Kn] D. E. Knuth, *Fundamental Algorithms: Vol. 1 of the Art of Computer Programming,* Addison-Wesley, Reading, MA, 1968.

[Kr] H. B. Keller, *Numerical Methods for Two-Point Boundary-Value Problems,* Blaisdell, Waltham, Mass., 1968.

[Ks] F. Kaiser, "The Climb of Jet Propelled Aircraft. Part 1. Speed Along the Path in Optimum Climb," *Ministry of Supply (Great Britain),* RTP/TIB Translation No. GDC/15/14ST, 1944.

[KS] H. Kwakernaak and R. Sivan, *Linear Optimal Control Systems,* Wiley-Interscience, New York, 1972.

[KT] H. Kuhn and A. W. Tucker, "Nonlinear Programming," *Second Berkeley Symposium of Mathematical Statistics and Probability,* University of California Press, Berkeley, CA, 1951.

[La] D. F. Lawden, *Optimal Trajectories for Space Navigation,* Butterworths, London, 1963.

[Lb] A. J. Laub, "A Schur Method for Solving Algebraic Riccati Equations," *IEEE Trans. Auto. Ctrl.,* Vol. AC-24, pp. 913–921, 1979.

[Le] G. Leitmann (Ed.), *Optimization Techniques,* Academic Press, New York, 1962.

[Ln] Luenberger, D. C., *Introduction to Linear and Nonlinear Programming,* Addison-Wesley, Reading, MA, 1973.

[Lt] A. M. Letov, "Analytical Synthesis of Regulators," *Automation and Telemechanics,* Vol. 21, 1960; also Editor of *Optimal Automatic Control Systems,* Moscow, 1967.

[Lu] K. J. Lush, "A Review of the Problem of Choosing a Climb Technique with Proposals for a New Climb Technique for High Performance Aircraft," *Aero. Res. Council,* RM No. 2557, 1951.

[Ly] U. L. Ly, "A Design Algorithm for Robust Low-Order Controllers," Ph.D. Thesis, Stanford University Dept. of Aero./Astro. Rpt. 536, 1982.

[Lz] C. Lanczos, *The Variational Principles of Mechanics,* University of Toronto Press, Toronto, Canada, 1949.

[Ma] A. G. J. MacFarlane, "An Eigenvector Solution of the Linear Optimal Control Problem," *Jour. Elect. and Control,* 14, 643–654, 1963.

[MAG] D. McRuer, I. Ashkenas, and D. Graham, *Aircraft Dynamics and Automatic Control,* Princeton University Press, Princeton, NJ, 1973.

[Mc]   G. McCormick, "Second Order Sufficient Conditions for a Constrained Minimum," *SIAM Jour. Appl. Math.,* Vol. 15, No. 3, 1967.

[MC]   G. Meyer and L. Cicolani, "Application of Nonlinear System Inverses to Automatic Flight Control," *Theory and Applications of Optimal Control in Aerospace Systems,* AGARDograph 251, 1981.

[McR]   S. R. McReynolds, "The Successive Sweep Method and Dynamic Programming," *J. Math. Anal. & Appl.,* Vol. 19, p. 565, 1967.

[McS]   E. J. McShane, "On Multipliers for Lagrange Problems," *Amer. J. Math.,* Vol. LXI, pp. 809–818, 1939.

[MD]   R. K. Mehra and R. E. Davis, "A Generalized Gradient Method for Optimal Control Problems with Inequality Constraints and Singular Arcs," *IEEE Trans. Auto. Control,* Vol. AC-17, pp. 69–78, 1972.

[Me]   E. Meier, "An Efficient Algorithm for Bang-Bang Control Systems Applied to a Two-Link Manipulator," Ph.D. Thesis, Stanford University Dept. of Aero./Astro., 1986.

[Mf]   A. G. J. MacFarlane, "An Eigenvector Solution of the Optimal Linear Regulator," *Jour. Electron. Control,* Vol. 14, pp. 643–54, 1963.

[Mi]   A. Miele (Ed.), *Theory of Optimum Aerodynamic Shapes,* Academic Press, New York, 1965.

[MP]   H. G. Moyer and G. Pinkham, "Several Trajectory Optimization Techniques, Part II: Application," pp. 91–105 of Ref. BN.

[MS]   C. Moler and Stewart, Vol. 10, No. 2, *SIAM J. Numer. Anal.,* pp. 241–256.

[MWM]   A. Miele, Wang, and Melvin, in "Piloting Strategies for Optimum Performance in a Wind Shear, Part I, Take-Off," AIAA Aerospace Sciences Meeting, Reno, NV, January 1986.

[NGK]   G. Newton, L. A. Gould, and J. F. Kaiser, *Analytical Design of Linear Feedback Controls,* Wiley, New York, 1957.

[NS]   S. G. Nash and A. Sofer, *Linear and Nonlinear Programming,* McGraw-Hill, New York, 1996.

[OE]   D. E. Okhotsimskii and T. M. Eneev, "Some Variational Problems Connected with the Launching of Artificial Satellites of the Earth," *Jour. Brit. Interplanet. Soc.,* Vol. 16, No. 5, 1958.

[OG]   H. J. Oberle and W. Grimm, "BNDSCO—A Program for the Numerical Solution of Optimal Control Problems," DLR Inter. Rpt. 515-89/22, Oberpfaffenhoven, Germany, 1989.

[PBGM]   L. S. Pontryagin, V. G. Boltyanskii, R. V. Gamkrelidze, and E. F. Mishchenko, *The Mathematical Theory of Optimal Processes,* Moscow, 1961, translated by K. N. Trirogoff, L. W. Neustadt (Ed.), Interscience, New York, 1962.

[Ru]   E. S. Rutowski, "Energy Approach to the General Aircraft Performance Problem," *Jour. Aero. Sci.,* Vol. 21, No. 3, 1954.

[SB1]   J. L. Speyer and A. E. Bryson, "Optimal Progamming Problems with a Bounded State Space," *AIAA Jour.,* Vol. 6, pp. 1488–1492, 1968.

[SB2]   J. L. Speyer and A. E. Bryson, "A Neighboring Optimum Feedback Control Scheme Based on Estimated Time-to-Go with Application to Re-Entry Flight Paths," *AIAA Jour.,* Vol. 6, No. 5, 1968, pp. 769–776.

[SMB]   J. L. Speyer, R. Mehra, and A. E. Bryson, "The Separate Computation of Arcs for Optimal Flight Paths with State Variable Inequality Constraints," *Proc. Colloquium on Advanced Problems and Methods for Space Flight Optimization,* Liège, Belgium, Pergamon Press, June 1967.

[Sc]   E. Schmitz, "Experiments on the End-Point Position Control of a Very Flexible One-Link Manipulator," Ph.D. Thesis, Stanford University, Dept. of Aero./Astro., Stanford, CA, 1985.

[Se]   H. Seywald, "Trajectory Optimization Based on Differential Inclusion," *Jour. Guidance, Control and Dynamics,* Vol. 17, No. 3, pp. 480–487, 1994.

[Sm]   G. Smuck, M.Sc. Thesis, M.I.T., Cambridge, MA, 1966.

[TE]   H. S. Tsien and R. C. Evans, "Optimum Thrust Programming for a Sounding Rocket," *Amer. Rocket Soc. Jour.,* Vol. 21, No. 5, pp. 99–107, 1951.

[TB]   T. L. Trankle and A. E. Bryson, "Control Logic to Track Outputs of a Command Generator," *Jour. Guidance, Control and Dynamics,* Vol. 1, No. 2, pp. 130–135, 1978.

[Va]   D. R. Vaughan, "A Nonrecursive Algebraic Solution for the Discrete Riccati Equation," *IEEE Trans. Auto. Ctrl.,* Vol. 15, No. 5, 1970.

[Vg]   J. Vagners, Personal Communication, 1995.

[VL]   C. F. Van Loan, "Computing Integrals Involving the Matrix Exponential," *IEEE Trans. of Automatic Control,* Vol. AC-23, No. 3, 1978.

[Vt]   F. A. Valentine, "The Problem of Lagrange with Differential Inequalities as Added Side Conditions," *Contributions to the Calculus of Variations 1933–37,* University of Chicago Press, Chicago, IL.

[VYC]   N. X. Vinh, C.-Y. Yang, and J.-S. Chern, "Optimal Trajectories for Maximum Endurance Gliding in a Horizontal Plane," *Jour. Guidance, Control and Dynamics,* Vol. 7, No. 2, pp. 246–248, 1984.

[WB]   L. J. Wood and A. E. Bryson, "Second-Order Optimality Conditions for Variable End-Time Terminal Control Problems," *AIAA Journal,* Vol. 11. No. 9, pp. 1241–1246, 1973.

[Wh]   E. T. Whittaker, *A Treatise on Analytical Dynamics of Particles and Rigid Bodies,* 4th Edition, Dover, New York, pp. 248–253, 1944 (1st Edition Cambridge University Press, Cambridge, England, 1904).

[Wi]   D. Winfield, Tech. Report 507, Div. Engineering and Appl. Physics, Harvard University, Cambridge, MA, 1966.

[WMP]   J. H. Wilkinson, R. S. Martin, and G. Peters, "The QR Algorithm for Real Hessenberg Matrices," *Numer. Mat.,* Vol. 14, pp. 219–231, 1970.

[Wo]   L. J. Wood, "Perturbation Guidance for Minimum Time Flight Paths of Spacecraft," Ph.D. Thesis, Stanford University Dept. of Aero./Astro., Stanford, CA, 1973.

[Wr]   J. L. Wright, *Space Sailing,* Gordon and Breach, Philadelphia, 1992.

[Ze]   E. Zermelo, "Uber das Navigationsproblem bei Ruhender oder Veranderlicher Windverteilung," *Z. Angew. Math. Mech.,* Vol. 11, No. 2, pp. 114–124, 1931.

[Zh]   Y. Zhao, "Optimal Control of an Aircraft Flying through a Downburst," Ph.D. Thesis, Stanford University Dept. of Aero./Astro., Stanford, CA, 1989.

[Zo]   G. Zoutendijk, *Method of Feasible Directions,* Elsevier Pub. Co., London, 1961.

Further Reading

- V. M. Alekseev, V. M. Tikhomirov, and S. V. Fomin, *Optimal Control,* Consultants Bureau, New York, 1987.
- B. D. O. Anderson and J. B. Moore, *Optimal Control—Linear-Quadratic Methods,* Prentice-Hall, Englewood Cliffs, NJ, 1989.
- J. Gregory, *Constrained Optimization in the Calculus of Variations and Optimal Control Theory,* Van Nostrand Reinhold, New York, 1992.
- M. J. Grimble and M. A. Johnson, *Optimal Control and Stochastic Estimation,* Wiley, New York, 1988.
- L. M. Hocking, *Optimal Control; an Introduction to the Theory and Applications,* Oxford University Press, London, 1991.
- F. L. Lewis and V. L. Syrmos, *Optimal Control,* Wiley, New York, 1995.
- K. L. Teo, C. J. Goh, and K. H. Wong, *A Unified Computational Approach to Optimal Control Problems,* Longman Scientific & Technical, White Plains, N. Y., 1991.

Index